Textbooks in Telecommunication Engineering

Era of the Internet and Network Science/Engineering

Series Editor
Tarek S. El-Bawab, Ph.D., Boston, MA, USA

Telecommunication and networks have evolved to embrace all aspects of our everyday life, including smart cities and infrastructures, healthcare, banking and businesses, manufacturing, space and aviation, meteorology and climate change, oceans and marine life, Internet of Things, defense, homeland security, education, research, social media, entertainment, and many others. Network applications and services continue to expand, virtually without limits. Therefore, specialized telecommunication and network engineering programs are recognized as a necessity to accelerate the pace of advancement in this field, and to prepare a new generation of engineers for imminent needs in our modern life. These programs need curricula, courses, labs, and textbooks of their own.

The IEEE Communications Society's Telecommunication Engineering Education (TEE) movement, led by Tarek S. El-Bawab- the editor of this Series, resulted in recognition of this field of engineering by the Accreditation Board for Engineering and Technology (ABET) on November 1, 2014. This Springer Series was launched to capitalizes on this milestone. The Series goal is to produce high-quality textbooks to fulfill the education needs of telecommunication and network engineering, and to support the development of specialized undergraduate and graduate curricula in this regard. The Series also supports research in this field and helps prepare its scholars for global challenges that lay ahead. The Series have published innovative textbooks in areas of network science and engineering where textbooks have been rare. It is producing high-quality volumes featuring innovative presentation media, interactive content, and online resources for students and professors.

Book proposals are solicited in all topics of telecommunication and network engineering including, but not limited to: network architecture and protocols; traffic engineering; network design, dimensioning, modeling, measurements, and analytics; network management and softwarization; cybersecurity; synchronization and control; applications of artificial intelligence in telecommunications and networks; applications of data science in telecommunications and networks; network availability, reliability, protection, recovery and restoration; wireless communication systems; cellular technologies and networks (through 5G, 6G, and beyond); satellite and space communications and networks; optical communications and networks; heterogeneous networks; broadband access and free-space optical communications; MSO/cable networks; storage networks; optical interconnects; and data centers; social networks; transmission media and systems; switching and routing (from legacy to today's paradigms); network applications and services; telecom economics and business; telecom regulation and policies; standards and standardization; and laboratories.

Proposals of interest shall be for textbooks that can be used to develop university courses, either in full or in part. They should include recent advances in the field while capturing whatever fundamentals that are necessary for students to understand the bases of the topic and appreciate its evolution trends. Books in this series will provide high-quality illustrations, examples, end-of-chapters' problems/exercises and case studies.

This series is indexed in Scopus. For further information and to submit proposals, please contact the Series Editor, Dr. Tarek S. El-Bawab, telbawab@ieee.org; or Mary James, Executive Editor at Springer, mary.james@springer.com.

Ivo Maljević · Faris Alfarhan · Raviraj Adve

Cellular Radio Access Networks

RF Fundamentals and Protocols

Springer

Ivo Maljević
TELUS
Toronto, ON, Canada

Faris Alfarhan
InterDigital Canada
Montreal, QC, Canada

Raviraj Adve
University of Toronto
Toronto, ON, Canada

ISSN 2524-4345   ISSN 2524-4353   (electronic)
Textbooks in Telecommunication Engineering
ISBN 978-3-031-76454-7   ISBN 978-3-031-76455-4   (eBook)
https://doi.org/10.1007/978-3-031-76455-4

© The Editor(s) (if applicable) and The Author(s), under exclusive license to Springer Nature Switzerland AG 2025

This work is subject to copyright. All rights are solely and exclusively licensed by the Publisher, whether the whole or part of the material is concerned, specifically the rights of translation, reprinting, reuse of illustrations, recitation, broadcasting, reproduction on microfilms or in any other physical way, and transmission or information storage and retrieval, electronic adaptation, computer software, or by similar or dissimilar methodology now known or hereafter developed.
The use of general descriptive names, registered names, trademarks, service marks, etc. in this publication does not imply, even in the absence of a specific statement, that such names are exempt from the relevant protective laws and regulations and therefore free for general use.
The publisher, the authors and the editors are safe to assume that the advice and information in this book are believed to be true and accurate at the date of publication. Neither the publisher nor the authors or the editors give a warranty, expressed or implied, with respect to the material contained herein or for any errors or omissions that may have been made. The publisher remains neutral with regard to jurisdictional claims in published maps and institutional affiliations.

This Springer imprint is published by the registered company Springer Nature Switzerland AG
The registered company address is: Gewerbestrasse 11, 6330 Cham, Switzerland

If disposing of this product, please recycle the paper.

Note from the Series Editor

This Series has its roots in the Telecommunication Engineering Education (TEE) movement in the USA, which I had the honor of leading from 2008 to 2015. The movement was a result of remarkable impacts telecom networks had already started to have on society, and many more that were looming on the horizon at that time. Major changes in the theoretical foundations of this field of engineering, its design strategies, and its practical applications were unfolding at the same time. The industries we had known as the Telecom, Cable, and Data-Communication industries through the twentieth century were going through technical innovations and were reshuffling, repositioning themselves in the market, and converging, in preparation for the global economy of the twenty-first century.

The TEE movement was prompted by a personal experience I went through as I moved from industry to academia. Telecom engineering, or Network Engineering, since the network is centric to the evolution telecom has been through for three to four decades so far, became a distinct multidisciplinary field of study. This field blends ingredients from electrical engineering, computer engineering, computer science, and other disciplines, but may not be taught merely as a branch of any one of them, per se. It became evident that society would be better served by adding specialized undergraduate/graduate courses and programs in telecom/network engineering to our higher-education system, especially in the USA where this was not the norm.

The story of the TEE movement has been told within broader contexts in the literature (e.g., IEEE Communications Magazine, Vol. 53, No. 11, pp. 35–9, November 2015). The movement had accomplished its main goal of adding telecom/network engineering to ABET's portfolio of recognized engineering disciplines. ABET is the Accreditation Board of Engineering and Technology, Inc., which accredits engineering programs in the USA and more than 40 other countries. A lot of work had to be done to capitalize on that milestone, however. The technical literature of this discipline was rich in scholarly research papers, but not as rich in textbooks designed to support the development of new university courses and curricula. These textbooks are needed to facilitate rigorous academic training for a new generation of network engineers. The idea of this Series came up subsequently to help the telecom/networking community overcome this shortcoming. Since its inception, Springer's TTE (Textbooks in Telecommunication Engineering) has been doing this job. The proximity of the acronyms herein—between TEE and TTE—is coincidental, but worth noting!

I am proud of the inroads TTE has made in publishing quality textbooks and in fulfilling the goals and objectives the Series has had since its beginning. By the end of 2024, we approach the milestone of having 35 books, including 20 published and about 15 in the making (whether in the writing, reviewing, or production phases). More than 50 authors have contributed to this Series so far and they come from about 20 different countries and cultures all over the world.

The topics covered by this Series so far scan the entire landscape of telecom and network engineering, including wireless, optical, satellite; access and core; devices, systems, and networks; transceiver systems; space-aerial-terrestrial integration; queuing theory and traffic engineering; network management; Internet and Internet of Things (IoT); signal processing; applied electromagnetics; and economy/business aspects of telecom. The Series includes a cellular-generations primer and has a Wi-Fi primer project under progress. It covers the most cutting-edge topics with textbooks that are carefully designed to strike the right balance between the need to create new courses thereto and the fact that progress is imminent in each of these topics, and in the short term. These include books that discuss the roles of edge and cloud computing; of blockchains, and of artificial intelligence and machine learning in network engineering. Special topics such as underwater communications, railways telecom, networked navigation systems, semantic communications, video analytics, and broadband to combat the digital divide are also covered by TTE. I have tried my best to make this Series a top de facto global home for quality telecom/network engineering textbooks, and am hoping that many in our community can start to consider it as such. The fruits of the hard work put by many people into fulfilling this goal are now handy to pick by our community.

I want to thank my wife Shahira and my daughter Nadine for their support of my work in this Series over nearly ten years. I have put a lot of time into this task and they have understood how this work is important to me. I owe a big deal to all the authors of this Series. Words cannot describe my appreciation of their contributions and talent. I would like to also express my thanks to Springer in general, and to Mary James in particular. Mary has been my main partner throughout this journey, and I couldn't have thought of a better partner than her. Finally, I thank all my colleagues in the field of telecom/network engineering and our readership for their continued support of the mission and the endeavor of this great project.

Boston, MA, USA Tarek El-Bawab
January 2025

Preface

Combining the strengths of industry and academia, this textbook leverages the knowledge of three authors with over 60 years of combined wireless industry and academic experience. As such, the material presented herein is based on the knowledge obtained through practical engineering, but also comes in part from the materials used in two graduate-level courses at the University of Toronto, ECE 1551 (Mobile Broadband Radio Access Network) and ECE 1552 (Modern Mobile Air Interfaces), and from a fourth year (senior level) undergraduate course ECE 464 (Wireless Communication). By combining these experiences, this textbook aims to provide fundamental theoretical concepts, describe practical aspects on how networks are implemented, and identify and bridge the gaps between the two. The practical descriptions follow the work of third generation partnership project (3GPP) in the context of the fourth (4G LTE) and fifth (5G New Radio) generation wireless networks. Furthermore, it covers some key practical issues that are, generally, ignored in most wireless textbooks, including practical problems faced by modern wireless engineers. Examples are inter-modulation and spectrum leakage issues.

This textbook is tailored to senior undergraduate and graduate students and newly practicing engineers. It covers Radio Access Network (RAN) aspects of the modern cellular communication networks. Important radio frequency (RF) parameters like power flux density, electrical field, and various power definitions are introduced and their relationship to regulatory requirements and standards-based usage are covered in detail. Also, various RF impediments such as the noise figure, out of band emissions, and adjacent channel selectivity/adjacent channel leakage ratio and intermodulation products are introduced. The link budget, receiver sensitivity, channel models, and how they relate to 3GPP based systems are explained. Spectrum and RF characteristics of cellular networks are an important part of the book. Theoretical and practical developments related to multiple input multiple output (MIMO) systems are also treated within the first half of the textbook. In the last segment of the first half of the textbook, we introduce multicarrier transmission and set the basis for understanding the air interface used in modern cellular technologies.

In the second half of the textbook, the fundamentals of RAN are introduced first by going over the architectural solutions. Next, an in-depth coverage of principles of modern mobile air-interfaces is provided, based on the examples from both 4G and 5G networks. The key elements of layers 1, 2, and 3 of air interfaces are covered in detail. Finally, the last segment of the book covers some more advanced topics, such as carrier aggregation, and system performance and optimization.

The materials presented herein assume basic understanding of digital modulation schemes and probability theory. Chapter 1, for example, covers these concepts, but at a high level; a course instructor may wish to add additional material covering digital modulation.

For an undergraduate level course, the authors recommend the first six chapters, as they provide a solid basis for understanding modern cellular communication systems like 4G and 5G networks. Advanced undergraduate, or a graduate course, may cover the remaining seven chapters, or Chaps. 7–12. Practicing engineers will find different parts of the book useful, depending on their expertise and areas of focus. For an RF network planning engineer, Chaps. 1–3 will be most instructive, as well as Chaps. 11 and 13. For RAN engineers, Chaps. 8–12 will be most useful. Spectrum strategy engineers will find Chaps. 4, 7, and 12 of particular interest.

Despite many years of experience between the three authors, anybody's knowledge has its limitations, and we would like to extend our gratitude to many colleagues from the industry and academia for their reviews, valuable comments, or simply words of encouragement. Our special thanks for numerous helpful suggestions and additions go to Frank Qing, Adam Tenenbaum, Rainer Iraschko, Matthew Mulvihill, Greg Thompson, Navaneetha Madan Gopal, and Jordan Melzer from TELUS, Ali Arad from Bell Mobility, Javad Jafarian and Marc Draper from Rogers, Colin Frank from Motorola, James Wang from Apple, Oumber Teyeb from Interdigital, Antti Toskala and the RAN4 team from Nokia, Rahim Nathoo and Christian Bergljung from Ericsson, David Mazzarese from Huawei, Masoud Ardakani from the University of Alberta, student Quinn Merkl from Simon Fraser University, Alex Bladenis from Telstra, Jad Berberi from Nokia, Scott Townley from Verizion, Bill Shvodian from T-Mobile, and Frank Rayal from Xonapartners. From the University of Toronto, our thanks go to Maryam Rezvan, a Ph.D. student, for many valuable inputs to Chap. 3, teaching assistant Ahmed Al-Mehdhar for help on the problem set solutions, and MEng student Xuyang Sun for valuable comments. Our thanks also go to Ljiljana Milic from the University of Belgrade, and Milica Pejanovic from University of Podgorica, for useful suggestions and words of encouragement. To the best of our knowledge, these individuals provided their personal inputs and do not represent their respective institutions. We would like to also acknowledge Josette Gallant from Innovation, Science and Economic Development Canada (ISED) and Narine Martel from Health Canada for their help in obtaining permissions to reproduce specific regulatory sections, and for their Safety Code 6 suggestions. Furthermore, our thanks go to Luka Maljevic, who helped with many drawings and made sure they are consistent throughout the textbook. Last, but not least, our gratitude goes to Tarek El-Bawab, Springer textbook series editor, and to Mary James, executive editor, whose patience and guidance have helped us in preparing this material.

While one of the authors works at InterDigital, it is disclaimed that views in this book represent the author's views and ideas, and should not be understood as supported or reviewed by InterDigital.

Toronto, ON, Canada	Ivo Maljević
Montreal, QC, Canada	Faris Alfarhan
Toronto, ON, Canada	Raviraj Adve
August 2024	

Contents

1	**Introduction to Wireless Communications and Networks**	1
1.1	Introduction	1
1.2	Electrical Signals in Communication Systems, Energy, and Average Power	1
	1.2.1 Linear and Logarithmic Scales for the Average Power	2
	1.2.2 Important Signals	4
	1.2.3 Frequency Representation of Signals	5
1.3	Power Spectral Density	7
1.4	Signal Bandwidth(s)	10
	1.4.1 Null and Null-to-Null Bandwidth	11
	1.4.2 3-dB Bandwidth	12
	1.4.3 Essential and Occupied Bandwidth	13
	1.4.4 RMS Bandwidth	13
1.5	Electric and Magnetic Fields and Power Flux Density	14
	1.5.1 Power Flux Density	15
	1.5.2 Electric Field in dBµV/m Scale	17
1.6	Cellular Antenna Basics	18
	1.6.1 N-Element Linear Array	19
	1.6.2 Important Antenna Parameters	23
	1.6.3 EIRP and ERP	33
1.7	Radiofrequency (RF) Modulation and Digital Communication	34
	1.7.1 BPSK Modulation	36
	1.7.2 QPSK	37
	1.7.3 Higher-Order Modulation Schemes	38
1.8	Cellular Networks	39
1.9	Multiple Access Schemes	41
	1.9.1 Frequency Division Multiple Access (FDMA)	42
	1.9.2 Time Division Multiple Access (TDMA)	43
	1.9.3 Code Division Multiple Access (CDMA)	44

	1.10	Duplexing Schemes	45
	Problems		47
	References		51

2 Pathloss and Channel Models — 53

- 2.1 Introduction — 53
- 2.2 Distance-Based Pathloss and Friis' Formula — 55
- 2.3 Empirical Pathloss Models — 58
  - 2.3.1 Penetration Through Materials and Foliage Loss — 62
- 2.4 Shadowing and Fade Margin — 63
- 2.5 Channel Impulse Response, Multipath, and Small-Scale Fading — 67
  - 2.5.1 Slow and Fast Fading — 67
  - 2.5.2 Statistical Description of a Wireless Channel — 69
  - 2.5.3 Overall Channel Model — 72
  - 2.5.4 Frequency Domain Analysis of the Channel Impulse Response — 73
  - 2.5.5 Power Delay Profile — 75
  - 2.5.6 Frequency Correlation and Coherence Bandwidth — 78
  - 2.5.7 Coherence Time — 79
- Problems — 80
- Appendix: Selected 3GPP Defined Pathloss Models — 82
- References — 84

3 Noise and Interference — 85

- 3.1 Introduction — 85
- 3.2 Noise and Interference Sources — 85
  - 3.2.1 Thermal Noise — 86
  - 3.2.2 Antenna Noise and System Temperature — 94
  - 3.2.3 Interference and Signal-to-Interference-Plus-Noise Ratio in AWGN Channels — 95
- 3.3 Link Budget — 100
  - 3.3.1 Coverage Difference — 105
- 3.4 Passive and Active Intermodulation — 106
  - 3.4.1 Narrowband Intermodulation — 107
  - 3.4.2 Wideband Intermodulation — 114
- Problems — 116
- References — 122

4 Spectrum and RF Characteristics of Cellular Systems — 125

- 4.1 Introduction — 125
- 4.2 Spectrum Licenses — 126
- 4.3 Role of ITU-R and 3GPP in Definition of Spectrum Bands and Standards — 127
  - 4.3.1 ITU Band Nomenclature — 128
  - 4.3.2 ITU-R Defined Geographic Regions — 129

Contents xiii

		4.3.3	WRC Conferences	130
		4.3.4	IMT Standardization	131
		4.3.5	The Role of 3GPP	131
		4.3.6	Frequency Bands and Characteristics of Each Frequency Range (sub-7GHz, mmWave, THz)	134
	4.4	Mobile Spectrum Allocation, Tier Organization, and Spectrum Auctions		136
		4.4.1	The Canadian Example	137
		4.4.2	The Australian Example	140
		4.4.3	The US Example	142
	4.5	Spectrum Cost Comparison		144
	4.6	Conducted and Radiated RF Requirements for 5G BS		146
	4.7	Device Output Power and Dynamic Range		149
	4.8	Unwanted Emission Requirements		151
		4.8.1	Unwanted Emissions at the BS	151
		4.8.2	Unwanted Emissions at the UT	154
	4.9	Reference Sensitivity		155
	4.10	Adjacent Channel Selectivity, Adjacent Channel Leakage Ratio, and Adjacent Channel Interference Ratio		156
	Problems			159
	References			162
5	**Multiple-Input, Multiple-Output Systems**			**165**
	5.1	Introduction		165
	5.2	Receive Diversity		166
		5.2.1	Selection Combining	170
		5.2.2	Maximum Ratio Combining	173
		5.2.3	Equal Gain Combining	176
		5.2.4	Impact of Spatial Correlation	177
	5.3	Transmit Diversity		178
		5.3.1	With CSIT	179
		5.3.2	Without CSIT	180
	5.4	MIMO Information Theory and Rate Considerations		185
		5.4.1	Parallel Channels	185
		5.4.2	General MIMO Channels	188
	5.5	Trade-Off Between Rate and Reliability		193
		5.5.1	Simultaneous Multiplexing and Space-Time Coding	193
		5.5.2	Diversity-Multiplexing Trade-Off	196
	5.6	Multiuser MIMO Systems		198
		5.6.1	Interference Mitigation	198
		5.6.2	Performance Metrics	202
		5.6.3	Massive MIMO Systems	203
		5.6.4	Impact of Massive MIMO on the Link Budget	205
	Problems			205
	References			210

6	**Introduction to Multicarrier Transmission**		**211**
	6.1	Orthogonal Frequency Division Multiplexing (Multicarrier Transmission)	211
	6.2	Cyclic Prefix and Channel Impact	219
	6.3	OFDM Challenges	225
		6.3.1 Peak-to-Average Power Ratio	225
		6.3.2 Phase Noise	228
		6.3.3 Carrier Frequency Offset	229
	6.4	Single-Carrier FDMA	232
		6.4.1 Localized Subcarrier Mapping	233
	Problems		235
	References		238
7	**Radio Access Network Architecture**		**239**
	7.1	Radio Access and Core Network	239
	7.2	Basestation Architecture	242
		7.2.1 Basestation Protocol Stack	244
		7.2.2 CPRI Line Rates	245
	7.3	Basestation Splits and the Midhaul	247
	7.4	eCPRI and a Physical Layer Split	248
	7.5	Backhaul Dimensioning	252
	7.6	Open RAN	253
		7.6.1 The O-RAN Working Groups	255
	Problems		256
	References		257
8	**3GPP-Based Air Interface, Layer 1**		**259**
	8.1	Time Domain Numerology	259
		8.1.1 4G Time Domain Numerology	259
		8.1.2 5G Time Domain Numerology	262
		8.1.3 4G Frequency Domain Numerology	265
		8.1.4 5G Frequency Domain Numerology	268
		8.1.5 5G Deployment Scenarios Based on Subcarrier Spacing	271
	8.2	5G DL and UL Frame Alignment and Slot Structures	272
	8.3	5G DL PHY Processing Chain	275
		8.3.1 Cyclic Redundancy Check (CRC) Insertion and Forward Error Control (FEC) Coding	275
		8.3.2 Rate Matching	278
		8.3.3 Scrambling	278
		8.3.4 Modulation	279
		8.3.5 MIMO Layer Mapping	279
	8.4	5G DL Physical Signals and Channels	280

Contents   xv

    8.5   5G UL Physical Signals and Channels 282
            8.5.1   Random Access and PRACH 283
            8.5.2   Other UL Physical Channels and Signals 289
    Problems 289
    References 294

9   3GPP-Based Air Interface: Layer 2 295
    9.1   Protocol Architecture 295
    9.2   The Medium Access Control Sublayer 296
            9.2.1   Mapping Between Logical Channels and
                        Transport Channels 298
            9.2.2   Random Access Procedure 299
            9.2.3   Scheduling 303
            9.2.4   Uplink Scheduling 304
            9.2.5   UL TB Construction and Logical Channel
                        Prioritization 308
            9.2.6   Buffer Status Reporting 312
            9.2.7   Scheduling Requests 314
            9.2.8   Power Headroom Reporting 315
            9.2.9   Downlink Scheduling 317
            9.2.10   Discontinuous Reception (DRX) and UE Power
                        Savings 318
            9.2.11   Beam Failure Detection and Recovery 319
            9.2.12   Timing Advance 321
            9.2.13   HARQ 321
            9.2.14   Wideband Carrier Operation 323
    9.3   The Radio Link Control Sublayer 324
            9.3.1   Error Correction Through ARQ 324
    9.4   Packet Data Convergence Protocol 325
    9.5   Service Data Adaptation Protocol 325
    Problems 326
    References 328

10   3GPP Based Air Interface, Layer 3 329
    10.1   The Radio Resource Control 329
            10.1.1   RRC States 331
    10.2   Initial Access Procedure 336
            10.2.1   Synchronization Procedure 336
            10.2.2   System Information 337
            10.2.3   Minimum System Information 338
            10.2.4   Additional System Information 339
            10.2.5   Cell Selection and Reselection 340
    10.3   Connected Mode Mobility 344
            10.3.1   Conditional Handover and CPC in NR 347
            10.3.2   Measurement Event Configurations for Handover 349
            10.3.3   Measurement Gaps 350

	10.4	Paging	352
		10.4.1 Paging Occasion and Paging Frames	353
		10.4.2 Radio Link Monitoring and Failure	353
	10.5	NAS Message Transfer	356
	Problems		356
	References		357
11	**Modeling and Performance Analysis of Cellular Systems**		**359**
	11.1	Cellular Network Performance Metrics and Modeling	359
		11.1.1 Reference Signal Received Power (RSRP)	359
		11.1.2 Received Signal Strength Indication (RSSI)	362
		11.1.3 Reference Signal Received Quality (RSRQ)	362
		11.1.4 Reference Signal SINR	365
		11.1.5 Synchronization Signal SINR	366
		11.1.6 PDSCH SINR	366
		11.1.7 PDSCH Maximum User Achievable MCS and Spectral Efficiency	366
		11.1.8 PUSCH SINR	366
		11.1.9 Maximum Achievable User Data Rate	367
	11.2	IEEE 802.11 Peak Data Rates	370
	11.3	Network Simulations	371
		11.3.1 Dynamic Cellular System Simulations	371
		11.3.2 Monte Carlo Simulations	371
		11.3.3 Scheduling Algorithms	373
	Problems		376
	References		380
12	**Carrier Aggregation with Licensed and Unlicensed Bands**		**381**
	12.1	LTE Carrier Aggregation	381
		12.1.1 Bandwidth Classes	383
		12.1.2 LTE CA Notation and the Number of Supported CCs	383
		12.1.3 LTE CA and Channel Spacing	385
		12.1.4 Primary and Secondary Serving Cell	385
		12.1.5 Peak Data Rate Calculations in LTE CA	387
	12.2	NR Carrier Aggregation	388
		12.2.1 5G/NR Bandwidth Classes	388
		12.2.2 NR CA Notation	389
		12.2.3 Supplemental Uplink	389
		12.2.4 NR Channel Spacings for CA	390
		12.2.5 Peak Data Rate Calculation in NR CA	391
	12.3	Dual Connectivity	391
		12.3.1 DC Notation and Peak Data Rate Calculation	393
	12.4	License-Assisted Access	394
		12.4.1 Interference Mitigation	396
		12.4.2 LAA Combinations	397

	12.5	NR-Unlicensed (NR-U)	397
		12.5.1 NR-U Carrier Aggregation	399
	Problems		400
	References		401

13 Cellular Network Planning, Design, and Optimization ... 403

- 13.1 Minimization of Drive Tests and Call Trace Data Collection 403
- 13.2 Positioning Methods in Cellular Networks 405
  - 13.2.1 Positioning by RF Finger Printing and Geolocations of Collected MDT Data 407
  - 13.2.2 Generation of Traffic Demand Maps from Geolocated MDT Data 408
- 13.3 Small Cell Placement 409
- 13.4 Automatic Cell Planning and Optimization 410
  - 13.4.1 Optimization and Planning of Network Configuration Parameters by Mixed Integer Programming Optimization 411
- 13.5 Self-Organizing Wireless Networks 421
- 13.6 Self-Configuration and Installation 422
  - 13.6.1 Automatic Physical Cell ID Configuration 423
  - 13.6.2 Autonomous Load Balancing in Wireless Networks 423
  - 13.6.3 Mobility Robustness Optimization 424
  - 13.6.4 Automatic Neighbor Relations 425
  - 13.6.5 Coverage Optimization and Self-Healing 425
  - 13.6.6 PRACH Optimization 426
- Problems 426
- References 426

Index 429

Acronyms

The list of acronyms used throughout the book is provided here.

3GPP	Third generation partnership project
5GC	5th Generation Core
A/D	Analog to digital
AAS	Active antenna systems
ACIR	Adjacent channel interference ratio
ACK	Acknowledgment
ACLR	Adjacent channel leakage ratio
ACP	Automatic cell planning
ACQ	Acknowledgment
ACS	Adjacent channel selectivity
AFC	Automated frequency coordination
AM	Amplitude modulation, analog technology; Acknowledged mode also
AMF	Access and mobility management
ANR	Automatic neighbor relation
APT	Asia-Pacific Telecommunity
ARQ	Automatic Repeat reQuest
AS	Access stratum
ASMG	Arab Spectrum Management Group
ATU	African Telecommunications Union
AWGN	Additive white Gaussian noise
BC	Bandwidth class
BCCH	Broadcast control channel
BCS	Bandwidth combination set
BFD	Beam failure detection
BFI	Beam failure instance
BLAST	Bell labs layered space-time
BLER	Block error rate
BPSK	Binary phase shift keying

BS	Basestation
BSD	Bucket size duration
BSR	Buffer status reporting
BWP	Bandwidth part
C-DRX	Connected mode DRX
C-RNTI	Cell Radio Network Temporary Identifier
CA	Carrier aggregation
CAPEX	Capital expenditures
CAZAC	Constant Amplitude Zero Autocorrelation
CBRA	Contention based random access
CBRS	Citizen's Broadband Radio Service
CC	Component carrier
CCA	Clear channel assessment
CCCH	Common control channel
CD-SSB	Cell-Defining SSB
CDF	Complementary distribution function
CDMA	Code division multiple access
CEPT	European Conference of Postal and Telecommunications Administrations
CFO	Carrier frequency offset
CFRA	Contention free random access
CG	Configured grants
CHO	Conditional handover
CITEL	Inter-American Telecommunication Commission
CMAS	Commercial Mobile Alert System
CN	Core network
CoMP	Coordinated multipoint
CP-OFDMA	Cyclic prefix orthogonal frequency division multiple access
CP	Cyclic prefix
CPD	Cross-polarization discrimination, see also XPD
CPE	Common phase error
CPRI	Common public radio interface
CQI	Channel quality indicator
CRC	Cyclic redundancy check
CSI-RS	Channel state information reference signal
CSI	Channel state information
CSIR	Channel state information at the receiver
CSIT	Channel state information at the transmitter
CU	Central unit
D/A	Digital to analog
dB	Decibel
dBc	Decibel relative to the carrier (power)
dBd	Decibel relative to half-wavelength dipole
dBi	Decibel relative to isotropic antenna
dBm	Decibel relative to power in milliwatts

Acronyms

dBW	Decibel relative to power in watts
DC	Dual connectivity
DCCH	Dedicated control channel
DCI	Downlink control information
DFT	Discrete Fourier transform
DL-AoD	Downlink Angle of Departure
DL-TDOA	Downlink Time Difference Of Arrival
DL	Downlink, transmission from the basestation toward the UE
DM-RS	Demodulation reference signal
DRB	Data radio bearer
DRX	Discontinuous reception
DTCH	Dedicated traffic channel
DU	Distributed unit
E-UTRAN	Enhanced universal terrestrial radio access network
eCPRI	Enhanced CPRI
EGC	Equal gain combining
EHF	Extremely high frequency, 30 to 30 GHz
eICIC	Enhanced Inter Cell Interference coordination
EIRP	Effective isotropic radiated power
eLAA	Enhanced license assisted access
EM	Electromagnetic (wave)
eMBMS	Evolved multimedia broadcast multicast service
EN-DC	E-UTRA-NR dual connectivity
eNB	Enhanced node B, see eNodeB
eNodeB	Enhanced node B, a 3GPP term for a 4G basestation
EPC	Enhanced packet core
ERP	Effective radiated power
ETWS	Earthquake and Tsunami Warning System
FCC	Federal Communications Commission
FDD	Frequency division duplex
FDMA	Frequency division multiple access
FEC	Forward error control
FeLAA	Further enhanced LAA
FFT	Fast Fourier transform
FM	Frequency modulation, analog technology
FRC	Fixed reference channels
FSS	Fixed satellite services
FSTD	Frequency switched transmit diversity
FT	Fourier Transform
GAA	General authorized access
GNSS	Global navigation satellite system
GPS	Global Positioning System
GSM	Global system for mobile communications
H-ARQ	Hybrid automatic repeat request
HAPS	High-Altitude Platform Stations

HF	High frequency, 3 to 30 MHz
HIBS	HAPS as IMT basestations
HO	Handover
HPBW	Half-power beamwidth
HRNN	Human-Readable Network Names
HSS	Home subscriber server
IAB	Integrated access and backhaul
ICIC	Inter Cell Interference coordination
IDF	Inverse discrete Fourier transform
IEEE	Institute of Electrical and Electronics Engineers
IFFT	Inverse Fourier transform
IM2	Second order intermodulation
IM3	Third order intermodulation
IMD	Intermodulation distortion
IMSI	International mobile subscriber identity
IMT	International mobile telecommunications
IoT	Internet of things
IPsec	Internet protocol security
IRAT	Inter radio access technology
ISED	Innovation, Science and Economic Development Canada
ISI	Inter-symbol interference
ITU-R	International telecommunication union, radiocommunications sector
ITU-R	TU Radio-communications sector
ITU	International telecommunication union
ITU	International Telecommunications Union
KPI	Key performance indicators
L1-RSRP	Layer 1 RSRP (see RSRP)
L1	Layer 1
L2	Layer 2
L3	Layer 3
L3-RSRP	Layer 3 RSRP (see RSRP)
LAA	License assisted access
LBT	Listen before talk
LCP	Logical channel prioritization
LDPC	Low density parity check
LF	Low frequency, 30 to 300 kHz
LMF	Location and mobility function
LO	Local oscillator
LOS	Line-of-sight
LPWA	Low-power wide area networks
LTE	Long term evolution, 4G technology
MAC	Medium access control
MADR	Maximum achievable data rate
MAPL	Maximum acceptable pathloss

MCG	Master cell group
MCOT	Maximum channel occupancy time
MCS	Modulation coding scheme
MDT	Minimization of drive test
MF	Medium frequency, 300 kHz to 3 MHz
MIB	Master information block
MILP	Mixed Integer Linear Program
MIMO	Multiple input, multiple output (antenna technology)
MIP	Mixed integer program
MIQP	Mixed Integer Quadratic Program
MME	Mobility management entity
MMSE	Minimum mean squared error
MN	Master node
MPE	Maximum permissible exposure
MR-DC	Multi-radio dual connectivity
MRC	Maximum ratio combining
MRT	Maximum ratio transmission
MSI	Minimum system information
MU-MIMO	Multi-user MIMO
NAS	Non-access stratum
NB-IoT	Narrowband IoT
NCL	Neighbor cell list
NCL	Non-Competitive Local Licensing
NG-AP	Next Generation Application Protocol
NG-RAN	Next Generation Radio Access Network
NGMN	Next Generation Mobile Networks Alliance
NLOS	Non line-of-sight
NPN	Non-public network
NR-DC	New radio dual connectivity
NR-U	New radio, unlicensed
NR-U	NR unlicensed
NR	New Radio, 5G technology
NSA	Non-standalone
NTIA	National Telecommunications and Information Administration
NTN	Non-terrestrial network
OBUE	Operating band unwanted emission
OCB	Occupied channel bandwidth
OFDM	Orthogonal frequency division multiplexing
OFDMA	Orthogonal frequency division multiple access
OH	Overhead
OOBE	Out of band emission
OPEX	Operational expenditures
OSTBC	Orthogonal space-time block code
OTA	Over the air
OTDOA	Observed Time Difference Of Arrival

P-GW	Packet data network gateway
P-MPR	Power management power reduction
P/S	Parallel to serial
PAL	Priority access licenses
PBCH	Physical broadcast channel
PCC	Primary component carrier
PCCH	Paging control channel
PCell	Primary cell
PCF	Policy control function
PCG	Project Coordination Group
PCH	Paging channel
PCI	Physical cell identifier
PCRF	Policy and charging rules function
PCS	Personal communications system; also, primary serving cell
PDCCH	Physical downlink control channel
PDCP	Packet Data Convergence Protocol
PDF	Probability density function
PDSCH	Physical downlink shared channel
PEI	Paging early indicator
PF	Proportional fair
PFD	Power flux density
PHR	Power headroom reports
PHY	Physical (layer)
PIM	Passive intermodulation
PLMN	Public land mobile network
PRACH	Physical random access channel
PRB	Physical resource block
PSCell	Primary secondary cell
PSD	Power spectral density
PSS	Primary synchronization signal
PT-RS	Phase tracking reference signal
PUCCH	Physical uplink control channel
PUSCH	Physical uplink shared channel
QAM	Quadrature amplitude modulation
QFI	QoS flow ID
QoS	Quality of service
QPSK	Quadrature phase shift keying
RA	Random access
RACH	Random access channel
RAN	Radio access network
RAR	Random access response
RAT	Radio access technology
RB	Resource block
RCC	Regional Commonwealth in the Field of Communications
RE	Resource element

REFSENS	Reference sensitivity
RET	Remote electrical tilting
RF	Radio frequency
RH	Radio head
RIB	Radiated interface boundary
RLF	Radio link failure
RLM	Radio link monitoring
RMS	Root mean-squared
RMSI	Remaining minimum system information
RNA	RAN-based notification area
RNAU	RAN-based notification area update
ROHC	Robust header compression
RR	Radio regulation
RRC	Radio resource control
RRH	Remote radio head
RRU	Remote radio unit
RSRP	Reference signal received power
RSRQ	Reference Signal Received Quality
RSS	Radio Standards Specification
RSSI	Received Signal Strength Indication
RSTD	Reference Signal Time Difference
RTO	Regional telecommunication organization
RTT	Round trip time
RV	Redundancy version
S-GW	Serving gateway
S/P	Serial to parallel
SA	Standalone (5G network)
SAS	Spectrum access system
SC-FDMA	Orthogonal frequency division multiplex access
SCC	Secondary component carrier
SCell	Secondary cell
SCG	Secondary cell group
SCS	Subcarrier spacing
SDAP	Service Data Adaptation Protocol
SDMA	Space-division multiple access
SDO	Standards developing organization
SDT	Small data transmission
SDU	Service data unit
SFBC	Space-frequency block coding
SFN	System frame number
SHF	Super high frequency, 3 to 30 GHz
SI	System information
SIBn	System information block $n, n = 1, 2, 3, \ldots$
SIC	Successive interference cancellation
SIMO	Single input, multiple output

SINR	Signal to interference plus noise ratio
SISO	Single input, single output
SLA	Side lobe attenuation, maximal; used in antenna gain modelling
SMF	Session management function
SMTC	SS/PBCH Block Measurement Timing Configuration
SN	Secondary node
SNR	Signal-to-noise ratio
SON	Self-organizing networks
SPS	Semi-persistent scheduling
SRB	Signaling radio bearers
SRS	Sounding reference signal
SS-RSRP	Synchronization signal RSRP (see RSRP)
SS-RSRQ	Synchronization signal RSRQ (see RSRQ)
SSS	Secondary synchronization signal
SU-MIMO	Single-user MIMO
SUL	Supplemental uplink
TAB	Transceiver array boundary
TABC	TAB connectors
TAC	Tracking area code
TAG	Timing advance group
TAU	Tracking area update
TB	Transport block
TBS	Transport block size
TDD	Time division duplex
TDMA	Time division multiple access
TM	Transparent mode
TN	Terrestrial network
TOI	Third order intercept
TR	Technical report, 3GPP
TRP	Transmission reception point
TS	Technical specification, 3GPP
TSG	Technical specification groups
TTI	Transmit time interval
UDM	Unified data management
UE	User equipment, synonymous with UT
UHF	Ultra high frequency, 300 MHz to 3 GHz
UL RTOA	Uplink Round Trip Time of Arrival
UL-AoA	Uplink Angle of Arrival
UL	Uplink, transmission from the UE towards the basestation
UPF	User plane function
UT	User terminal, synonymous with UE
UTC	Coordinated Universal Time
UTDOA	Uplink Time Difference Of Arrival
UTRAN	Universal terrestrial radio access network
VHF	Very high frequency, 30 to 300 MHz

VLF	Very low frequency, 3 to 30 kHz
VoIP	Voice over IP
VPLMN	Visiting public land mobile network
WRC	World radio-communication conference
XPD	Cross-polarization discrimination, see also CPD
ZC	Zadoff-Chu
ZF	Zero-forcing

Chapter 1
Introduction to Wireless Communications and Networks

1.1 Introduction

The introductory material covered in this chapter provides an overview of the material that forms the background of this textbook. In this chapter, we cover the signals, communications, and radiofrequency (RF) concepts that will be used throughout the book. If completely new to this material, the reader is encouraged to consult undergraduate texts in the appropriate subject area.

1.2 Electrical Signals in Communication Systems, Energy, and Average Power

In communication systems, electrical signals are represented as functions of time t: $x(t), g(t), s(t)$, etc. These signals may represent information data such as voice, text, video, music, or any other media or the carrier signal modulated to *carry* the information signals.

The *size* of a signal $x(t)$ is measured as a squared norm:

$$E_x = \int_{-\infty}^{\infty} |x(t)|^2 \, dt, \tag{1.1}$$

Supplementary Information The online version contains supplementary material available at (https://doi.org/10.1007/978-3-031-76455-4_1).

© The Author(s), under exclusive license to Springer Nature Switzerland AG 2025
I. Maljević et al., *Cellular Radio Access Networks*, Textbooks in Telecommunication Engineering, https://doi.org/10.1007/978-3-031-76455-4_1

which, if finite, represents the energy in the signal. An alternate measure is the average signal power defined as

$$P_x = \lim_{T \to \infty} \frac{1}{T} \int_{-T/2}^{T/2} |x(t)|^2 \, dt. \tag{1.2}$$

Often, in practical systems, there is a finite time segment of interest. In this case, the parameter T covers this time segment only. One example is periodic signals of period T since the average power in any period is the same. Signals with finite energy are known as energy signals; such signals have zero average power. Conversely, signals with nonzero average power are called power signals; such signals have infinite energy.

> **Note** Since all real-world signals have finite duration, theoretically, all real-world signals are energy signals. However, in many cases, signals of interest last a very long time (e.g., the carrier signal) and so are better treated as power signals. In general, we will treat signals of short durations as energy signals and long durations as power signals.

> **Example 1.1** Determine the average power for the complex exponential signal defined as $x(t) = Ae^{j2\pi f_0 t}$.

Solution The average power is

$$P_x = \lim_{T \to \infty} \frac{1}{T} \int_{-T/2}^{T/2} |x(t)|^2 \, dt \tag{1.3}$$

$$= \lim_{T \to \infty} \frac{1}{T} A^2 T = A^2. \tag{1.4}$$

∎

1.2.1 Linear and Logarithmic Scales for the Average Power

Wireless communication signal powers usually vary over a very wide range of values, typically from below 1 pW to tens of kW. Given such a wide range,

1.2 Electrical Signals in Communication Systems, Energy, and Average Power

comparisons of power levels can be challenging. For this reason, it is more convenient to compress the range of values using logarithmic compression. Using a log scale also makes the modeling of signal attenuation easier.

The logarithmic scale of power levels is called the decibel or dB scale:

$$P_{|\text{dBW}} = 10\log_{10}(P) \qquad (1.5)$$

and is expressed in dBW to indicate that the decibel scale relates to power level in watts. For instance, 1 watt is 0 dBW, 1 mW is -30 dBW, 1 µW is -60 dBW, etc.

The dB scale, in fact, is based on the ratio of a measurement to a reference:

$$P_{|\text{dBW}} = 10\log_{10}(P/P_{\text{ref}}) \qquad (1.6)$$

In the dBW scale, the reference is $P_{\text{ref}} = 1$ W. Due to the low power levels in wireless communication, a widely used choice for the reference power is $P_{\text{ref}} = 1$ mW, called the dBm scale. Since 1 W = 1000 mW, we have

$$P_{|\text{dBm}} = 10\log_{10}\left(\frac{P}{1\,\text{mW}}\right) = 10\log_{10}\left(\frac{P \times 1000}{1\,\text{W}}\right) = P_{|\text{dBW}} + 30$$

Example 1.2 Determine the dBm value for the signal whose average power is $P = 20$ W.

Solution 20 W of power is equivalent to $P_{|\text{mW}} = 20{,}000$ mW, so we can determine the dBm either directly:

$$P_{|\text{dBm}} = 10\log_{10}(20) + 30 = 43\,\text{dBm} \qquad (1.7)$$

or by using the mW value

$$P_{|\text{dBm}} = 10\log_{10}(20000) = 43\,\text{dBm} \qquad (1.8)$$

Conversely, we can calculate the linear domain power level from the dB scale by reversing the expression in (1.6):

$$P_x = 10^{\frac{P_{|\text{dBm}} - 30}{10}} = 20\,(\text{W}). \qquad (1.9)$$

■

> **Note** The maximum transmit power level for most cellular phones, as defined by the 3rd Generation Partnership Project (3GPP) is 23 dBm or 200 mW (see Chap. 4 for more details). These are the so-called Power Class 3 devices. Other power level classes are defined in 3GPP technical specification TS 38.101-1 [1] (for 5G technology) or in TS 36.101 [2] (for 4G technology).
>
> We should note here that in subsequent text, we will use the term Long-Term Evolution (LTE) interchangeably with the 4G term to align the terminology with 3GPP specifications (similarly, we will use the term New Radio (NR) to indicate 5G). Also, terms user terminal (UT) and user equipment (UE) will be used interchangeably to indicate user devices (e.g., smartphones).

Generally, determining the average power of the signal, or its energy, is the extent of the usage of time domain representation of the wireless communication signals. The rest of signal analysis relies on the frequency domain representation of the signal (e.g., signal bandwidth, in-band power spectral density, out-of-band emissions, etc.), and we will discuss the basics after introducing several important signals.

1.2.2 Important Signals

Before continuing further, it is worth defining three signals that will play important roles in our discussion. The first is rect(t), defined as[1]

$$\text{rect}(t) = \begin{cases} 1, & |t| \leq 1/2 \\ 0, & |t| > 1/2. \end{cases} \quad (1.10)$$

A time-scaled version of rect(t) is shown in Fig. 1.1.

A function that plays a crucial role in theoretical analyses—and, hence, system design—is the δ-function, $\delta(t)$, defined as

$$\delta(t) = 0 \quad \forall t \neq 0$$
$$\int_{-\infty}^{\infty} \delta(t) dt = 1 \quad (1.11)$$

For our purposes, it suffices to say that the δ-function "operates" at a single point.

---

[1] Alternative definitions, where rect($\pm \frac{1}{2}$) is 0 or $\frac{1}{2}$, are also used in the literature.

1.2 Electrical Signals in Communication Systems, Energy, and Average Power

Fig. 1.1 Rectangular pulse shape $x(t) = \text{rect}(t/T)$

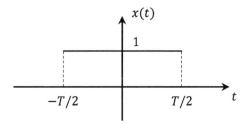

The third and fundamental signal is the complex exponential function $e^{j2\pi ft}$, ($j = \sqrt{-1}$), a pure sinusoid at frequency f. We have

$$e^{j2\pi ft} = \cos(2\pi ft) + j \sin(2\pi ft). \tag{1.12}$$

1.2.3 Frequency Representation of Signals

The frequency domain representation of communication signals is based on the time-to-frequency domain signal transformation. This *to-* and *from-* transformation between time and frequency domain is obtained via the Fourier transform (FT) and the inverse Fourier transform.

Let $x(t)$ be a time domain signal. Its FT is defined as

$$\mathcal{FT}[x(t)] = X(f) = \int_{-\infty}^{\infty} x(t)e^{-j2\pi ft}\, dt \tag{1.13}$$

and the inverse FT is defined as

$$\mathcal{FT}^{-1}[X(f)] = x(t) = \int_{-\infty}^{\infty} X(f)e^{j2\pi ft}\, df \tag{1.14}$$

We will denote Fourier transform pairs as $x(t) \leftrightarrow X(f)$.

Example 1.3 Determine the FT of the signal $x(t) = \text{rect}(t/T)$ as shown in Fig. 1.1.

Solution Using (1.13), the FT of the rectangular signal is

$$X(f) = \int_{-\infty}^{\infty} x(t)e^{-j2\pi ft}\, dt$$

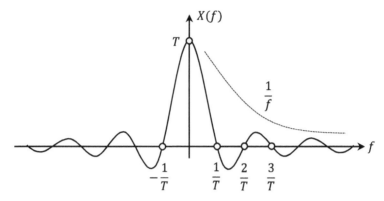

Fig. 1.2 Frequency domain version of the rectangular pulse is a *sinc* pulse, which decays at the $1/f$ rate

$$= \int_{-T/2}^{T/2} 1 \times e^{-j2\pi ft} \, dt = \frac{1}{-2\pi jf} e^{-j2\pi ft} \Big|_{-T/2}^{T/2}$$

$$= \frac{T}{\pi fT} \frac{e^{j\pi fT} - e^{-j\pi fT}}{2j} = T \frac{\sin(\pi fT)}{\pi fT}$$

$$= T \operatorname{sinc}(fT), \tag{1.15}$$

where we have defined $\operatorname{sinc}(x) = \sin(\pi x)/(\pi x)$. We, therefore, have $\operatorname{rect}(t/T) \leftrightarrow T\operatorname{sinc}(fT)$. ∎

> As illustrated below, the nulls of the *sinc* pulse are spaced $1/T$ apart in frequency, and the decay of the amplitude of *sinc* pulse is relatively slow, at a $1/f$ rate.

This signal is known as a *baseband* signal since, as seen in Fig. 1.2, the strongest frequency components are near zero frequency.

An important property of the Fourier transform is the so-called shifting property, which states that

$$x(t) \leftrightarrow X(f) \quad \Rightarrow \quad x(t)e^{2j\pi f_0 t} \leftrightarrow X(f - f_0) \tag{1.16}$$

This property enables signal up-conversion (or signal modulation) and down-conversion (demodulation). Up-conversion is illustrated below, which plots the Fourier transform of $\operatorname{rect}(t/T)e^{j2\pi f_0 t}$ after *up-conversion* of the baseband signal in Fig. 1.2. This is an example of a *passband* signal (Fig. 1.3).

1.3 Power Spectral Density

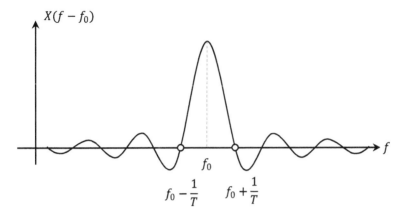

Fig. 1.3 Frequency domain *sinc* pulse shifted to the right by f_0 Hz

As we will see in Chaps. 6 and 8, in, e.g., 4G and 5G systems, rectangular pulse shapes, combined with frequency shifting, are the key elements of the air interface.

Another useful property of the Fourier transform is Parseval's theorem, which establishes the relationship between the time domain signal energy and frequency domain spectral contributions. If $x(t) \leftrightarrow X(f)$, then

$$E_x = \int_{-\infty}^{\infty} |x(t)|^2 \, dt = \int_{-\infty}^{\infty} |X(f)|^2 \, df. \tag{1.17}$$

1.3 Power Spectral Density

While (1.17) is appropriate for energy signals, in communications, we must often deal with power signals; the power spectral density (PSD) provides a convenient framework to analyze the signal's *power distribution across frequencies*. This description is particularly convenient for *random or stochastic* signals. We begin with deterministic signals.

If we truncate the signal $x(t)$ so that $x_T(t) = x(t)\text{rect}(t/T)$, the truncated version of the signal is an energy signal as long as T is finite. A finite signal $x_T(t)$ has a Fourier transform, denoted as $X_T(f)$. We can now analyze the average power of the signal and its power as

$$P_x = \lim_{T \to \infty} \frac{1}{T} \int_{-T/2}^{T/2} |x(t)|^2 \, dt = \lim_{T \to \infty} \frac{1}{T} \int_{-\infty}^{\infty} |x_T(t)|^2 \, dt$$

$$= \lim_{T \to \infty} \frac{1}{T} \int_{-\infty}^{\infty} |X_T(f)|^2 \, df = \int_{-\infty}^{\infty} \lim_{T \to \infty} \frac{|X_T(f)|^2}{T} \, df$$

$$= \int_{-\infty}^{\infty} S_x(f) \, df \qquad (1.18)$$

where we have defined the PSD as $S_x(f) = \lim_{T \to \infty} \frac{|X_T(f)|^2}{T}$.

Importantly, as (1.18) shows, the power in the original signal is the integral of the PSD. The PSD can, therefore, be interpreted as the *distribution of power across frequencies*.

In Example 1.1, we saw that the average power of a complex sinusoid, $x(t) = Ae^{j2\pi f_0 t}$ is A^2. Using the definition of $x_T(t)$, the FT of the rect($\cdot$) function, and the definition of the PSD above, we can show that $S_x(f) = A^2 \delta(f - f_0)$, i.e., all the power of this sinusoid is concentrated at a single frequency, $f = f_0$.

While the PSD of deterministic signals is based on the Fourier transform, analyzing random signals requires a different approach. We consider a random signal $x(t)$ such that for every value of time t the signal is a random variable and define $\mathbb{E}[\cdot]$ as the statistical expectation operator. We can now define the mean signal $\mu_x(t) = \mathbb{E}[x(t)]$. For most signals of our interest, the random signals are zero mean, i.e., $\mu_x(t) = 0$.

We also define the autocorrelation function $R_x(t_1, t_2) = \mathbb{E}[x^*(t_1)x(t_2)]$. A wide-sense stationary signal satisfies

$$\mu_x(t) = \text{constant} \qquad (1.19)$$

$$R_x(t_1, t_2) = R_x(\tau), \quad \tau = t_2 - t_1, \qquad (1.20)$$

i.e., the correlation between any two samples of the random signal only depends on the time difference between the two samples. Unless otherwise specified, the random signals in this text will be considered to be wide-sense stationary.[2] For these processes, we can apply the Wiener-Khinchin theorem that establishes the relationship between the PSD of a process $x(t)$ and its autocorrelation:

---

[2] A detailed discussion of random processes and stationarity is outside the scope of this text. We refer the reader to a relevant textbook such as [3].

1.3 Power Spectral Density

$$S_x(f) = \int_{-\infty}^{\infty} R_x(\tau) e^{-j2\pi f \tau} d\tau \quad (1.21)$$

Example 1.4 An important stochastic signal in communications is the so-called white-noise process. This zero-mean signal satisfies $R_x(\tau) = \frac{N_0}{2}\delta(\tau)$, i.e., any two samples of the process are uncorrelated.

Consequently, $S_x(f) = \frac{N_0}{2}$, i.e., on average, all frequencies contribute equally (and, hence, the name *white noise*).

It is worth noting that noise is present in all electrical equipment. The constant PSD is a valid model up to frequencies in the range of THz, i.e., higher in frequency than of interest here. We will, therefore, use the white-noise model.

Example 1.5 The random binary process

$$x(t) = \sum_{k=-\infty}^{\infty} a_k \text{rect}\left(\frac{t - kT - \alpha}{T}\right) \quad (1.22)$$

where α is a uniformly distributed random variable in the range $(0, T)$ and a_k takes on values of $+1$ or -1 in every interval of T seconds. For this signal, the autocorrelation function is $R_x(\tau) = (1 - |\tau|/T)\text{rect}(t/2T)$, and the power spectral density is [4]:

$$S_x(f) = T\text{sinc}^2(fT). \quad (1.23)$$

Using (1.23), the average power of this process is given by

$$P_x = \int_{-\infty}^{\infty} S_x(f) df = 1. \quad (1.24)$$

Figure 1.4 plots the PSD in (1.23). The figure illustrates that most of the power is contributed by the lower frequencies. Furthermore, while, in theory, all frequencies contribute, if we integrate the PSD between the first nulls (from $-1/T$ to $1/T$), we get the value of 0.90282, that is, on average, 90.282% of the total signal power there is contained within the main lobe of the PSD.

With these preliminaries established, we can now define the signal bandwidth.

1.4 Signal Bandwidth(s)

In theory, the bandwidth of a signal is the width of its PSD on the positive frequency axis. For example, in the case of

$$S_x(f) = \text{rect}\left(\frac{|f| - f_c}{W}\right) \Rightarrow \text{Bandwidth} = W,$$

since $S_x(f)$ is nonzero only in the range $f_c - W/2 < |f| < f_c + W/2$. Note that $S_x(f)$ is nonzero on both the positive and negative frequency axes, but only the positive axis is used to calculate bandwidth.

Unfortunately, most signals are not as nicely behaved as in this example. As we see in Fig. 1.4, a signal may have a theoretically infinite bandwidth, but only a limited range of frequencies make significant contributions. In these situations, the bandwidth is the effective range of frequencies that contribute significantly to the overall signal. However, what is significant can depend on the application, i.e., there is no single definition of signal bandwidth; several definitions are used in practice, and one needs to specify which definition is used. The definition chosen may even change the measurement technique used to determine the signal bandwidth.

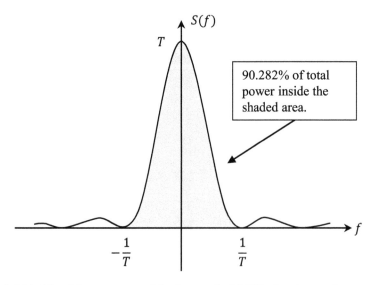

Fig. 1.4 Main lobe of the power spectral density contains 90.28% of total energy/power

1.4.1 Null and Null-to-Null Bandwidth

This is the simplest definition and states that for a baseband signal, the signal bandwidth is the frequency that corresponds to the first null of the power spectral density. In the example of Fig. 1.4, the first null bandwidth would be at $1/T$. For instance, if the pulse duration is $T = 1\mu s$, null bandwidth of such signal would be $BW_{null} = 1/T = 1\,\text{MHz}$.

In the case the signal is not baseband (e.g., after carrier modulation or up-conversion), the bandpass signal power is half of the power of the baseband signal [4]. This is because

$$y(t) = x(t)\cos(2\pi f_c t) \Rightarrow S_y(f) = \frac{1}{4}[S_x(f - f_c) + S_x(f + f_c)] \quad (1.25)$$

We now use the null-to-null bandwidth, defined as the range of positive frequencies between the two spectral nulls of the PSD closest to the center frequency, as illustrated in Fig. 1.5, for a frequency-shifted rectangular pulse shape of $1\mu s$ duration, the null-to-null bandwidth $BW_{\text{null-to-null}} = 2/T = 2\,\text{MHz}$.

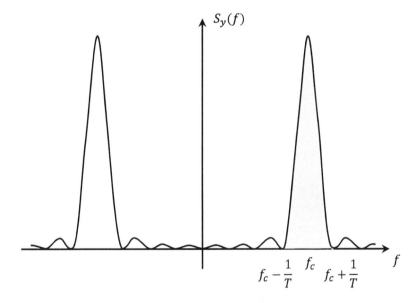

Fig. 1.5 PSD of frequency-shifted rectangular pulse

1.4.2 3-dB Bandwidth

Another simple measurement method is the so-called *3-dB bandwidth*, where the bandwidth is the determined by the frequency point where the PSD is 3 dB lower than peak (or half in linear scale). As in the previous definitions, only positive frequencies are counted.

In the case of a baseband rectangular pulse shape, we have the power spectral density $S_x(f) = T\text{sinc}^2(fT)$, and the reference point is at the frequency $f = 0$. Therefore, $S_x(0) = T$, and $S_x(f = BW_{3\,\text{dB}}) = T/2$. After numerical calculation, we have

$$BW_{3\,\text{dB}} = \frac{0.443}{T} \quad (1.26)$$

As can be seen from the Fig. 1.6, clearly, even less energy is contained within the 3-dB bandwidth when compared to the null bandwidth.

> **Homework** Numerically determine the percentage of energy within the 3-dB bandwidth of the rectangular pulse shape. *Result: 72.2%.*

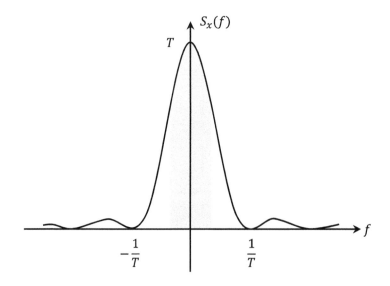

Fig. 1.6 Shaded area from $-0.443/T$ to $0.443/T$ has an area equal to 0.772

1.4.3 Essential and Occupied Bandwidth

The definitions so far, while relatively simple to use, do not suffice for all purposes. Specifically, a significant portion of the signal power (energy) falls outside the defined bandwidths (e.g., for the rectangular pulse case, almost 10% for the null bandwidth and 30% for the 3 dB bandwidth). Communication standards and government regulations often use a stricter definition of bandwidth.

In 3GPP-based cellular specifications, the occupied bandwidth is defined as the bandwidth that contains 99% of the total signal's power. A related definition is the $X\%$ essential bandwidth, where $X\%$ of the power (or energy) within a frequency range limited by that bandwidth. For the rectangular pulse shape, the 99% essential bandwidth (occupied bandwidth) is numerically determined by solving the following equation:

$$\int_{-BW_{occ}}^{BW_{occ}} T\operatorname{sinc}^2(fT)\,df = 0.99, \tag{1.27}$$

which results in

$$BW_{occ} = \frac{10.286}{T}, \tag{1.28}$$

i.e., more than ten times greater than the null-to-null bandwidth!

> As an aside, it is evident that the rectangular pulse shape is not spectrally efficient as the spectrum content is not contained nicely within a narrow range (ideally, the whole energy should be within $1/T$). Instead, it spreads over a wide range of frequencies.

1.4.4 RMS Bandwidth

The *root mean square* or RMS bandwidth of a baseband (low-pass) signal $x(t)$ is defined as following:

$$BW_{\text{RMS}} = \sqrt{\frac{\int_{-\infty}^{\infty} f^2 |X(f)|^2\,df}{\int_{-\infty}^{\infty} |X(f)|^2\,df}} \tag{1.29}$$

where $X(f) = \mathcal{FT}[x(t)]$. For the rectangular pulse shape, $BW_{\text{RMS}} = \infty$.

While generally useful only in theoretical analyses, the one important consequence of this definition is that

$$BW_{RMS}T_{RMS} \geq 1/2,$$

where T_{RMS} is the RMS duration defined, in the time domain, analogously to the definition of RMS bandwidth. Our interpretation is that signals of short duration generally occupy large bandwidth and vice versa.

> As is clear, in practice, engineers use multiple definitions of bandwidth. Each has its own purpose; for example, the 3-dB bandwidth is often used to define the passband of a filter, whereas a wireless communications engineer will often use the 99% bandwidth to define the range of operationally useful power. Importantly, these bandwidths are operationally *consistent*, e.g., if we speed up the signal by a timescale factor of two, all measures of bandwidth will double.

1.5 Electric and Magnetic Fields and Power Flux Density

Wireless communication signals are transmitted by means of electromagnetic waves. These electromagnetic waves are composed of electric and magnetic components.

At distances sufficiently far from the transmit antenna, the so-called *far-field*, the electric and magnetic field vectors $\vec{E}$ and $\vec{H}$ form a plane wave, and, as shown in Fig. 1.7, these vectors are perpendicular to each other and to the direction of propagation.

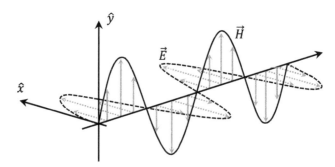

Fig. 1.7 Electromagnetic wave with perpendicular field vectors $\vec{E}$ and $\vec{H}$ in a 3D space

1.5 Electric and Magnetic Fields and Power Flux Density

Measurement equipment can detect the field strength, that is, the magnitude of the electrical or magnetic component of the RF signals. For a wave propagating in free space, field strengths of the electric and magnetic components, $E = |\vec{E}|$ and $H = |\vec{H}|$, are linearly related as $E = Z_0 H$, where Z_0 is an intrinsic impedance of free space measured in ohms (Ω) and is equal to

$$Z_0 = \sqrt{\frac{\mu_0}{\epsilon_0}} \approx 120\pi \qquad (1.30)$$

where $\mu_0 = 1.25663706212 \times 10^{-6}$ H/m is the magnetic permeability in vacuum and $\epsilon_0 = 8.8541878128 \times 10^{-12}$ F/m is the vacuum permittivity. The electric field strength E is measured in V/m (volts per meter).

There is a relationship between the electric field and the signal power, and we will establish that relationship by introducing the notion of power flux density next.

1.5.1 Power Flux Density

The power flux density (PFD) measured in W/m^2 is given by $PFD = |\vec{E} \times \vec{H}|$, which, in the far-field, is just $PFD = EH = E^2/Z_0$ since $\vec{E}$ and $\vec{H}$ are perpendicular. Alternatively, we can define the PFD by considering a point radiation source emitting power P_T uniformly in all directions in a 3D space, at distance d:

$$PFD = \frac{P_T}{4\pi d^2}, \qquad (1.31)$$

i.e., the transmitted power spreads uniformly over the surface of the sphere of radius d. We can use this equation to measure the field strength given a power measurement:

$$PFD = \frac{E^2}{Z_0} = \frac{P_T}{4\pi d^2} \qquad (1.32)$$

$$E^2 = \frac{30 P_T}{d^2} \qquad (1.33)$$

$$E = \frac{\sqrt{30 P_T}}{d} \qquad (1.34)$$

Example 1.6 For health considerations, Health Canada has set recommended exposure limits for RF energy that are considered safe for human exposure [5].

(continued)

Example 1.6 (continued)
For the extracted exposure table below, verify the relationship between the electric field and PFD for the frequency range from 300 to 6000 MHz.

Solution From Table 1.1, we have that the limit on E for the frequency range from 300 to 6000 MHz is expressed as

$$E = 3.142 f^{0.3417} \tag{1.35}$$

and by using the conversion from above, we show that

$$PFD = \frac{E^2}{Z_0} = \frac{3.142^2 f^{2\times 0.3417}}{120\pi} = 0.02619 f^{0.6834}, \tag{1.36}$$

which matches the value in Table 1.1. For further details on health considerations, you can refer to *What we do to keep you safe from radiofrequency EMF exposure on the Understanding Safety Code 6* [6] webpage or to the section on *How does the Government of Canada protection you?* on the ISED[3] Radiofrequency Energy and Safety webpage [7]. ∎

Table 1.1 Reference levels for electric field strength, magnetic field strength, and power density in uncontrolled environments, Safety Code 6, specified by Health Canada. Reproduced here by written permission from Health Canada

Frequency (MHz)	Electric field strength; rms (V/m)	Power density (W/m^2)	Reference period (minutes)
10–20	27.46	2	6
20–48	$58.07/f^{0.25}$	$8.944/f^{0.5}$	6
48–300	22.06	1.291	6
300–6000	$3.142 f^{0.3417}$	$0.02619 f^{0.6834}$	6
6000–5000	61.4	10	6
15000–150000	61.4	10	$616000/f^{1.2}$
150000–300000	$0.158 f^{0.5}$	$6.67 \times 10^{-5} f$	$616000/f^{1.2}$

Frequency f is in MHz

---

[3] Innovation, Science, and Economic Development Canada.

1.5.2 Electric Field in dBµV/m Scale

In the previous subsection, we have seen that the power is proportional to the squared value of the electrical field. Because the dB scale relates to powers, the dB scale of the electrical field (measured in V/m) is

$$E_{|\text{dBV/m}} = 10\log_{10}\left(E^2\right) = 20\log_{10}(E), \tag{1.37}$$

where the reference voltage is 1 V/m.

However, the electrical field values are quite small and are typically expressed in µV/m. Consequently, the dB scale for the electrical field is defined as

$$E_{|\text{dB}\mu\text{V/m}} = 20\log_{10}\left(E_{|\mu\text{V/m}}\right) = 20\log_{10}\left(E_{|\text{V/m}} \times 10^6\right) = 20\log_{10}(E) + 120 \tag{1.38}$$

and is expressed in dBµV/m.

> **Homework** Find the expression for $PFD_{|\text{dB(W/m}^2)}$ as a function of the electric field in $E_{|\text{dB}\mu\text{V/m}}$ scale.

For example, if $E = 0.00224$ V/m, the dBµV/m value is $20\log_{10}(0.00224) + 120 = 67$ dBµV/m.

> **Example 1.7** The Canadian regulation for Advanced Wireless Services Equipment operating in the bands 1710–1780 MHz and 2110–2200 MHz, specified in RSS-139 [8], states that the unwanted emission shall not exceed the limits of -13 dBm/MHz when the measurement is conducted more than 1 MHz away from the band edge. In the radio certification process, this value may be verified by measuring the electrical field at a distance $d = 3$ m from the transmitter. Determine the transmitted power (P_T) if the measured electrical field strength is $E_{|\text{dB}\mu\text{V/m}} = 57.36$ dBµV/m.

Solution To find the transmit power, we first find the electric field strength using (1.38):

$$E_{|dB\mu V/m} = 57.36 \Rightarrow E_{|\mu V/m} = 10^{\frac{E_{|dB\mu V/m}}{20}} = 737.90\,\mu\text{V/m}$$

$$\Rightarrow E_{|V/m} = \frac{E_{|\mu V/m}}{10^6} = 0.0007379\,\text{V/m}$$

$$\tag{1.39}$$

Now, using (1.33)

$$P_T = \frac{(E_{|V/m})^2 d^2}{30} = 1.633 \times 10^{-7} \text{ W}$$

$$\Rightarrow P_{T|\text{dBm}} = 10\log_{10}(P_T \times 1000) = -37.87 \text{ dBm} \quad (1.40)$$

Given that the value of -37.87 dBm/MHz is smaller than the required limit of -13 dBm/MHz (note that the field strength measurements was conducted using the 1 MHz measurement bandwidth), the transmitted power of the radio meets the regulatory requirement by a large margin. ∎

Homework Verify that a transmit power limit of -27 dBm/MHz translates into an electric field strength value of 68.2 dBuV/m at 3 m.

1.6 Cellular Antenna Basics

Antennas are used as a means for transmission or reception of radio waves. In this section, we provide a basic introduction to the antennas used in cellular communication systems and key terminology terms related to dB-scale parameters associated with antennas.

Practical antennas are compared to an idealized theoretical antenna called an isotropic antenna, which serves as a reference point for practical antennas. Isotropic (omnidirectional) antennas radiate uniformly in every direction, thus forming a radiation pattern that is spherical around the source of radiation (transmitter), as shown in Fig. 1.8. In Sect. 1.5.1, we discussed the power flux density of such antennas. If we slice the sphere vertically across the center, we get the *elevation pattern* which, in this case, is a perfect circle. Similarly, a horizontal slice through the center gives an *azimuth pattern* which, in this case of an isotropic antenna, is also circular.

Practical antennas, however, do not radiate equally in every direction. Instead, they focus the transmitted energy in the radio waves in a certain direction. This radiation focusing is accomplished without amplification of the input signal, i.e., focusing the signal in one direction must attenuate the signal in other(s). Compared to the isotropic case, practical antennas have a *gain* in a certain direction. This gain, relative to an isotropic antenna, is usually given in logarithmic scale and is expressed in dBi (decibels relative to isotropic).

In order to understand how the gain is accomplished in practical antennas, we will examine an array of idealized antenna elements and derive the expression for the gain after combining the elements.

1.6 Cellular Antenna Basics

Fig. 1.8 Illustration of an isotropic antenna that radiates equally in 3D space over all angles

Fig. 1.9 Vertical N-element linear array of antenna elements

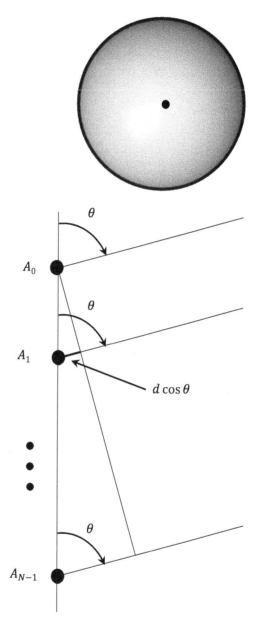

1.6.1 N-Element Linear Array

To consider the simplest case, we consider N isotropic antenna elements, as shown in Fig. 1.9, stacked vertically in a straight line. Assume that the same signal $s(t) = e^{j2\pi f_0 t}$ is transmitted from each antenna elements and received at some distant point.

Denote as D_n, $n = 0, 1, \ldots, N-1$ the distance from each antenna element to this observation point, which means the signal takes time D_n/c to propagate from the transmitter to the observation point (c is the speed of light). The combined received signal can be written as

$$y(t) = \sum_{n=0}^{N-1} s\left(t - \frac{D_n}{c}\right) = \sum_{n=0}^{N-1} e^{j2\pi f_0(t - D_n/c)}$$

$$= e^{j2\pi f_0 t} \sum_{n=0}^{N-1} e^{-jk_0 D_n}$$

$$= e^{j2\pi f_0 t} e^{-jk_0 D_0} \sum_{n=0}^{N-1} e^{-jk_0(D_n - D_0)} \tag{1.41}$$

where $k_0 = 2\pi f_0/c = 2\pi/\lambda_0$ is the called the *wave number*.

From Fig. 1.9, we have $D_n - D_0 = nd\cos(\theta)$. Therefore, the received signal is

$$y(t) = e^{j2\pi f_0 t} e^{-jk_0 D_0} M(\theta) \tag{1.42}$$

where $M(\theta)$ is the complex amplitude gain that is expressed as

$$M(\theta) = 1 + e^{-jk_0 d \cos(\theta)} + \cdots + e^{-jk_0(N-1)d\cos(\theta)} \tag{1.43}$$

We can simplify this expression as the following:

$$M(\theta) = \sum_{n=0}^{N-1} e^{-jk_0 nd\cos(\theta)} \tag{1.44}$$

$$= e^{-jk_0 d \frac{N-1}{2}\cos(\theta)} \frac{\sin\left(k_0 \frac{N}{2} d\cos(\theta)\right)}{\sin\left(k_0 d \frac{1}{2}\cos(\theta)\right)} \tag{1.45}$$

If we ignore the complex phase rotation part, the real part is known as the *array factor* or *radiation pattern*. Therefore, for the array in Fig. 1.9, the array factor is

$$AF(\theta) = \frac{\sin\left(k_0 \frac{N}{2} d\cos(\theta)\right)}{\sin\left(k_0 \frac{d}{2}\cos(\theta)\right)}. \tag{1.46}$$

Figure 1.10 below illustrates this array factor, in two ways, for the case of $N = 12$ antennas and the inter-element distance is $d = \lambda_0/2$. From the figure, we see that the maximum received power, called the *main beam*, is in the direction of $\theta = 90°$, called the boresight direction. Since the main beam is in the boresight direction, such an arrangement is known as a broadside array. The figure also shows that no

1.6 Cellular Antenna Basics

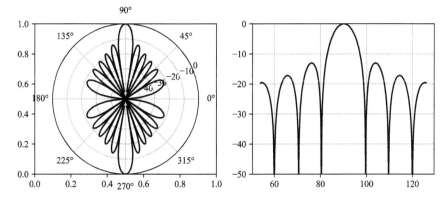

Fig. 1.10 Array factor for $N = 12$ and $d = \lambda_0/2$, in polar and log-linear scale

signal is received in some directions—these are the *nulls* in the array factor. There are also local peaks, but not as strong as the main peak; these are the *sidelobes*.

Due to the periodicity of the $\sin(\cdot)$ function, the radiation pattern can have multiple peaks of equal strength. This can happen if the distance between the antenna elements is larger than or equal to λ_0. An example where $d = 2\lambda_0$ is shown below; the multiple peaks are known as *grating lobes* (Fig. 1.11).

Beam Steering The discussion so far has assumed we transmitted the same signal from all N elements in the array. If, instead, we applied a linear shift across the array, i.e., if the transmit signal from the nth element were set to $s_n(t) = e^{j(2\pi f_0 t + n\beta)} = e^{jn\beta} e^{j2\pi f_0 t}$, the expression for the array factor would change to

$$AF(\theta) = \frac{\sin\left(\frac{N}{2}(k_0 d \cos(\theta) - \beta)\right)}{\sin\left(\frac{1}{2}(k_0 d \cos(\theta) - \beta)\right)}. \tag{1.47}$$

Crucially, now, the peak of the array factor shifts to θ_0 where the phase shift (β) is related to θ_0 by

$$\beta = k_0 d \cos(\theta_0) = \frac{2\pi}{\lambda_0} d \cos(\theta_0). \tag{1.48}$$

This result implies that we can direct the transmitted energy toward any desired direction by choosing appropriate phase shifts at each transmit element.

We can illustrate this concept using the following example that often happens in real-world deployments (Fig. 1.12).

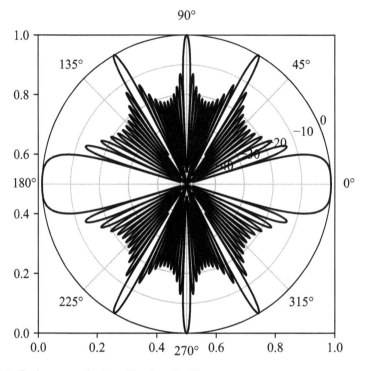

Fig. 1.11 Grating pattern for $N = 12$, when $d = 2\lambda_0$

Example 1.8 The base station antenna is placed vertically at the 50 m of height, and the user terminal (UT) is at 1.5 m of height, located at 300 m of distance. Find a phase shift β such that the maximum radiation pattern occurs at the given UT distance.

Solution The angle $\theta_0 = \pi/2 + \alpha$, where

$$\alpha = \arctan\left(\frac{50 - 1.5}{300}\right) = 0.16 \text{ radians}. \tag{1.49}$$

Therefore, we have

$$\theta_0 = \frac{\pi}{2} + 0.16 = 99.18° \tag{1.50}$$

$$\beta = \frac{2\pi}{\lambda_0} \frac{\lambda_0}{2} \cos(\theta_0) = -0.16\pi = -28.8°. \tag{1.51}$$

1.6 Cellular Antenna Basics

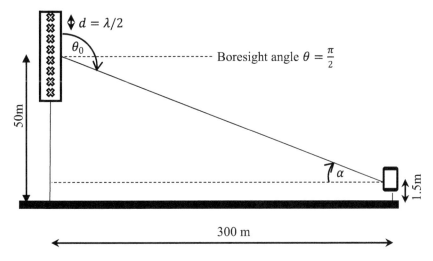

Fig. 1.12 Example of electrical downtilt of the base station antenna

At the base station, after applying a progressive phase shift of β between successive antenna elements, the radiation pattern has maximal gain at the desired $\theta_0 = 99.18°$. ∎

The adjustment of phase between individual elements of the antenna can be done mechanically through rods or screws, usually located at the bottom of the antenna, which, when moved, change the phases of individual antenna elements, thus resulting in a beam downtilt. Modern antenna arrays also use electronic steering, providing greater agility in beam steering. The peak array gain at a given distance away from the base station can also be implemented by physically downtilting the antenna by α, so that the receiver is at the boresight angle. In practice, a combination of electrical and physical downtilt (or uptilt) is used.

1.6.2 Important Antenna Parameters

1.6.2.1 Directivity

The directivity of an antenna measures its relative preference for a specific direction. The directivity is defined as the maximum radiation intensity (power gain) divided by the average radiation intensity:

$$D_0 = \frac{F^{\max}(\theta, \varphi)}{\frac{1}{4\pi} \int_0^{2\pi} \int_0^{\pi} F(\theta, \varphi) \sin(\theta) d\theta d\varphi} \tag{1.52}$$

where θ is the elevation angle, φ is the azimuth angle, and the radiation intensity is

$$F(\theta, \varphi) = |M(\theta, \varphi)|^2.$$

The denominator in (1.52) is proportional to what an isotropic antenna would transmit. An isotropic antenna, therefore, has a directivity of 1. Let us illustrate directivity calculation with a few examples. These are important results as they show what can we expect in terms of peak antenna gain for a given number of antenna elements.

Example 1.9 A short Hertzian dipole has the antenna radiation pattern $M(\theta, \varphi) = \sin(\theta)$ and is uniform over azimuth φ. Determine the value for directivity.

Solution The radiation intensity is

$$F(\theta, \varphi) = |M(\theta, \varphi)|^2 = \sin^2(\theta) \tag{1.53}$$

with the maximum radiation intensity of $F_{\max} = 1$, and the directivity is

$$D_0 = \frac{4\pi}{2\pi \int_0^\pi \sin^3(\theta) d\theta} = \frac{4\pi}{2\pi \int_0^\pi \left[\frac{3}{4}\sin(\theta) - \frac{1}{4}\sin(3\theta)\right] d\theta}$$

$$= \frac{2}{\frac{4}{3}} = 1.5. \tag{1.54}$$

∎

Example 1.10 A $\lambda/2$ dipole has the antenna radiation pattern:

$$M(\theta, \varphi) = \frac{\cos\left(\frac{\pi}{2}\cos(\theta)\right)}{\sin(\theta)}$$

and is uniform over azimuth φ. Determine the value for directivity.

Solution The radiation intensity is

$$F(\theta, \varphi) = |M(\theta, \phi)|^2 = \frac{\cos^2\left(\frac{\pi}{2}\cos(\theta)\right)}{\sin^2(\theta)} \tag{1.55}$$

1.6 Cellular Antenna Basics

with the maximum radiation intensity of $F_{\max} = 1$, and the directivity is

$$D_0 = \frac{4\pi}{2\pi \int_0^\pi \frac{\cos^2\left(\frac{\pi}{2}\cos(\theta)\right)}{\sin^2(\theta)} \sin(\theta) d\theta} = 1.64, \quad (1.56)$$

where the integral has been solved numerically. In the dB scale, the directivity is $10\log_{10}(1.64) = 2.15$ dBi. This is the gain a $\lambda/2$ dipole has over an ideal isotropic radiator with the same input power. ∎

Example 1.11 A linear antenna array with N elements and inter-element distance d has the antenna radiation pattern:

$$|M(\theta, \varphi)| = \left| \frac{\sin\left(k_0 \frac{N}{2} d \cos(\theta)\right)}{\sin\left(\frac{1}{2} k_0 d \cos(\theta)\right)} \right|$$

and is uniform over azimuth φ. Determine the value for directivity if $d = \lambda/2$.

Solution The radiation intensity is

$$F(\theta, \varphi) = |M(\theta, \phi)|^2 = \left| \frac{\sin\left(\frac{N}{2} k_0 d \cos(\theta)\right)}{\sin\left(\frac{1}{2} k_0 d \cos(\theta)\right)} \right|^2$$

with the maximum radiation intensity of $F_{\max} = N^2$ if we do not normalize the array factor.[4] The directivity is (see the problem set for the derivation)

$$D_0 = N. \quad (1.57)$$

∎

So far, we have considered the directivity of a single element or an array of N isotropic elements. If we create an array of N non-isotropic elements, each with an *element factor* of $M_{\text{elem}}(\theta, \varphi)$, the overall transmit pattern is given by the product of the array factor in (1.46) and the element factor, i.e., $M(\theta, \varphi) = M_{\text{elem}}(\theta, \varphi) M_{\text{array}}(\theta, \varphi)$ and the overall radiation pattern, as before, $F(\theta, \varphi) = |M(\theta, \varphi)|^2$. This overall radiation pattern determines the directivity, as in (1.52). For

---

[4] A common practice is to apply a $1/N$ normalization, in which case the directivity result does not change, but with normalization $F_{\max} = 1$..

example, an antenna system with $N = 64$ antenna elements and with per element antenna gain of $G_{elem} = 7$ dBi would have a total directivity in dB scale:

$$D_{0|dBi} = G_{elem} + 10 \times \log_{10}(N) = 25 \text{ dBi}. \tag{1.58}$$

1.6.2.2 Antenna Gain and Efficiency

In practice, we use the term directivity interchangeably with the term antenna gain. However, the actual antenna gain is somewhat reduced relative to directivity due to implementation losses. If P_{IN} is the power input into the antenna, and P_{rad} is the total radiated power, then the antenna gain is

$$G = \eta D_0 \tag{1.59}$$

where η is an efficiency measure that is the measure that accounts for antenna losses:

$$\eta = \frac{P_{rad}}{P_{IN}} \tag{1.60}$$

In a wireless network, at base stations, the efficiency η is usually close to 1; however, at the device, it can be much lower due to the reduced antenna size (in relation to the wavelength).

In the previous section, we defined the power flux density (see (1.32)), which was appropriate for an isotropic radiation. For a transmit antenna with a gain of G_T, the power flux density is given by

$$PFD = \frac{E^2}{Z_0} = \frac{G_T P_T}{4\pi d^2}, \tag{1.61}$$

and the relationship between the field strength and PFD, as stated in (1.33) is now given by

$$E = \frac{\sqrt{30 P_T G_T}}{d} \tag{1.62}$$

We emphasize that, given the definition of gain, this is the PFD and electric field strength in the direction of the maximum power transmission.

1.6 Cellular Antenna Basics

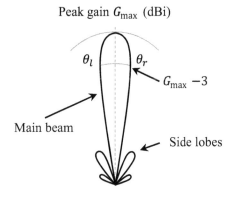

Fig. 1.13 Illustration of the main beam and symmetric half-power points to the left (θ_l) and to the right (θ_r) that determine the half-power beamwidth

This relationship between the electrical field strength, the transmit power, and transmitter's gain is useful in practice, where by measuring the electrical field with an instrument at some distance d from the transmitter, we can determine the transmitted power:

$$P_T = \frac{E^2 d^2}{30 G_T}.$$

This type of measurement is often used in radio or device certification if the transmitter does not have an accessible port (e.g., systems with radio and antenna integrated do not have an accessible measurement port).

1.6.2.3 Half-Power Beamwidth

The half-power beamwidth (HPBW) is similar to the 3-dB bandwidth stated earlier for signals. The HPBW of a radiation pattern is determined by finding the half powers located symmetrically with respect to the maximum direction or main beam (Fig. 1.13).

Example 1.12 A linear antenna array has a normalized antenna radiation pattern:

$$|M(\theta, \varphi)| = \left| \frac{\sin\left(k_0 \frac{N}{2} d \cos(\theta)\right)}{N \sin\left(\frac{1}{2} k_0 d \cos(\theta)\right)} \right|$$

(continued)

Example 1.12 (continued)
and is uniform over φ. Determine the HPBW for the angle θ when $N = 12$ and $d = \lambda/2$.

Solution The radiation intensity is

$$F(\theta, \varphi) = |M(\theta, \phi)|^2 = \left| \frac{\sin\left(\frac{N}{2} k_0 d \cos(\theta)\right)}{N \sin\left(\frac{1}{2} k_0 d \cos(\theta)\right)} \right|^2 \quad (1.63)$$

and the half-power points can be found numerically:

$$F(\theta_r) = 0.5 \quad (1.64)$$

$$\theta_r = 1.497 \text{ rad} = 85.75° \quad (1.65)$$

and the HPBW angle is $\theta_{3dB} = 2(\pi/2 - \theta_r) = 8.49°$ due to symmetry.

Alternatively, we can approximate the term in the denominator by the linear term:

$$\sin\left(\frac{1}{2} k_0 d \cos(\theta)\right) \approx \frac{1}{2} k_0 d \cos(\theta) \quad (1.66)$$

because the argument of sin() function in (1.66) is small for angles close to $\theta = \pi/2$, which is valid for relatively long arrays. Consequently, we can write

$$\left| \frac{\sin\left(\frac{N}{2} k_0 d \cos(\theta)\right)}{\frac{N}{2} k_0 d \cos(\theta)} \right|^2 = \frac{1}{2} \quad (1.67)$$

$$\frac{N}{2} k_0 d \cos(\theta) = \pm 1.3915 \quad (1.68)$$

$$\theta_{r,l} = \cos^{-1}\left(\pm \frac{2.783}{N} \frac{\lambda}{2\pi d}\right) \quad (1.69)$$

where θ_l and θ_r are half-power left and right points, respectively. The HPBW then becomes

$$\theta_{3dB} = |\theta_l - \theta_r| \quad (1.70)$$

$$= \left| \cos^{-1}\left(-\frac{2.783}{N} \frac{\lambda}{2\pi d}\right) - \cos^{-1}\left(\frac{2.783}{N} \frac{\lambda}{2\pi d}\right) \right| \quad (1.71)$$

$$\approx \left| \frac{\pi}{2} + \frac{2.783}{N} \frac{\lambda}{2\pi d} - \left(\frac{\pi}{2} - \frac{2.783}{N} \frac{\lambda}{2\pi d}\right) \right|$$

1.6 Cellular Antenna Basics

$$= \frac{2.783}{N} \frac{\lambda}{\pi d} = \frac{0.886\lambda}{Nd}$$

where we approximated the $\cos^{-1}(x)$ function with its linear term in the Taylor series expansion around $x = 0$. For $d = \lambda/2$, the HPBW simplifies to

$$\theta_{3dB} = \frac{1.772}{N} \quad (1.72)$$

and in our example, it becomes $\theta_{3dB} = 8.46°$, which is almost the same value as the numerically calculated value without approximation. This illustrates that the term *relatively long array* applies even for $N = 12$.

We can also calculate the location of the first nulls immediately next to the main lobe:

$$F(\theta) = 0$$

$$\frac{N}{2} k_0 d \cos(\theta) = \pm \pi$$

$$\theta = \cos^{-1}\left(\frac{1}{6}\right) = \pi/2 \pm 0.167448 \text{ rad}$$

and the first null-to-null beamwidth is $2 \times 0.167448 = 0.335$ rad or $19.2°$. ∎

If we consider the vertical beamwidth, as illustrated in Fig. 1.14, it is clear that points closer than the inner radius and points further away than the outer radius will have lower received powers relative to the main beam direction. In practice, areas well inside the inner radius are known as dead zones as there is effectively no signal reception.

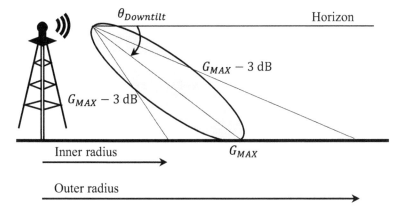

Fig. 1.14 Illustration of typical sector design with dowtilted antenna. Areas within the inner radius, especially closer to the base station, have very low signal level

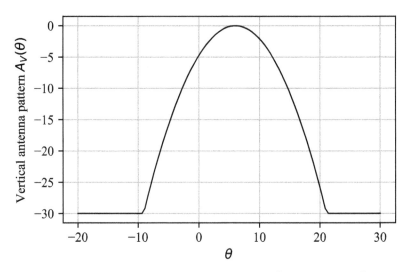

Fig. 1.15 Illustration of a vertical antenna pattern for $\theta_{\text{Dtilt}} = 7^0$ and, $\theta_{\text{3dB}} = 9.55^0$, based on a 3GPP model

In 3GPP studies, an antenna pattern used in system level simulations has a simple analytic form. It combines 3-dB beamwidth (HPWB) and downtilt and is specified in TS 38.901 [9], in International Telecommunication Union, Radiocommunications sector (ITU-R) report M.2412-0 [10] and several previous versions of these documents. The horizontal antenna pattern is given as

$$A_H(\varphi) = -\min\left[12\left(\frac{\varphi}{\varphi_{\text{3dB}}}\right)^2, SLA\right], \quad (1.73)$$

where SLA is the maximum side lobe attenuation, φ_{3dB} is the horizontal HPBW, and the vertical pattern is

$$A_V(\theta) = -\min\left[12\left(\frac{\theta - \theta_{\text{Dtilt}}}{\theta_{\text{3dB}}}\right)^2, SLA\right], \quad (1.74)$$

where $\theta_{\text{Dtilt}} = 0$ points toward horizon. Both $A_H(\varphi)$ and $A_V(\theta)$ are in dB. An example of a vertical antenna pattern is shown in Fig. 1.15.

Combined, they provide a 3D radiation pattern, where we can use the gain $G(\theta, \varphi)$ notation instead of radiation intensity $F(\theta, \varphi)$:

$$G(\theta, \varphi) = G_{\max} - \min\{-[A_V(\theta) + A_H(\varphi)], SLA\} \quad (1.75)$$

where all terms are in the dB scale. For example, in [10], the value for evaluation is $SLA = 30\,\text{dB}$.

1.6 Cellular Antenna Basics

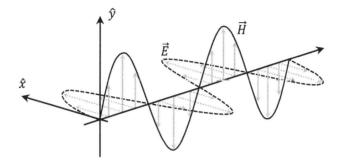

Fig. 1.16 Illustration of a linearly polarized electric field

1.6.2.4 Antenna Polarization

Figure 1.16 illustrates an electric field that changes over time t, where the direction of the propagation of the electromagnetic wave is along the z-axis.

The electrical field, in the horizontal plane, is given by

$$\vec{E}_H(z,t) = E_0 e^{j2\pi f_c t} e^{-jk_0 z} \hat{x}, \tag{1.76}$$

i.e., the electric field is a vector in the x-direction with magnitude given by (1.76). The magnetic field is in the direction of $\hat{y}$, but, in wireless communications, we normally focus our discussion on the E-field.

When the electric field lies on a single axis, it is said to have *linear polarization* (as shown in Fig. 1.16). If the receive antenna is oriented the same way as the transmit orientation ($\hat{x}$ in our example), it will pick up the signal due to the current induced on the antenna; the signal is then fed into a receiver. On the other hand, if the antenna is orthogonal to the direction of polarization, it will not pick up any signal.

Let us now generate another signal that is vertically oriented:

$$\vec{E}_V(z,t) = E_0 e^{j2\pi f_c t} e^{-jk_0 z} \hat{y} \tag{1.77}$$

The two waves combine into a single composite signal:

$$\vec{E}_{VH}(z,t) = \vec{E}_V(z,t) + \vec{E}_H(z,t) = E_0 e^{j2\pi f_c t} e^{-jk_0 z} \hat{x} + E_0 e^{j2\pi f_c t} e^{-jk_0 z} \hat{y} \tag{1.78}$$

This way, we can transmit horizontally and vertically polarized signals (H and V).

Importantly, since an $\hat{x}$-oriented ($\hat{y}$-oriented) receive antenna picks up signals from the $\hat{x}$ ($\hat{y}$) component, but not the $\hat{y}$ ($\hat{x}$) component, one could transmit two different message signals, $m_x(t)$ and $m_y(t)$:

$$\vec{E}_{VH}(z,t) = E_x m_x(t) e^{j2\pi f_c t} e^{-jk_0 z} \hat{x} + E_y m_y(t) e^{j2\pi f_c t} e^{-jk_0 z} \hat{y} \quad (1.79)$$

A receive antenna with an $\hat{x}$ orientation would receive only $m_x(t)$, while one oriented in the $\hat{y}$ direction would only receive $m_y(t)$.

In practice, it is hard to align transmit and receive antennas. However, we can exploit the fact that two perpendicular receive antennas will receive different linear combinations of the two messages; some simple linear processing will then disentangle the two messages.

Circular Polarization Now, let us consider a case where the second signal is transmitted with a 90° phase shift:

$$\vec{E}_{circ}(z,t) = E_0 e^{j2\pi f_c t} e^{-jk_0 z} \hat{x} + E_0 e^{j2\pi f_c t} e^{-jk_0 z} e^{j\frac{\pi}{2}} \hat{y} \quad (1.80)$$

This way, the direction of the electrical field changes over time—the tip of the E-field array rotates over a circle (an ellipse if the amplitudes of the two components are unequal) on the $\hat{x} - \hat{y}$ plane. We call this a circularly polarized signal.

Cross-Polarization Discrimination As we will see in the next chapter, wireless signals almost always propagate in a multipath environment in which a transmitted signal gets reflected and refracted before arrival, from multiple directions, at the receiver. In such an environment, a portion of energy leaks from one polarization to another. The measure of this leakage is called cross-polarization discrimination (XPD) and is defined as

$$XPD_V = \frac{\mathbb{E}[|h_{V,V}|^2]}{\mathbb{E}[|h_{V,H}|^2]}, \quad (1.81)$$

for the vertical plane, and

$$XPD_H = \frac{\mathbb{E}[|h_{H,H}|^2]}{\mathbb{E}[|h_{H,V}|^2]}, \quad (1.82)$$

for the horizontal plane. Here, $\mathbb{E}[|h_{V,V}|^2]$ denotes the average power in the vertically polarized wave, while $\mathbb{E}[|h_{V,H}|^2]$ is the leakage from the vertical to horizontal polarization. Typically, these two values are different.

1.6 Cellular Antenna Basics

> In cellular systems, typical XPD (the term CPD is also used) values are around 20 dB or better, which limits the energy leakage from one polarization plane onto another to less than 1%.

1.6.3 EIRP and ERP

The product of the transmit power P_T and the antenna gain G_T is known as the effective isotropic radiated power (EIRP):

$$EIRP = P_T G_T = \frac{E^2 d^2}{30}. \tag{1.83}$$

As the name suggests, in the direction of maximum transmitted power, the transmitter effectively behaves as if it were transmitting the power of $P_T G_T$ watts. Since an isotropic antenna has unit gain, in the dB scale, the antenna gain is the difference between the EIRP and the transmitted power and is expressed in dBi (dB relative to an isotropic antenna) scale:

$$G_{|dBi} = EIRP_{|dBX} - P_{T|dBX}, \tag{1.84}$$

where X could be W or m (the reference could be watts or milliwatts). This represents an antenna gain relative to an ideal (theoretical) isotropic antenna.

A closely related term to the EIRP is the effective radiated power (ERP), where the antenna gain is expressed relative to an ideal half-wave dipole antenna. An ideal half-wave dipole antenna gain has gain $G_T = 1.64$, and the ERP, in linear scale, is related to EIRP as

$$ERP = \frac{EIRP}{1.64}. \tag{1.85}$$

The antenna gain, relative to half-wave dipole, is

$$G_{|dBd} = ERP_{|dB} - P_{T|dB} = EIRP_{|dB} - 10\log_{10}(1.64) - P_{T|dB} = G_{|dBi} - 2.15. \tag{1.86}$$

Practical antennas, especially those used at base stations, have much higher gains, in the range of 15–25 dBi. An illustration of various antenna gains is shown below (Fig. 1.17).

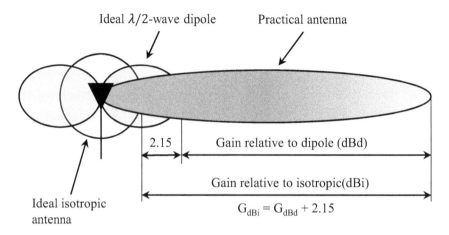

Fig. 1.17 Illustration of antenna radiation patterns for an ideal isotropic radiator, $\lambda/2$ dipole, and a typical practical antenna used in base stations

1.7 Radiofrequency (RF) Modulation and Digital Communication

Modulation is the process of mapping data bits to symbols fit for transmission. Before modulation, the raw data bits are protected using an error control code. Error control codes provide a degree of protection from reception errors to the data bits. To do so, the coding scheme maps n data bits to k coded bits ($k > n$), i.e., the coding rate

$$R = \frac{n}{k}$$

is below one and represents an overhead to the transmission system. While error control coding is outside the scope of this text, it is worth emphasizing that every communication system uses error control coding. The benefits are worth the price of the associated overhead. Chapter 8 provides a brief overview of the use of coding in wireless networks. Here, we present a brief review of digital communication and modulation concepts. The reader is referred to texts such as [4, 11] for a more thorough development.

Although communication signals are real, mathematically it is convenient to represent them as complex signals. Specifically, these signals comprise two message signals, one representing the real part of the signal and the other the imaginary component of the overall complex message signal:

$$m(t) = m_I(t) + jm_Q(t) \text{ (baseband signal)} \quad (1.87)$$

1.7 Radiofrequency (RF) Modulation and Digital Communication

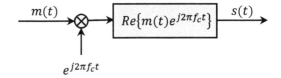

Fig. 1.18 Illustration of quadrature amplitude modulation using a carrier frequency f_c

that carries some information. To transmit such a signal via electromagnetic (radio) waves, we typically apply some form of quadrature amplitude modulation (a process of frequency translation from baseband to some RF frequency range), as illustrated above (Fig. 1.18).

Mathematically, the propagating RF signal can be expressed as

$$s(t) = \mathcal{R}\left\{m(t)e^{j2\pi f_c t}\right\} = m_I(t)\cos(2\pi f_c t) - m_Q(t)\sin(2\pi f_c t), \quad (1.88)$$

where f_c is a carrier frequency and $\mathcal{R}\{\cdot\}$ denotes the real part of a complex number.

This expression is most relevant for communications using digital modulation; to create such a signal, long sequences of bits are mapped to appropriately long sequences of *symbols*. The most commonly used mappings, also called digital modulation schemes, are binary or quadrature phase shift keying (BPSK or QPSK) and quadrature amplitude modulation (or M-QAM where $\log_2(M)$ bits are mapped to one of M symbols). The mapping from bits to symbols, represented on the complex plane, is called the *constellation diagram*. In 4G and 5G systems, possible values are $M = 16, 64, 256, 1024$.

In each of these cases, the message signal can be written in pulse amplitude modulated (PAM) form:

$$m(t) = \sum_k a_k g(t - kT_s),$$

where a_k is the symbol in the k-th symbol period and T_s denotes the symbol period. If the symbols can take M possible values, we have $N_{bits} = \log_2(M)$ per symbol interval, and $1/T_s$ is the symbol rate.

> The pulse shape $g(t)$ plays a crucial role in wireless networks: since, in practice, $1/T_s \ll f_c$, this pulse shape determines the bandwidth used. Conceptually, the easiest pulse shape to consider is the rect($\cdot$) signal of duration T_s, but as we have seen, this implies a very large occupied bandwidth.
>
> Since duration and bandwidth are inversely proportional to each other, ideally, the bandwidth would be approximately $1/T_s$. Mathematically, one can show that the pulse $g(t) = \text{sinc}(t/T_s)$ would satisfy this relation, it raises practical difficulties in terms of large delay. A popular balance between
>
> (continued)

bandwidth, while enabling practical delay, is the *raised cosine pulse* given by

$$g(t) = \frac{\cos(\pi \alpha T_s)}{1 - (2\alpha t/T_s)^2} \operatorname{sinc}\left(\frac{t}{T_s}\right), \qquad (1.89)$$

which results in an effective bandwidth of $(1+\alpha)/T_s$ for $s(t)$. The parameter, α, is referred to as the fractional excess bandwidth.

1.7.1 BPSK Modulation

In this simple form of digital modulation, the message signal $m(t)$ consists of a real sequence of binary symbols, with a pulse shape $g(t)$ applied to each of symbols:

$$m_I(t) = \sum_k a_k g(t - kT_b), \quad a_k \in \{\pm 1\}, \qquad (1.90)$$

$$m_Q(t) = 0, \qquad (1.91)$$

where T_b is a binary symbol (bit) duration. Binary symbols a_k are usually obtained from a binary stream of bits b_k using the following mapping:

b_k	a_k
0	1
1	-1

which can be also written as $a_k = 1 - 2b_k$. The RF version of the BPSK signal is

$$s_{BPSK}(t) = \mathcal{R}\left\{m(t)e^{j2\pi f_c t}\right\} = \sum_k a_k g(t - kT_b) \cos(2\pi f_c t) \qquad (1.92)$$

An illustration of three binary symbols transmitted using BPSK is shown in Fig. 1.19.

A more commonly used visualization method for displaying digitally modulated signals is the so-called constellation diagram shown in Fig. 1.20, where the amplitude and phase of the signal are shown only. The BPSK signal over a duration of one symbol T_b, assuming a rectangular pulse shape for $g(t)$, can be written as

$$s_{BPSK}(t) = \sqrt{\frac{2E_b}{T_b}} \cos(2\pi f_c t + \phi_k) \qquad (1.93)$$

1.7 Radiofrequency (RF) Modulation and Digital Communication

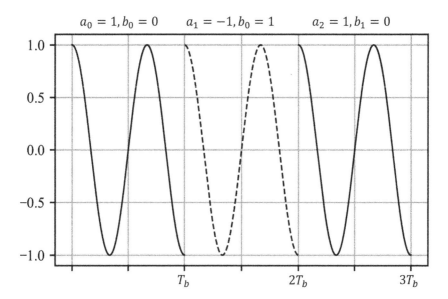

Fig. 1.19 Example of 3 bits represented using BPSK modulation

Fig. 1.20 BPSK constellation diagram

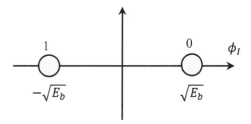

where[5] $\phi_k = b_k \pi$, and the constellation diagram is as follows:

1.7.2 QPSK

While the BPSK approach uses only one of the two terms in the QAM framework, it is possible to transmit simultaneously 2 bits in a two-dimensional signal space using a BPSK-like approach. In QPSK, we have

$$m_I(t) = \sum_k a_k g(t - kT_s), \quad a_k \in \{\pm 1\}, \tag{1.94}$$

---

[5] Note that this representation is consistent with the alternative above, i.e., $a_k = 1 - 2b_k$.

Fig. 1.21 QPSK constellation diagram

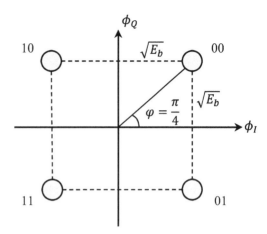

$$m_Q(t) = \sum_k b_k g(t - kT_s), \quad b_k \in \{\pm 1\} \tag{1.95}$$

where $T_s = T_b$ is the symbol duration. The QPSK signal can therefore be expressed as

If the chosen pulse shape $g(t)$ has the energy of E_b, the signal has four possible constellation points, each at the $\sqrt{2E_b} = \sqrt{E_s}$ distance from the signal space origin, where E_s is the symbol energy (kept the same for all four constellation points). Figure 1.21 illustrates the symbol constellation showing the two bits per constellation point. The QPSK signal can also be expressed as a signal with constant envelope and four different phase values φ_k:

$$s_{QPSK}(t) = \sum_k \sqrt{a_k^2 + b_k^2}\, g(t - kT_s) \cos(2\pi f_c t + \varphi_k), \tag{1.96}$$

where $\varphi_k = \arctan\left(\frac{b_k}{a_k}\right) \in \{\pm \pi/4, \pm 3\pi/4\}$ is the information-dependent signal phase (and hence the term "phase shift keying").

1.7.3 Higher-Order Modulation Schemes

M-ary Quadrature Amplitude Modulation (QAM) has the same form as QPSK, except that symbols a_k and b_k are no longer binary:

$$m_I(t) = \sum_k a_k g(t - kT_s), \quad a_k \in \{\pm 1, \pm 3, \ldots, \pm(L-1)\}, \tag{1.97}$$

Fig. 1.22 16-QAM constellation points and binary sequence to symbols mapping

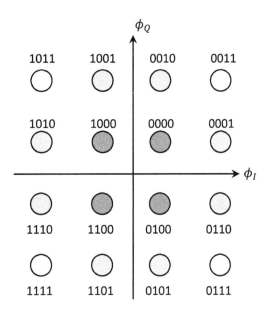

$$m_Q(t) = \sum_k b_k g(t - kT_s), \quad b_k \in \{\pm 1, \pm 3, \ldots, \pm(L-1)\}. \tag{1.98}$$

The QAM modulation schemes are designed such that $M = L^2$ (e.g., $M = 16 \Longrightarrow L = 4$), where M is the number of constellation points.

Figure 1.22 illustrates the constellation diagram for the 16-QAM modulation. It has $M = 16$ constellation points (see below), and the amplitudes are selected from $a_k, b_k \in A \times \{\pm 1, \pm 3\}$. There are three distinct signal magnitudes and 12 different phases as indicated by three different shades of gray.

1.8 Cellular Networks

In order to cover a larger geographic area and to provide service to as many users as needed, a partition of the coverage area into smaller, hexagonal cells is typically used to visually represent contiguous coverage. This type of coverage is associated with the so-called macrocell, generally a large geographical area. A single serving node, a base station, serves users in each cell. Other smaller types of cells are also used: microcells, with base stations typically deployed on light poles or sides of building walls along street corridors, covering small areas, and picocells, typically deployed indoors. The serving node in a picocell may be called an access point. Rather than by the size of the hexagon, a more appropriate classification of base stations is based on their transmit powers, as shown in Table 1.2 (Fig. 1.23).

Table 1.2 Base station classification defined by 3GPP, TS 38.104 [12]

BS class	Max transmit power
Wide area BS	No upper limit specified by 3GPP
Medium-range BS	≤ 38 dBm
Local area BS	≤ 24 dBm

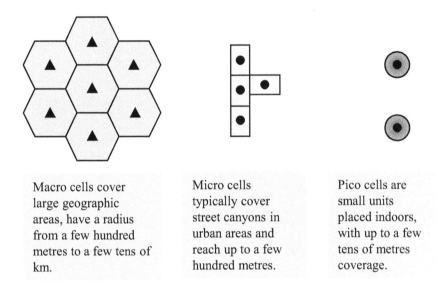

Macro cells cover large geographic areas, have a radius from a few hundred metres to a few tens of km.

Micro cells typically cover street canyons in urban areas and reach up to a few hundred metres.

Pico cells are small units placed indoors, with up to a few tens of metres coverage.

Fig. 1.23 Illustration of macro, micro, and pico deployment use cases

While the 3GPP standardization group does not define the power limit for macro base stations, the limit is set by national regulators. For example, the Canadian regulator, ISED, specifies the maximum limit for base stations operating in the frequency range from 3450 to 3900 MHz in radio specification standard RSS-192 and allows up to 40 dBm/MHz of the total transmit power before the antenna gain is applied. Medium-range base stations, used in microcells, typically transmit up to 40 watts (includes built-in antenna gain), whereas picocell (local area base stations) transmit powers are < 1 watts. These transmit powers will be discussed again, in some detail, in Chap. 4.

To further increase availability of resources to multiple users, cell sectorization is used, where a coverage area (e.g., hexagon) is divided in three or six equal areas. Figure 1.24 shows three sectors, oriented 120 degrees from each other. While the actual coverage shape of a cell is not hexagonal (ideally, it is a circle or an arc in case of sectorization, but local propagation changes the shape), a hexagonal shape has been adopted to illustrate cells (the terms sector and cell are used interchangeably) because they fill up the entire space leaving no holes. In reality, sector coverage areas overlap.

1.9 Multiple Access Schemes

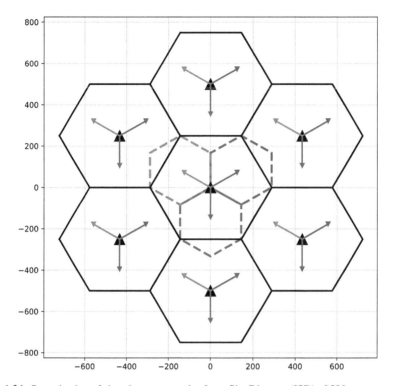

Fig. 1.24 Sectorization of sites that are spaced at Inter-Site Distance (ISD) of 500 m

1.9 Multiple Access Schemes

The wireless communication infrastructure is designed to service many users simultaneously. The available radio resources must, therefore, be shared between different users, allowing for *multiple access*:[6]

FDMA   frequency domain (Frequency Division Multiple Access)
TDMA   time domain (Time Division Multiple Access)
CDMA   code space domain (Code Division Multiple Access)

as illustrated below (Fig. 1.25).

---

[6] In the previous section, we described cells and sectorization to enable simultaneous service to multiple users—a form of spatial multiple access.

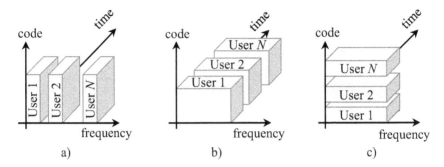

Fig. 1.25 Block diagram of different multiple access schemes. Combinations of these are also possible. (**a**) Frequency Division Multiple Access (FDMA). (**b**) Time Division Multiple Access (TDMA). (**c**) Code Division Multiple Access (CDMA)

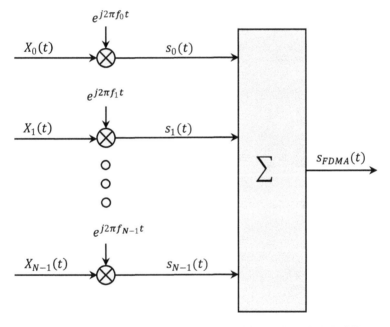

Fig. 1.26 Analog FDMA transmitter with N parallel signals/channels carried via different carrier frequencies

1.9.1 Frequency Division Multiple Access (FDMA)

In frequency division multiple access schemes, N data streams, belonging to N different users, are transmitted at different carrier frequencies, as illustrated above (Fig. 1.26).

The combined transmit signal can be expressed as

1.9 Multiple Access Schemes

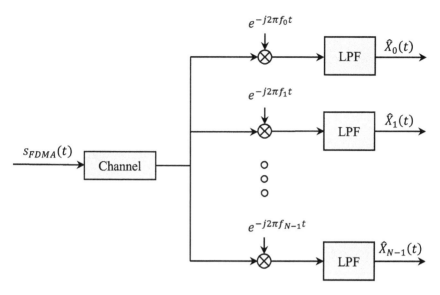

Fig. 1.27 Analog FDMA receiver with N parallel signals/channels carried via different carrier frequencies

$$s_{FDMA}(t) = \sum_{k=0}^{N-1} X_k(t) e^{j2\pi f_k t} \qquad (1.99)$$

Typically, the spacing between the carriers is kept constant at $\Delta f = f_2 - f_1 = f_3 - f_2 = \ldots = f_{N-1} - f_{N-2}$. The receiver structure for this type of transmission is shown in Fig. 1.27.

This type of multiple access scheme was typical for the first generation of cellular systems and is similar to how the analog amplitude modulation/frequency modulation (AM/FM) radio station operate.

> 4G and 5G systems also use a form of frequency division multiple access, but a far more efficient one called *orthogonal frequency division multiplexing/multiple access* (OFDM/OFDMA). This scheme will be covered in some detail in Chap. 6.

1.9.2 Time Division Multiple Access (TDMA)

In TDMA, the data stream is divided into frames, which are further subdivided into time slots. Each user is allocated one slot. The Global System for Mobile

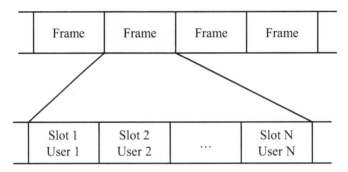

Fig. 1.28 TDMA frame and slot structure illustration

Communications (GSM), a second-generation cellular technology, is TDMA based, where there are eight slots shared dynamically among different users. The carrier frequency is common for all users. Slot assignment on the transmit and receive side is synchronized to avoid inter-user interference and can be dynamically assigned for increased capacity (Fig. 1.28).

1.9.3 Code Division Multiple Access (CDMA)

CDMA was introduced in the second generation of cellular systems. In CDMA-based schemes, N data streams, belonging to N different users, are multiplied by different mutually orthogonal codes $c_k(t)$, where $k = 0, 1, \ldots, N-1$, as illustrated in Figs. 1.29 and 1.30; this is a simplified view and is more representative of transmission that would happen at the base station.

The orthogonality condition that allows us to detect the signals at the receiver is given by

$$\int_0^{T_s} c_k(t) c_l(t) \, dt = \begin{cases} 1, & k = l \\ 0, & k \neq l. \end{cases} \quad (1.100)$$

We note that this requires the length of the code to be at least N which, in turn, means the bandwidth of the transmission expands by a factor of N (at least). The receiver structure is similar to that of FDMA system:

1.10 Duplexing Schemes

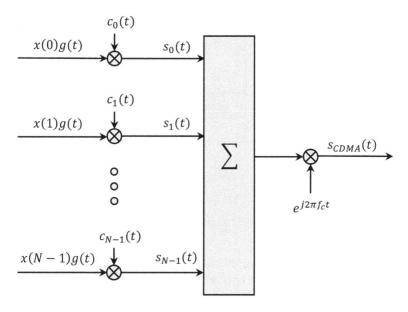

Fig. 1.29 Block diagram of a CDMA transmitter, where $x(0), x(1), \ldots$ are the symbols being transmitted

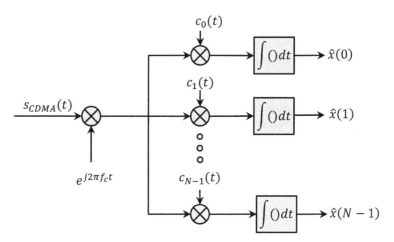

Fig. 1.30 Block diagram of a CDMA receiver

1.10 Duplexing Schemes

Our final topic in this introductory chapter is how to separate the downlink (DL or the link from the base station to users) from the uplink (UL or from users to the base station). Communication in both directions, known as duplexing, is the norm

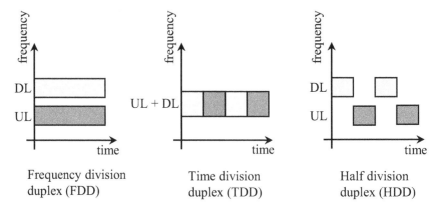

Fig. 1.31 Duplexing schemes in wireless communication

in modern cellular systems (the alternative is the so-called simplex mode, where the transmission only goes in one direction as in, for instance, TV or AM/FM radio transmission.) Since antennas, the basis for RF wave transmission and reception, generally cannot transmit and receive simultaneously on the same frequency band, the uplink and downlink must be separated.[7]

The duplex mode can be implemented in several ways, as illustrated in Fig. 1.31:

TDD: DL and UL signals share the same frequency but occur at different times (time division of resources). The majority of 5G bands at higher carrier frequencies will operate in TDD mode.

FDD: DL and UL signals occur at the same time but use different frequencies (frequency division of resources). The majority of 4G bands and lower frequency 5G bands operate in FDD mode (e.g., DL frequency at 2.1 GHz and UL frequency at 1.7 GHz).

HDD: DL and UL signals occupy different frequencies like in FDD but also occur at different times like in TDD. This mode is useful for low complexity implementation such as Internet of Things (IoT).

These three duplexing schemes are illustrated below. The Time Division Duplex (TDD) mode has a number of advantages over FDD mode:

- Channel reciprocity between the DL and UL direction due to the same transmit and receive frequency allows for more accurate knowledge of the DL channel at the base station (more accurate than heavily quantized channel state information transmitted from the UT to BS in FDD mode).
- Better channel knowledge enables the implementation of large antenna arrays, the so-called massive multiple input multiple output systems, at the BS.

---

[7] It is worth noting that full-duplex communications, i.e., enabling simultaneous transmit and receive on the same band, is an active research area. There are enormous implementation challenges, solutions for which are actively being investigated [13].

1.10 Duplexing Schemes

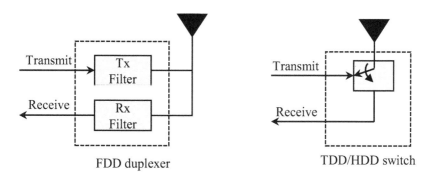

Fig. 1.32 Differences between FDD mode (with a duplexing filter) and TDD or HDD mode (without a duplexing filter)

- A flexible duty cycle allows for better matching of the traffic pattern. Typically, DL data volume is much larger than the UL data volume, and by setting a larger percentage of available time for the DL, we can better match traffic needs.
- There is no need for a duplexing filter (Fig. 1.32).

A significant drawback is the need for a guard-time to ensure all communications in one direction are complete before communication in the other direction can begin. This guard can represent a significant overhead to the systems.

The Frequency Division Duplex (FDD) mode is mostly used in low (<1 GHz) and midrange frequencies (2–3 GHz). Systems using these frequency ranges have been operating in this mode since the first generation. It is, therefore, a mature technology that is advantageous if very low latency is needed as it does not require switching between the DL and UL transmission periods and does not impose a guard-time.

The Half Duplex Division (HDD) scheme is intended for the Internet of Things (IoT) devices that operate in FDD bands but require low complexity. Unlike the proper FDD devices (e.g., smartphones), the low-cost IoT devices operate in FDD mode but lack the duplexing filter, as illustrated above (Fig. 1.32).

Problems

1. Let $x(t)$ be a signal represented in the time domain and $X(f) = \mathcal{FT}[x(t)]$ be its Fourier transform. This pair is written as

$$x(t) \Longleftrightarrow X(f)$$

   Prove the following:

   a. The time shifting property is given by

$$x(t-t_0) \iff X(f)e^{-j2\pi f t_0}$$

b. The frequency shifting property is given by

$$x(t)e^{j2\pi f_0 t} \iff X(f-f_0)$$

2. Let $x(t)$ be a signal represented in the time domain and $X(f) = \mathcal{F}\mathcal{T}[x(t)]$ be its Fourier transform. Prove Parseval's theorem which is given by

$$\int_{-\infty}^{\infty} |x(t)|^2 \, dt = \int_{-\infty}^{\infty} |X(f)|^2 \, df$$

3. Calculate the linear value of the received power expressed in the following units:

   a. -50 dBm
   b. -90 dBm
   c. -40 dBW

4. Determine the power of the sum of three mutually independent signals whose individual powers, expressed in different units, are -72 dBm, -106 dBW, and 100 pW.

5. Two independent signals, with a dB scale power levels of $P_{1|\text{dBm}}$ and $P_{2|\text{dBm}}$, are added together. Determine by how much is power increased if:

   a. $P_{2|\text{dBm}} = 6 + P_{1|\text{dBm}}$.
   b. $P_{1|\text{dBm}} = P_{2|\text{dBm}}$.

6. If $x(t) = Ae^{j2\pi f_0 t}$, determine its power spectral density $S_x(f)$.

7. For a rectangular pulse shape, whose power spectral density is expressed as $S_x(f) = T\,\text{sinc}^2(fT)$, verify that the average power, when integrated over a range defined by a multiple of n/T (multiple of nulls) results in the following values:

n	Integrated Power
1	0.902823
2	0.949939
3	0.96641
4	0.974748
5	0.979776
6	0.983137
7	0.98554
8	0.987345
9	0.988749
10	0.989873

1.10 Duplexing Schemes

8. Derive the expression for complex array factor for an N-element linear array (verify equation (45)).
9. If $s_n(t) = e^{j(2\pi f_0 t - n\beta)}$, where β is a phase shift applied to nth antenna element in a linear N-element array, determine the expression for the array factor.
10. A base station antenna is located at the top of a building and the antenna height from the ground is 45 meters.

    a. If a UT at the height of 1.5 m from the ground and at a distance of 300 m from the base station (as shown in the Fig. 1.10) is to experience the peak antenna gain, what should the value of angle α_{tilt} be?
    b. If the vertical antenna gain is expressed as

    $$G(\alpha) = G_{max} - \min\left(12\left(\frac{\alpha - \alpha_{tilt}}{\alpha_{3dB}}\right)^2, A_{max}\right)$$

    where angle α is expressed in degrees, calculate the vertical antenna gain for the UT at half the distance (150 m) if $G_{max} = 16.4$ dBi, $A_{max} = 25$ dB, and $\alpha_{3\,dB} = 9.5$ degrees?

11. Determine the antenna directivity if the radiation intensity is given by $F(\theta, \varphi) = \sin(\theta)$ and is uniform over φ.
12. Determine the antenna directivity if the radiation intensity is given by $F(\theta, \varphi) = \cos(\theta)$ and is uniform over φ. Assume that the radiation intensity exists only in the upper hemisphere ($0 \leq \theta \leq \pi/2$, $0 \leq \varphi \leq 2\pi$).

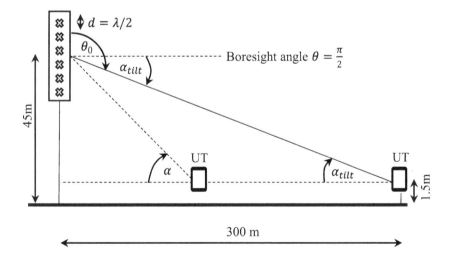

13. Determine the antenna directivity of the linear N-element array with the array factor $AF(\theta) = \frac{\sin(N\psi)}{\sin(\psi)}$. Assume that the radiation intensity is uniform over φ and that the element spacing is $d = \lambda/2$.

14. As a part of radio certification requirements, unwanted emission cannot exceed $P_{t|dBm} = -13$ dBm. Determine the electrical field strength measured at $d = 3$ m of distance assuming an ideal half-wave dipole antenna with a received gain $G_r = 1.64$ if the unwanted emission is exactly -13 dBm/MHz.

15. An N-element linear array has an array factor $AF(\theta) = \frac{\sin(N\psi)}{\sin(\psi)}$, where $\psi = \frac{1}{2}k_0 d \cos(\theta)$. Show that for $d = \lambda/2$, the first null-to-null beamwidth is $4/N$.

16. A normalized N-element linear array has an array factor $AF(\theta) = \frac{\sin(N\psi)}{N\sin(\psi)}$, where $\psi = \frac{1}{2}k_0 d \cos(\theta)$. This array factor has its maximum values at $\psi = 0, \pm\pi, \pm 2\pi, \ldots$. Show that if the spacing between the antenna elements $d = a\lambda$, where $a \geq 1$, grating pattern occurs (more than one maximum).

17. Two transmit signals, m_1 and m_2, are sent simultaneously using linearly polarized electrical fields:

$$\vec{E}_1 = m_1 e^{-jk_0 z}\vec{x} + m_1 e^{-jk_0 z}\vec{y} \qquad (1.101)$$

$$\vec{E}_2 = m_2 e^{-jk_0 z}\vec{x} - m_2 e^{-jk_0 z}\vec{y} \qquad (1.102)$$

Determine m_1 and m_2 at the receiver.

18. Two transmit signals, m_1 and m_2, are sent simultaneously using circular polarization:

$$\vec{E}_1 = m_1 e^{-jk_0 z}\vec{x} + m_1 e^{-jk_0 z} e^{j\frac{\pi}{2}}\vec{y} \qquad (1.103)$$

$$\vec{E}_2 = m_2 e^{-jk_0 z}\vec{x} - m_2 e^{-jk_0 z} e^{j\frac{\pi}{2}}\vec{y} \qquad (1.104)$$

Determine m_1 and m_2 at the receiver.

19. Find the energy and average power of the BPSK signal specified over one bit duration:

$$s_{BPSK}(t) = \sqrt{\frac{2E_b}{T_b}} \cos(2\pi f_c t). \qquad (1.105)$$

Assume that $f_c T_b \gg 1$.

20. Two functions, $\phi_I(t) = \sqrt{\frac{2}{T_s}} \cos(2\pi f_c t)$ and $\phi_Q(t) = -\sqrt{\frac{2}{T_s}} \sin(2\pi f_c t)$, are defined over a time period $(0, T_s)$. Prove the following:

   a. $\int_0^{T_s} \phi_I(t)\phi_I(t)\, dt = 1$
   b. $\int_0^{T_s} \phi_Q(t)\phi_Q(t)\, dt = 1$
   c. $\int_0^{T_s} \phi_I(t)\phi_Q(t)\, dt = 0$

21. For the given 16-QAM mapping table shown below

Binary sequence	16QAM symbol
0000	$1+j$
0001	$1+3j$
0010	$1-j$
0011	$1-3j$
0100	$3+j$
0101	$3+3j$
0110	$3-j$
0111	$3-3j$
1000	$-1+j$
1001	$-1+3j$
1010	$-1-j$
1011	$-1-3j$
1100	$-3+j$
1101	$-3+3j$
1110	$-3-j$
1111	$-3-3j$

and an input integer sequence is $i = [5, 7, 13, 10]$, determine the following:

a. If each of the integers given above is represented as a 4-bit binary number with the leftmost bit being the MSB and the rightmost the LSB, what is the total bit sequence to be transmitted?
b. 16-QAM mapped output sequence $X(k)$ (constellation points) using the mapping table shown above. Draw the constellation diagram indicating the points being transmitted.
c. Draw the time domain representation of the 16-QAM modulated signal after the carrier modulation. Assume that there are two full periods of the carrier frequency over one symbol duration.
d. What is the order of an M-QAM modulation that would transmit the binary sequence determined in (a) using 2 transmit symbols?

References

1. User Equipment (UE) radio transmission and reception, Part 1: Range 1 Standalone. https://www.3gpp.org/ftp/Specs/archive/38_series/38.101-1/38101-1-i30.zip. Cited Oct 25, 2023
2. Evolved Universal Terrestrial Radio Access (E-UTRA), User Equipment (UE) radio transmission and reception. https://www.3gpp.org/ftp/Specs/archive/36_series/36.101/36101-eq0.zip. Cited Oct 25, 2023
3. Alberto L.G, *Probability, Statistics, and Random Processes for Electrical Engineering*, 3rd edn. (Pearson Prentice Hall, New York, 2007)
4. B.P. Lathi, Z. Ding, *Modern Digital and Analog Communication Systems*, 4th edn. (Oxford University, Oxford, 2009)

5. Limits of Human Exposure to Radiofrequency Electromagnetic Energy in the Frequency Range from 3 kHz to 300 GHz. https://www.canada.ca/en/health-canada/services/publications/health-risks-safety/limits-human-exposure-radiofrequency-electromagnetic-energy-range-3-300.html. Cited Oct 27, 2023
6. Understanding Safety Code 6: Health Canada's radiofrequency exposure guidelines. https://www.canada.ca/en/health-canada/services/health-risks-safety/radiation/occupational-exposure-regulations/safety-code-6-radiofrequency-exposure-guidelines.html. Cited Apr 28, 2024
7. Radiofrequency Energy and Safety. https://ised-isde.canada.ca/site/spectrum-management-telecommunications/en/safety-and-compliance/facts-about-towers/radiofrequency-energy-and-safety. Cited Apr 28, 2024
8. RSS-139—Advanced Wireless Services Equipment Operating in the Bands 1710-1780 MHz and 2110-2200 MHz. https://ised-isde.canada.ca/site/spectrum-management-telecommunications/en/devices-and-equipment/radio-equipment-standards/radio-standards-specifications-rss/rss-139-advanced-wireless-services-equipment-operating-bands-1710-1780-mhz-and-2110-2200-mhz. Cited Oct 27, 2023
9. Study on channel model for frequencies from 0.5 to 100 GHz. https://www.3gpp.org/ftp/Specs/archive/38_series/38.901/38901-h00.zip. Cited Oct 27, 2023
10. Guidelines for evaluation of radio interface technologies for IMT-2020. https://www.itu.int/pub/R-REP-M.2412-2017. Cited Oct 27, 2023
11. S. Haykin, M. Moher, *Communication Systems* (Wiley, New York, 2009)
12. NR; Base Station (BS) radio transmission and reception. https://www.3gpp.org/ftp/Specs/archive/38_series/38.104/38104-i30.zip. Cited Oct 27, 2023
13. M. Mohammadi, Z. Mobini, D. Galappaththige, C. Tellambura, A comprehensive survey on full-duplex communication: current solutions, future trends, and open issues. IEEE Commun. Surv. Tutorials **25**(4), 2190–2244 (Fourthquarter 2023)

Chapter 2
Pathloss and Channel Models

2.1 Introduction

Wireless communication systems suffer unique challenges brought about by the *channel*: the propagation between the transmitter and receiver. Specifically, RF communications invariably deal with obstacles in the path, each with a random impact on the propagation. Understanding the challenges posed by this randomness is crucial to understanding the design and analysis of wireless networks (Fig. 2.1).

The obstacles have three crucial impacts: one, the power in the propagating signal attenuates much faster than in free space; two, because user and obstacle locations are unknown a priori, this attenuation of power, called shadowing, is random; finally, since the obstacles reflect and refract signals, propagation from transmitter to receiver is over multiple paths, leading to yet another source of randomness, called small-scale fading, a term that varies rapidly over distance. Figure 2.2 illustrates these phenomena, which taken together are referred to as *fading*. It is this complex propagation environment that truly distinguishes cellular communications from other communication systems. It is worth noting that early RF designs, such as AM/FM radio, overcome these difficulties largely by increasing their transmit power; however, cellular communications does not have this luxury: not only is the transmit power strictly regulated to relatively low levels, but also, as we will see in the next chapter, increasing transmit power does not necessarily improve performance. Additionally, the trend in networks today is toward more energy efficient systems with lower power consumption equipment.

Supplementary Information The online version contains supplementary material available at (https://doi.org/10.1007/978-3-031-76455-4_2).

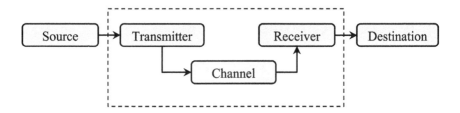

Fig. 2.1 Simplified model of a communication system

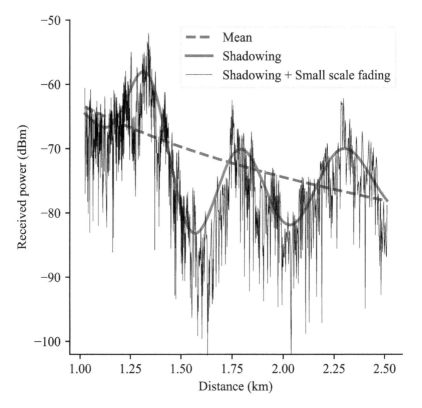

Fig. 2.2 Received signal power as a function of distance, impacted by shadowing (large-scale fading) and small-scale fading

An important consequence of this is that the overall fading is a local measure: two receivers at the same distance from the transmitter will likely not measure the same received signal power. This is because the obstacles each receiver sees are likely to be different. You may have observed that sometimes even a very small change of location in the range of few centimeters will have an impact on the received signal power (e.g., how many bars one sees on the phone). These small-scale impacts play a

2.2 Distance-Based Pathloss and Friis' Formula

crucial role in system design. Explaining the wireless channel and the corresponding models is the focus of this chapter.

2.2 Distance-Based Pathloss and Friis' Formula

Let us begin with the power attenuation model in free space. In Chap. 1, we introduced the power flux density (PFD) as a function of transmit power and the distance from the transmitter. We can use the PFD to calculate the received power at a given distance: assuming an isotropic transmitter with a transmit power P_T and a receiver, also using an isotropic antenna, that receives a signal power P_R, the ratio between the two powers is called the pathloss.

An isotropic antenna has an effective antenna size [1]:

$$A_e = \frac{\lambda^2}{4\pi}. \qquad (2.1)$$

Using the PFD for an isotropic antenna, $PFD = P_T/4\pi d^2$, the received signal power, at a receiver a distance d from the transmitter, is (Fig. 2.3)

$$P_R = PFD \times A_e = \frac{P_T}{4\pi d^2} \frac{\lambda^2}{4\pi} = P_T \left(\frac{\lambda}{4\pi d}\right)^2. \qquad (2.2)$$

The free space pathloss is then determined as

$$PL(d) = \frac{P_T}{P_R} = \left(\frac{4\pi d}{\lambda}\right)^2, \qquad (2.3)$$

i.e., the path loss is proportional to d^2, the square of distance between transmitter and receiver; we say that the pathloss exponent is $n = 2$.

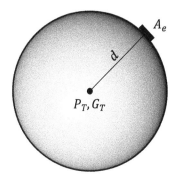

Fig. 2.3 Illustration of an antenna element's received power at distance d with effective area A_e

As mentioned in Chap. 1, real-world antennas are not isotropic and provide a power gain. Assuming the transmit and receive antennas are properly oriented, we can account the for the transmit and receive antenna, with gains G_T and G_R, respectively, and rewrite the received power as

$$P_R = P_T G_T G_R \left(\frac{\lambda}{4\pi d}\right)^2 = \frac{P_T}{PL(d)} G_T G_R. \quad (2.4)$$

This relationship, valid for free space propagation, is known as **Friis' formula**.

In free space conditions, Friis' formula provides a convenient approach for power calculations. For example, national regulatory bodies, such as ISED in Canada and the Federal Communications Commission (FCC) in the United States, specify a limit on the PFD level at the receiver. Given this limit, we can calculate how much received power is allowed in a given geographic area. We rewrite the received power as a function of the PFD:

$$P_R = PFD \frac{G_R \lambda^2}{4\pi}$$

where, as in Chap. 1, the PFD term includes the transmit antenna gain, G_T.

Expressing this equation in the dB scale and isolating the PFD term, we have

$$PFD_{|dB(W/m^2)} = P_{R|dBW} - 10\log_{10}(G_R) - 10\log_{10}\left(\frac{\lambda^2}{4\pi}\right), \quad (2.5)$$

Example 2.1 The border sharing arrangement between the Canadian regulator ISED and the FCC of the United States [2] concerning the use of the frequency bands 806–824 MHz and 851–869 MHz by the Land Mobile Service Along the Canada-United States Border, known as Arrangement F, specifies:

```
The maximum power flux density (PFD) of the signal at and
beyond the border of the primary user's country does not
exceed -124 dB(W/m2)/25 kHz.
```

Assuming the carrier frequency of $f_c = 850$ MHz and the receiver antenna gain of $G_{R|dB} = 0$ dBi, determine the maximal allowed received power over 5 MHz of bandwidth.

2.2 Distance-Based Pathloss and Friis' Formula

Solution The maximal received power level over 25 kHz of bandwidth is

$$P_{R|\text{dBW}} = PFD_{|\text{dB(W/m}^2)} + 10\log_{10}(G_R) + 10\log_{10}\left(\frac{\lambda^2}{4\pi}\right), \tag{2.6}$$

and for the carrier frequency is $f_c = 850$ MHz, the wavelength of the signal is

$$\lambda = \frac{c}{f_c} = 0.353 \text{ m}. \tag{2.7}$$

Given the receive antenna gain of $G_{R|\text{dB}} = 0$ dBi, the received power over 25 kHz is

$$P_{R|\text{dBW}} = -124 + 10\log_{10}\left(\frac{0.353^2}{4\pi}\right) = -144 \text{ dBW/25 kHz}. \tag{2.8}$$

Unless otherwise specified, we can assume a flat power spectral density (PSD), so that the total received power over 5 MHz is scaled by the ratio of 5 MHz/25 kHz. That is equivalent to finding the PSD (per 1 Hz) and then integrating a constant value over the 5 MHz:

$$S(f)_{|\text{dBW/Hz}} = P_{R|\text{dBW}} - 10\log_{10}\left(25 \times 10^3\right) = -188 \text{ dBW/Hz}, \tag{2.9}$$

resulting in total received power over 5 MHz:

$$P_{R|\text{dBW}} = -188 + 10\log_{10}\left(5 \times 10^6\right) = -121 \text{ dBW}. \tag{2.10}$$

In dBm scale, that is $P_{R|\text{dBm}} = P_{R|\text{dBW}} + 30 = -91$ dBm over 5 MHz. ∎

In the dB scale (log-domain), the free space pathloss can be written as

$$\begin{aligned} PL(d)_{|\text{dB}} &= 10\log_{10}\left(\frac{4\pi d}{\lambda}\right)^2 = 20\log_{10}\left(\frac{4\pi}{\lambda}\right) + 20\log_{10}(d), \\ &= b + 20\log_{10}(d), \end{aligned} \tag{2.11}$$

where the $b = 20\log_{10}(4\pi/\lambda)$ term does not depend on the distance. Importantly, it is the pathloss exponent of 2 that leads to the factor of 20 (10×2) in the second, distance-related, term.

2.3 Empirical Pathloss Models

In developing Friis' formula, we assumed free space propagation, which is rarely feasible in real-world (terrestrial) systems. Obstacles in the propagation path attenuate signals faster than the $1/d^2$ attenuation seen in free space.[1] In 1968, Okumura reported the results of field strength measurements in Tokyo [3]. More than a decade later, in 1980, Hata developed an empirical formula [1] that applied to the operating frequency range of 150 to 1500 MHz. This was further extended to cover the range up to 2000 MHz, which is known as the COST 231 Hata model [4]. The pathloss is calculated as

$$PL(d)_{|\text{dB}} = b + 10n \log_{10}(d) - K_r. \quad (2.12)$$

For a base station whose antenna is at a height of h_{BS} meters, a user terminal whose receiver is at a height of h_{UT} meters, and a carrier frequency f_c, expressed in MHz, is in the range from 150 MHz to 1500 MHz, the Hata model has the following model for the parameters in (2.12):

$$b = 69.55 + 26.16 \log_{10}(f_c) - 13.82 \log_{10}(h_{BS}) - a(h_{UT}, f_c) \quad (2.13)$$

$$n = 4.49 - 0.655 \log_{10}(h_{BS}) \quad (2.14)$$

where the $a(h_{UT}, f_c)$ correction value is calculated as

$$a(h_{UT}, f_c) = \begin{cases} (1.1 \log_{10}(f_c) - 0.7) h_{UT} - 1.56 \log_{10}(f_c) + 0.8, & \text{small to medium city} \\ 3.2 (\log_{10}(11.75 h_{UT}))^2 - 4.97, & \text{large city.} \end{cases} \quad (2.15)$$

The large city value for $a(h_{UT}, f_c)$ for frequencies below 400 MHz is omitted in (2.15) since that frequency range is not used for cellular communication. The small to medium size city area is also referred to as "urban," while the large city area is referred to as "dense urban."

For only suburban and rural areas, based on Okumura's measurements, there is a correction factor K_r [1, 5]:

---

[1] The case wherein tall buildings create a sort of tunnel effect, thereby reducing the attenuation—resulting in a path loss exponent below 2—is rare and generally out of the control of the system designer.

2.3 Empirical Pathloss Models

$$K_r = \begin{cases} 2\left[\log_{10}(f_c/28)\right]^2 + 5.4, & \text{Suburban} \\ 4.78\left[\log_{10}(f_c)\right]^2 - 18.33\log_{10}(f_c) + 40.94, & \text{Rural} \\ 4.78\left[\log_{10}(f_c)\right]^2 - 18.33\log_{10}(f_c) + 35.94, & \text{Rural, quasi-open} \end{cases} \quad (2.16)$$

and the correction factor $a(h_{UT}, f_c)$ for urban areas applies.

The extended frequency range spanning frequencies from 1500 to 2000 MHz, described by the *COST 231-Hata* model [6], has the following parameter values:

$$b = 46.3 + 33.9\log_{10}(f_c) - 13.82\log_{10}(h_{BS}) - a(h_{UT}, f_c) + C \quad (2.17)$$

$$n = 4.49 - 0.655\log_{10}(h_{BS}) \quad (2.18)$$

where

$$C = \begin{cases} 0, & \text{medium city and suburban} \\ 3, & \text{metropolitan centers} \end{cases} \quad (2.19)$$

and the $a(h_{UT}, f_c)$ value is calculated the same way as in the Hata model. The COST 231 Hata model is then calculated as

$$PL(d)_{|dB} = b + 10n\log_{10}(d) - K_r. \quad (2.20)$$

where d is a distance in kilometers and the same correction factor K_r for rural areas can be applied as for the Hata model.

Both Hata and COST 231 Hata models apply to the following parameters values:

$$30\,\text{m} \leq h_{BS} \leq 200\,\text{m} \quad (2.21)$$

$$1\,\text{m} \leq h_{UT} \leq 10\,\text{m} \quad (2.22)$$

$$1\,\text{km} \leq d \leq 20\,\text{km} \quad (2.23)$$

In (2.12) and (2.20), the variable n denotes the pathloss exponent. From (2.11), (2.12), and (2.20), we see that all three models, in fact, satisfy the same framework. In each case, in the dB scale, the *average* loss in power, as a function of distance, is the sum of a constant (which can depend on multiple parameters, like frequency, height, etc.) and a distance-related term. The pre-log factor is, by definition of the dB scale, 10× the pathloss exponent.

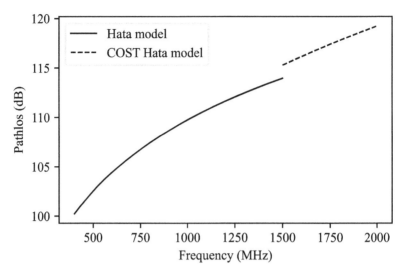

Fig. 2.4 Pathlos values for Hata and COST 231 Hata models, for $h_{BS} = 35$ m, $h_{UT} = 1.5$ m and $d = 0.5$ km, for medium size city model

It is worth noting that the COST Hata model was created as an extension of the PCS band around 1900 MHz. As a result, these two models do not meet at $f_c = 1500$ MHz, as illustrated in Fig. 2.4.

Example 2.2 Given the pathloss formula

$$PL(d)_{|dB} = 128.1 + 37.6 \log_{10}(d), \quad d \text{ is in km.} \qquad (2.24)$$

where the distance d is in kilometers, determine the following:

1. Signal attenuation (pathloss) at $d = 1$ km.
2. If the base station transmit power is $P_T = 48$ dBm and $G_T = G_R = 0$ dBi, what is the received signal level P_R at $d = 1$ km?
3. What is the distance you need to travel to measure an additional 10 dB of attenuation (relative to $d = 1$ km)?
4. What is the value of the pathloss exponent n?

Solution For the given pathloss formulas

$$PL(d)_{|dB} = 128.1 + 37.6 \log_{10}(d), \quad d \text{ is in km.} \qquad (2.25)$$

2.3 Empirical Pathloss Models

we obtain

1. Signal attenuation (pathloss) at $d = 1$ km

$$PL(d=1)_{|dB} = 128.1 + 37.6\log_{10}(1) = 128.1 \text{ dB} \quad (2.26)$$

It is worth noting the substantially larger pathloss as compared to the free space expression in (2.11). In free space, for a carrier frequency of 1 GHz and a transmitter-receiver distance of 1 km, the expression in (2.11) evaluates to a pathloss of 92.4 dB, i.e., the Hata model indicates an increased pathloss of 36 dB (a factor of approximately 4000)!

2. The received signal level

$$P_R = P_T + G_T + G_R - PL(d)_{|dB} = 48 + 0 + 0 - 128.1 = -80.1 \text{ dBm.} \quad (2.27)$$

3. Additional distance

$$PL(d+\Delta)_{|dB} = PL(d=1)_{|dB} + 10 = 138.1 \quad (2.28)$$
$$128.1 + 37.6\log_{10}(1+\Delta d) = 138.1 \quad (2.29)$$
$$1 + \Delta d = 10^{10/37.6} = 1.845 \text{ km} \quad (2.30)$$
$$\Delta d = 0.845 \text{ km.} \quad (2.31)$$

The rule of thumb is that for every doubling of distance we get additional 10 dB of attenuation, and this result is in a similar ballpark.

4. The pathloss exponent is determined by the pre-log factor:

$$n = \frac{37.6}{10} = 3.76.$$

∎

3GPP has modified the pathloss models and extended their range of applicability up to 100 GHz, as specified in TR 38.901, Section 7.4 [7]. Additionally, the distances between the base station and user terminals have been refined by introducing a ground and the actual 3D distance, as illustrated below (Fig. 2.5).

> It is worth emphasizing that these models are empirical and provide insights into system design. For example, these models are used in system level simulations and in link budget calculations. Furthermore, these models are not exhaustive; an example of non-line-of-sight (NLOS) and line-of-sight (LOS) models for rural and urban macro deployments are given in Table 2.3 at the end of this chapter.

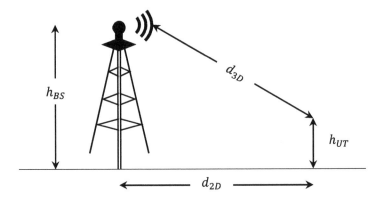

Fig. 2.5 Illustration of 3GPP specified distances involved in channel models

Table 2.1 Penetration loss through different materials

Material	Penetration loss [dB]
Standard multi-pane glass	$L_{glass} = 2 + 0.2f$
IRR glass	$L_{IIRglass} = 23 + 0.3f$
Concrete	$L_{concrete} = 5 + 4f$
Wood	$L_{wood} = 4.85 + 0.12f$

Note: f is in GHz

Table 2.2 High and low penetration loss formulas

Model	Penetration loss [dB]
Low-loss model	$5 - 10\log_{10}\left(0.3 \times 10^{-L_{glass}/10} + 0.7 \times 10^{-L_{concrete}/10}\right)$
High-loss model	$5 - 10\log_{10}\left(0.7 \times 10^{-L_{IIRglass}/10} + 0.3 \times 10^{-L_{concrete}/10}\right)$

2.3.1 Penetration Through Materials and Foliage Loss

Penetration through materials has been extensively studied by industry and academia. 3GPP has summarized the findings in the TR 38.901, Section 7 [7], where the following losses are listed (Table 2.1).

Houses are not uniformly made of concrete, or glass, or any other material, so a weighted function is applied to different penetration losses in link budget calculations or link level simulations, as listed in Table 2.2.

In addition to material penetration, there is an additional indoor loss that adds 0.5 dB for every meter of distance from the wall (see Table 7.4.3-2, TR 38.901).

For foliage loss, the penetration loss model through vegetation developed by CCIR report 236-6 is usually used [8]:

$$L_{foliage|dB} = 0.2 \times f_c^{0.3} \times d^{0.6} \text{ (dB)} \quad (2.32)$$

where f_c is the carrier frequency in MHz and d is the depth of foliage traversed in meters. For example, a signal at the carrier frequency of $f_c = 3500$ MHz that goes through 10 meters of foliage will experience $L_{foliage|\text{dB}} = 9.21$ dB of attenuation.

2.4 Shadowing and Fade Margin

In the previous section, we saw that, empirically, pathloss can be expressed in the dB scale as

$$PL(d)_{|\text{dB}} = b + 10n \log_{10}(d) \tag{2.33}$$

This model represents the *average* loss in power at a distance d. The pathloss exponent, n, is usually larger than 2 because of the obstacles between the transmitter and receiver. As seen in Fig. 2.6, an obstacle, such as a building, can significantly attenuate a signal. Specifically, such large obstacles are said to cause a *shadow*. These power variations are observable on a relatively large scale, on order of tens of meters.

Indeed, if the signal had to go through obstacles to arrive at a receiver, the resulting penetration losses would be too high for any effective communications. Luckily, the signal, at microwave frequencies, also reflects off, and refracts from, these obstacles. So, while each reflection/refraction results in some path loss, this is far smaller than penetration losses. Crucially, the number and kind of obstacles between the transmitter and receiver are a priori unknown. Over a significant propagation distance, the number of obstacles is unknown but generally large. In the dB scale, we could model the received signal as

$$P_{R|\text{dBW}} = P_{T|\text{dBW}} - \text{Free space pathloss} - L_1 - L_2 - \ldots L_M \tag{2.34}$$

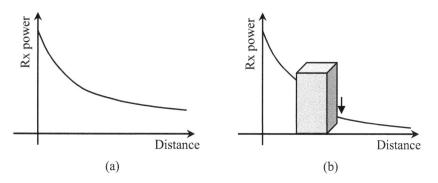

Fig. 2.6 Illustration of shadowing of the signal level caused by a building. (**a**) No shadowing. (**b**) With shadowing

where M denotes the number of obstacles (modeled as random) and L_m, $m = 1, \ldots, M$ denote the power loss due to obstacle m *in the dB scale* (also modeled as random). Denoting the total loss (in dB) as $L = \sum_m L_m$, since M is large, using the central limit theorem, we can model L as a Gaussian random variable with mean $\mathbb{E}[L] = \mu_L$ and variance σ_L^2. Equivalently

$$P_{R|\text{dBW}} = P_{T|\text{dBW}} - \text{Free space pathloss} - \mu_L - X_\sigma, \qquad (2.35)$$

where X_σ is a *zero-mean* Gaussian random variable with variance σ_L^2 (standard deviation of σ_L), often denoted as $X_\sigma \sim \mathcal{N}(0, \sigma_L^2)$.

We are now able to identify the source of the *average* power loss in the previous section. It is the sum of the free space pathloss and mean of X_σ that leads to the pathloss models mentioned. Physically, as the transmitter-receiver distance grows, we expect more obstacles and a greater value of μ_L. This explains why the effective pathloss exponent is greater than 2.[2]

Given this discussion, we can rewrite the pathloss expression as

$$PL(d)_{|\text{dB}} = b + 10n \log_{10}(d) + X_\sigma. \qquad (2.36)$$

Since the random term, X_σ, is caused by obstacles, it changes only over relatively large distances. This randomness is referred to as *large-scale fading*. Furthermore, because this randomness is modeled as additive and Gaussian in the dB scale, it is also referred to as "log-normal fading."

In practice, the standard deviation of X_σ, σ_L, ranges from 3 dB up to 12 dB, with a typical value of $\sigma_L = 6$ dB for macro base stations in urban areas.

The randomness in the received pathloss, and hence the resulting received signal power, requires a probabilistic analysis. Overall, the received signal power can be written as

$$P_{R|\text{dBW}} = P_{T|\text{dBW}} + G_T + G_R - PL(d) = EIRP + G_R - PL(d)_{|\text{dB}}. \qquad (2.37)$$

If we were to design the network using only the mean pathloss, the received power at any chosen distance would be lower than this mean in half the cases (and greater than this mean in half the cases). The network performance for cell-edge users would, therefore, be poor. Instead, we add a *fade margin* (also called shadowing

---

[2] It is worth noting that a path loss exponent of $n > 2$ can also arise from deterministic reflections. One scenario is when a single reflected path is added to the direct path between transmitter and receiver; depending on the distance travelled and the reflection coefficient, a path loss of $2.5 < n \leq 4$ can be achieved [9].

2.4 Shadowing and Fade Margin

margin) on the minimum allowed received power. For instance, instead of having minimum receive power of $P_S = -100$ dBm, we design the network for the minimum signal level of $P_S + M = -92$ dBm, thus using an $M = 8$ dB margin. This will guarantee that only a small percentage of cell-edge users actually receive a signal level lower than -100 dBm (due to the random X_σ component).

The question then is how to choose this fade margin. If chosen too low, many users will receive inadequate service; if chosen too high, we will be too conservative: wasting power or making cells too small. In choosing the fade margin, we define an acceptable *outage probability*, P_OUTAGE, essentially the percentage of cell-edge users who will receive a power too low for adequate service (are in *outage*).

From (2.37), the received power, in the dB scale, is a Gaussian random variable with mean $\mu_{P_r} = EIRP + G_R + b + 10n \log_{10}(d)$ and variance σ_L^2. If we denote the minimum required power for service as P_s, we have

$$P_\text{COVERAGE} = 1 - P_\text{OUTAGE} = \mathbb{P}[P_r > P_S] = \int_{P_S}^{\infty} \frac{1}{\sqrt{2\pi\sigma_L^2}} e^{-\frac{(x-\mu_{P_r})^2}{2\sigma_L^2}} dx,$$

$$= \frac{1}{2}\text{erfc}\left(\frac{P_S - \mu_{P_r}}{\sqrt{2}\sigma_L}\right) \qquad (2.38)$$

where $\mathbb{P}[\cdot]$ denotes the probability of an event and $\text{erfc}(\cdot)$ denotes the complementary error function given by

$$\text{erfc}(x) = \frac{2}{\sqrt{\pi}} \int_x^\infty e^{-t^2} dt. \qquad (2.39)$$

Figure 2.7 illustrates the probabilities involved in this calculation. We have the mean $\mu_{P_r} = P_S + M$, i.e., we can write (2.38) as

$$P_\text{COVERAGE} = \frac{1}{2}\text{erfc}\left(\frac{-M}{\sqrt{2}\sigma_L}\right),$$

and, using the convenient property of $\text{erfc}(-x) = 2 - \text{erfc}(x)$, we have

$$P_\text{OUTAGE} = 1 - P_\text{COVERAGE} = \frac{1}{2}\text{erfc}\left(\frac{M}{\sqrt{2}\sigma_L}\right). \qquad (2.40)$$

Inverting this equation provides the required fade margin, M, in dB. The coverage probability is also called the cell edge reliability or the contour reliability.

An important impact of using a fade margin is a reduction in cell size (as compared to when the fading is not random). This impact on cell edge radius is illustrated below (Fig. 2.8).

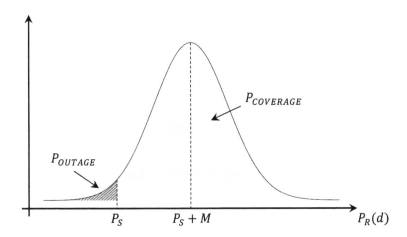

Fig. 2.7 Coverage outage probability with hatched area on the left caused by shadowing

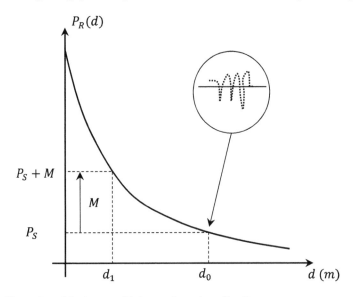

Fig. 2.8 Illustration of the impact of fade margin on the cell radius

Example 2.3 Let the cell edge reliability be 85% and the standard deviation of the shadowing fading is $\sigma = 8$ dB. Determine the required margin.

Solution Cell edge reliability of 85% means that 15% of the cell-edge users are in outage, i.e., $P_{\text{OUTAGE}} = 0.15$. The value for the shadow margin M is given by

$$M = \sqrt{2}\sigma \, \text{erfc}^{-1}(2P_{\text{OUTAGE}}) \qquad (2.41)$$
$$= 8\sqrt{2}\,\text{erfc}^{-1}(2 \times 0.15) = 8.29 \text{ dB}. \qquad (2.42)$$

■

It is worth mentioning that, due to fading, even regions inside a cell may experience low receive power. We can also average out all the points within a coverage area, called the area reliability. Since the probability of an inside region being in outage is lower than at the cell edge, this area reliability is higher than the cell edge reliability (e.g., a cell edge reliability of 85% may result in a 95% area reliability).

2.5 Channel Impulse Response, Multipath, and Small-Scale Fading

Our discussion, so far, has focused on received power without looking at the channel properties. As seen in Chap. 1, the transmitted signal is represented as a complex value. In this section, we develop the fundamental concept of multipath and investigate its impact on the complex transmitted signal.

2.5.1 Slow and Fast Fading

The communication channel is often modeled as a linear system characterized by an impulse response $h(t)$ such that, given a transmitted signal, $s(t)$, the received signal is given by the following convolution expression:

$$r(t) = \int_{-\infty}^{\infty} s(\tau)h(t-\tau)\,d\tau \qquad (2.43)$$

$$= \int_{-\infty}^{\infty} h(\tau)s(t-\tau)\,d\tau. \qquad (2.44)$$

We denote this convolution as $r(t) = s(t) * h(t)$.

If we denote H as a linear time-invariant (LTI) operator that represents the channel impulse response, then the time shifted input produces time shifted outputs:

$$y(t) = H[x(t)] \Rightarrow y(t-t_0) = H[x(t-t_0)]. \qquad (2.45)$$

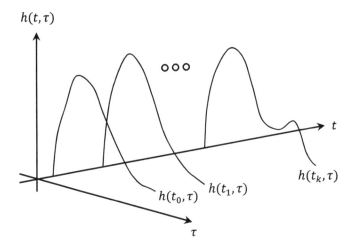

Fig. 2.9 Channel impulse response with time-variable delay profile showing three different channel impulse responses at time instances τ_0, τ_1 and τ_k

This is contrast to linear time-variant (LTV) systems; in such systems, the channel impulse response changes with time and, as illustrated in Fig. 2.9, can be written as a function of two time variables, absolute time t and delay-since-input τ: $h(t, \tau)$. For LTV systems, the channel response for an input signal $s(t)$ is written as

$$r(t) = \int_{-\infty}^{\infty} s(t-\tau) h(t, \tau) \, d\tau. \tag{2.46}$$

An important question for our purposes is which of the two models is valid for wireless communications. The answer, interestingly, is scenario dependent. At one extreme, if the transmitter and receiver are stationary—and so is every obstacle in between—the channel between the two is fixed and the LTI model is valid. However, in a wireless setting, a user (and obstacles such as cars) might be moving and the channel is not fixed anymore. In theory, then, we must use the LTV model.

Crucially, in most cases, the channel changes very slowly compared to the data rate, i.e., the channel is *effectively fixed* for many data symbols. This is called *slow fading* and an LTI model is valid for short time scales. On the other hand, if the channel is changing rapidly, on order of the symbol rate, the LTI model is invalid and the LTV model must be used. This scenario is referred to as *fast fading*.

(continued)

2.5 Channel Impulse Response, Multipath, and Small-Scale Fading

> The rate at which the channel changes is the Doppler shift in frequency due to motion. If $v_{\max}$ is the maximum velocity of an object in the channel (transmitter/receiver/obstacle), the highest Doppler shift is $f_D = v_{\max}/\lambda_c$, where λ_c is the operating wavelength. In Chap. 1, we denoted the symbol period as T_s. Essentially, if $f_D \ll 1/T_s$, we have slow fading.

Unless otherwise stated, our discussion will assume slow fading.

2.5.2 Statistical Description of a Wireless Channel

A fundamental characteristic of wireless channels is *multipath propagation*. In the previous section, we discussed how obstacles attenuate power; however, these obstacles also reflect and refract the propagating signal, and, as a consequence, the transmitted signal arrives at the receiver over many (different) paths. Each path has a different path length, and, hence, the propagation time is also different. To investigate the impact on the signal, consider the associated channel impulse response:

$$h(t) = \sum_\ell \alpha_\ell \delta(t - \tau_\ell) \Rightarrow r(t) = \sum_\ell \alpha_\ell s(t - \tau_\ell), \qquad (2.47)$$

where τ_ℓ is the delay in receiving the transmitted signal due to path distance. Crucially, the amplitudes of individual paths, denoted here as α_ℓ, are complex, indicating the attenuation and phase shift associated with the path. The number of paths is considered to be large.

> A crucial parameter is the spread of the path delays. The *channel delay spread* is given by $\Delta\tau = \max_\ell\{\tau_\ell\} - \min_\ell\{\tau_\ell\}$

As we will see in subsequent chapters, it is hard to overemphasize the importance of the channel delay spread in the design of wireless networks. Let us start with the simple case when the delay spread $\Delta\tau \ll T_s$, the symbol period. In this case, we could consider all the paths arrive at approximately the same time and the received signal is given by

$$r(t) = \alpha s(t - \tau), \quad \text{where} \quad \alpha = \sum_\ell \alpha_\ell. \qquad (2.48)$$

In the case of a short delay spread, therefore, the received signal is just a scaled and time-shifted version of the transmitted signal. The scaling term, or channel, α, is the sum of the amplitudes of each path. As in the previous section, the individual amplitudes are random and α is, therefore, the sum of many (complex) random variables. Importantly, the phase of the signal changes by 2π over one wavelength (which is on the order of 10s of cm), i.e., the phase of the individual paths, with random lengths, can be treated as a uniform random variable between 0 and 2π. The central limit theorem again plays a role in our model for this overall amplitude.

Rayleigh Fading The simplest case is when each individual path follows approximately the same pdf, and, writing the real and imaginary components, we have

$$h = \mathfrak{R}\{h\} + j\mathfrak{I}\{h\} = \sum_\ell \mathfrak{R}\{\alpha_\ell\} + j\mathfrak{I}\{\alpha_\ell\}. \tag{2.49}$$

Since the phase of α_ℓ is uniform, the real and imaginary components are zero mean and uncorrelated. Using the central limit theorem, we have that $\mathfrak{R}\{h\} \sim \mathcal{N}(0, \sigma_\alpha^2/2)$ and $\mathfrak{I}\{h\} \sim \mathcal{N}(0, \sigma_\alpha^2/2)$, i.e., individually, the real and imaginary parts are zero-mean Gaussian random variables with variance $\sigma_\alpha^2/2$. The factor of two arises because, from the definition of variance

$$\mathbb{E}[|\alpha|^2] = \mathbb{E}[\mathfrak{R}\{\alpha\}^2] + \mathbb{E}[\mathfrak{I}\{\alpha\}^2] = \sigma_\alpha^2. \tag{2.50}$$

We say that α is a *complex* zero-mean Gaussian random variable with variance σ_α^2, denoted as $\alpha \sim \mathcal{CN}(0, \sigma_\alpha^2)$.

This scenario, where the paths are all statistically similar and the overall channel is a zero-mean complex Gaussian random variable, is referred to as *Rayleigh fading*.[3] We note that if there is (relatively slow) movement in the channel, the value of α can change over long (in relation to the symbol period) time scales. Figure 2.10 plots such a case with a Doppler frequency of 20 Hz.

> In the previous section, we discussed large-scale fading—the shadowing effect changed over relatively large distances. The multipath components, on the other hand, change rapidly with distances. Specifically, as mentioned, the phase of each component covers the entire range of 0 to 2π over one wavelength (10 cm at 3 GHz). The overall channel value, therefore, can change quite quickly over small distances. This multipath fading is referred to as *small-scale fading*.

---

[3] Because the magnitude of the channel follows a Rayleigh probability density function. In turn, the channel power $|\alpha|^2$ is an exponential random variable with mean of σ_α^2.

2.5 Channel Impulse Response, Multipath, and Small-Scale Fading

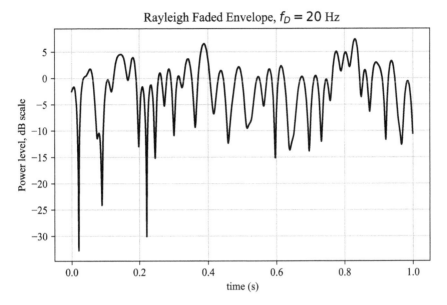

Fig. 2.10 Rayleigh distributed signal power variations, in dB scale

Rician Fading In the case where one path is dominant, e.g., due to a line-of-sight between the transmitter and receiver, the overall channel has a nonzero mean. Essentially, the situation is as if we have one dominant path plus Rayleigh fading. We model this scenario as $\alpha \sim \mathcal{CN}(\mu_\alpha, \sigma_\alpha^2)$, i.e., the fading has nonzero mean (which models the dominant path). This model is called *Rician fading*. The Rician factor measures the relative powers in the mean and random component (the variance):

$$K = \frac{|\mu_\alpha|^2}{\sigma_\alpha^2}. \tag{2.51}$$

It is worth noting that Rayleigh and Rician fading are just two (popular) models for the random channel. Other examples are Nakagami and Weibull; we do not explore these models here.

We now turn to the case where the delay spread $\Delta \tau > T_s$, the symbol period. We can extend our previous discussion by gathering all the paths that arrive within one symbol period into one term. Our model is now

$$h(t) = \sum_{\ell=0}^{L} \alpha_\ell \delta(t - \ell T_s), \tag{2.52}$$

where $L = \lceil \Delta\tau/T_s \rceil$ denotes the number of symbols for which the channel is nonzero. Our model for each individual term, h_ℓ is the same as before, e.g., Rayleigh.

> From Chap. 1, we know that the transmitted signal can be written as $s(t) = \sum_k a_k g(t - kT_s)$. Since the channel lasts for L symbols, L symbols *overlap* at the receiver—this is known as *inter-symbol interference* (ISI). ISI significantly complicates receiver design; indeed, 4G and 5G systems place dealing with ISI at the core of how the physical (signal) layer is designed.

Outage Probability Since the channels are random, the received power is random. Consequently, a receiver may be in outage if the power falls below some required threshold. For example, for a threshold of $\bar{p}$, the outage probability is

$$P_{\text{OUTAGE}} = \mathbb{P}[|\alpha|^2 < \bar{p}]. \tag{2.53}$$

Now, if the channel arises from Rayleigh fading, the power is exponentially distributed with mean σ_α^2, i.e.,

$$P_{\text{OUTAGE}} = \int_0^{\bar{p}} \frac{1}{\sigma_\alpha^2} e^{-p/\sigma_\alpha^2} dp = 1 - e^{-\bar{p}/\sigma_\alpha^2}. \tag{2.54}$$

Interestingly, if we were to set the threshold to the average power, i.e., $\bar{p} = \sigma_\alpha^2$, we obtain $P_{\text{OUTAGE}} = 1 - e^{-1} = 0.63$, i.e., 63% of the receivers do not receive adequate power!

2.5.3 Overall Channel Model

In Sects. 2.2 and 2.4, we had discussed the power loss due to distance (pathloss) and the randomness in the received signal power due to shadowing (large-scale or log-normal fading). In the previous section, we have discussed yet another source of randomness, namely, multipath propagation or small-scale fading. We are now able to combine these concepts into a single overall channel model. The channel is given by

$$\text{Channel} = \underbrace{\text{Pathloss}}_{\text{loss due to distance}} \times \underbrace{\text{Large-scale fading}}_{\text{Shadowing}} \times \underbrace{\text{Small-scale fading}}_{\text{Due to multipath}}.$$

Mathematically, gathering our models, we have for the overall channel (with a slight abuse of notation) as

2.5 Channel Impulse Response, Multipath, and Small-Scale Fading

$$h = \sqrt{PL(d) \times X_\sigma} \times \alpha, \qquad (2.55)$$

The abuse of notation in (2.55) arises from the fact that our previous analysis for $PL(d)$ and X_σ was in the dB scale, but here these terms are in the linear scale. Importantly, since $PL(d)$ accounts for the *average* power loss as a function of distance, and X_σ the fluctuation in the power level, we often set $\mathbb{E}[|\alpha|^2] = 1$, absorbing all power terms into $PL(d)$.

Another equivalent model, particularly convenient for Rayleigh fading, is to consider the overall channel h as a zero-mean complex Gaussian random variable with variance given by

$$\sigma^2_{h|\text{dB}} = PL(d) + X_\sigma, \qquad (2.56)$$

where $PL(d)$ and X_σ are in the dB scale.

2.5.4 Frequency Domain Analysis of the Channel Impulse Response

The impact of the delay spread and the potential for ISI is best interpreted in the frequency domain. Let us consider a simple two-path time-invariant channel impulse response:

$$h(\tau) = \alpha_1 \delta(\tau - \tau_1) + \alpha_2 \delta(\tau - \tau_2). \qquad (2.57)$$

where α_1 and α_2 are real. The frequency domain channel impulse response for such a channel is obtained by determining the Fourier Transform of $h(\tau)$:

$$\begin{aligned}
H(f) &= \int_{-\infty}^{\infty} h(\tau) e^{-j2\pi f \tau} d\tau \\
&= \int_{-\infty}^{\infty} \alpha_1 \delta(\tau - \tau_1) e^{-j2\pi f \tau} d\tau + \int_{-\infty}^{\infty} \alpha_2 \delta(\tau - \tau_2) e^{-j2\pi f \tau} d\tau \\
&= \alpha_1 e^{-j2\pi f \tau_1} + \alpha_2 e^{-j2\pi f \tau_2}
\end{aligned} \qquad (2.58)$$

The magnitude of the frequency domain impulse response is

$$|H(f)| = \sqrt{\alpha_1^2 + \alpha_2^2 + 2\alpha_1 \alpha_2 \cos(2\pi f \Delta \tau)}, \qquad (2.59)$$

where $\Delta \tau = \tau_2 - \tau_1$ is the channel delay spread.

Fig. 2.11 Frequency domain channel impulse response around 2100 MHz

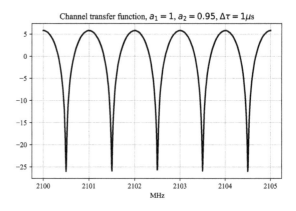

Fig. 2.12 Flat fading example, when the channel bandwidth $BW \ll 1/\Delta\tau$

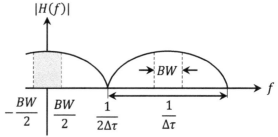

An illustration of a channel impulse response in frequency domain, for the frequency range from 2100 to 2105 MHz and two components arriving with $\tau_1 = 1\,\mu s$ and $\tau_2 = 2\,\mu s$, i.e., $\Delta\tau = 1\,\mu s$, is shown above (Fig. 2.11).

Now, consider the case of a relatively narrowband transmission. In this case, as illustrated in Fig. 2.12, the signal transmission occupies a small fraction of the channel frequency response; indeed, from the transmission *sees* a relatively flat spectrum, i.e., the received signal, in the frequency domain is given by

$$R(f) = \text{constant} \times S(f).$$

This is referred to as *flat fading*. From the figure and recognizing that the channel delay spread is $\Delta\tau$), we have flat fading when

$$\frac{BW}{2} \ll \frac{1}{2\Delta\tau} \implies BW \ll \frac{1}{\Delta\tau}. \tag{2.60}$$

We note that, in some sense, we have already analyzed the case of flat fading. The transmission bandwidth is inversely proportional to the symbol rate, i.e., $BW \propto 1/T_s$. The flat fading relation in (2.60) can be written as $\Delta\tau \ll T_s$. For this case, we had

$$r(t) = \alpha s(t - \tau) \implies h(t) = \alpha \delta(t - \tau), \quad \text{i.e.,} \quad H(f) = \alpha e^{j2\pi f\tau}. \tag{2.61}$$

2.5 Channel Impulse Response, Multipath, and Small-Scale Fading

Consequently, $|H(f)| = |\alpha|$, a constant.

The alternative case is when the transmission is broadband, i.e., T_s, the symbol period, is short. Here, $\Delta\tau > T_s$, leading to *frequency selective fading*. In this case, different frequency components of the transmitted signal see different fading levels. The resulting spreading in the signal causes ISI. The terms *frequency selective fading* and *ISI* are synonymous.

Example 2.4 The widest supported channel bandwidth in 5G for carrier frequencies below 7125 MHz is $BW = 100$ MHz. If the path-length difference between two received multipath components is 50 meters, determine if this signal is subjected to flat or frequency selective fading.

Solution The channel delay spread, the time delay between the two received signal components, is

$$\Delta\tau = \frac{\Delta d}{c} = \frac{50}{3 \times 10^8} = 166.67 \text{ ns}$$

and the inverse of the signal bandwidth is

$$\frac{1}{BW} = \frac{1}{100 \times 10^6} = 10 \text{ ns}$$

Therefore, we have the following inequality:

$$\Delta\tau > \frac{1}{BW}$$

and the fading is considered to be frequency selective. ∎

2.5.5 Power Delay Profile

As we have seen, the wireless channel impulse response spreads over time. For a given system, if we were to measure a large number of channel impulse responses and averaged the channel power, the squared magnitude of the channel, we would obtain a continuous function of the delay (the delay between the time of transmission and reception) called the power delay profile.

Another type of statistical analysis of wireless channels involves the average channel power which is obtained via integration of the square magnitude of the channel impulse response over variable t. Assuming wide-sense stationary wireless channels with uncorrelated scatterers, the power delay profile of the channel is defined as

$$P(\tau) = \lim_{T\to\infty} \frac{1}{T} \int_{-T/2}^{T/2} |h(t,\tau)|^2 \, dt. \qquad (2.62)$$

For instance, COST 207 report [10] specifies a power delay profile for typical urban (TU) environments. This model, often used in 2G GSM communications, is as follows:

$$P(\tau) = \begin{cases} e^{-\tau}, & 0 < \tau < 7\,\mu s, \\ 0 & \text{elsewhere.} \end{cases} \qquad (2.63)$$

Essentially, the maximum channel delay spread in this model is $7\,\mu s$.

Another related term is the RMS delay spread, which measures the delays over which most of the signal energy arrives. The total average channel power is given by

$$P_h = \int_{-\infty}^{\infty} P(\tau) \, d\tau \qquad (2.64)$$

Similarly, the mean delay is determined as

$$\overline{\tau} = \frac{1}{P_h} \int_{-\infty}^{\infty} \tau P(\tau) \, d\tau \qquad (2.65)$$

and the mean squared delay is

$$\overline{\tau^2} = \frac{1}{P_h} \int_{-\infty}^{\infty} \tau^2 P(\tau) \, d\tau \qquad (2.66)$$

The RMS delay is calculated as

$$\tau_{RMS} = \sqrt{\overline{\tau^2} - (\overline{\tau})^2} = \sqrt{\frac{1}{P_h} \int_{-\infty}^{\infty} \tau^2 P(\tau) d\tau - (\overline{\tau})^2} \qquad (2.67)$$

For discrete channels, with the tapped-delay form

2.5 Channel Impulse Response, Multipath, and Small-Scale Fading

$$h(t, \tau) = \sum_{\ell=1}^{L} \alpha_\ell(t)\delta(\tau - \tau_\ell), \tag{2.68}$$

where $\alpha(t)$ are complex channel gains, the power delay profile is

$$P(\tau) = \sum_{\ell=1}^{L} P_\ell \delta(\tau - \tau_\ell) \tag{2.69}$$

where

$$P_\ell = \lim_{T \to \infty} \frac{1}{T} \int_{-T/2}^{T/2} |\alpha_\ell(t)|^2 \, dt \tag{2.70}$$

is the i-th delay average power. The total channel power and the mean delays for the discrete model formulas are as follows:

$$P_h = \sum_{\ell=1}^{L} P_\ell, \quad \overline{\tau} = \frac{1}{P_h} \sum_{\ell=1}^{L} \tau_i P_\ell, \quad \overline{\tau^2} = \frac{1}{P_h} \sum_{\ell=1}^{L} \tau_i^2 P_\ell \tag{2.71}$$

Finally, the RMS delay for the discrete channel models, which captures the effective duration of channel impact, is calculated as

$$\tau_{RMS} = \sqrt{\overline{\tau^2} - (\overline{\tau})^2} = \sqrt{\frac{1}{P_h} \sum_{\ell=1}^{L} \tau_\ell^2 P_\ell - (\overline{\tau})^2} \tag{2.72}$$

The power delay profile plays a subtle role in modeling channels. In Sect. 2.5.3, we suggested we could absorb all the power terms into the pathloss term, leaving the small-scale fading with unit average power. This is acceptable for a frequency flat channel wherein the channel is characterized by one value. However, for a frequency selective channel, as in (2.52), the individual amplitudes must be consistent with the power delay profile. In this case, we have

$$h(t) = \sqrt{PL(d)} \times X_\sigma \times \sum_{\ell=0}^{L} \alpha_\ell \delta(t - \ell T_s), \tag{2.73}$$

but, now, we need $\mathbb{E}[|\alpha_\ell|^2] = P(\ell T_s)$. The remaining power terms can still be absorbed into $PL(d)$.

2.5.6 Frequency Correlation and Coherence Bandwidth

The *frequency correlation function* is the Fourier transform of the power delay profile:

$$\rho_f(\Delta f) = \int_{-\infty}^{\infty} P(\tau) e^{-j2\pi \Delta f \tau} \, d\tau, \qquad (2.74)$$

which measures the correlation between channel seen by two frequency components, within the transmission bandwidth, separated by Δf.

For a tapped delay-line channel, the frequency correlation is

$$\rho_f(\Delta f) = \int_{-\infty}^{\infty} P(\tau) e^{-j2\pi \Delta f \tau} \, d\tau \qquad (2.75)$$

$$= \int_{-\infty}^{\infty} \sum_{\ell=1}^{L} P_\ell \delta(\tau - \tau_\ell) e^{-j2\pi \Delta f \tau} \, d\tau \qquad (2.76)$$

$$= \sum_{\ell=1}^{L} P_i e^{-j2\pi \Delta f \tau_\ell} \qquad (2.77)$$

A related concept is the *coherence bandwidth* B_c, as the range of frequencies over which the channel response is relatively flat. Therefore, transmissions that have bandwidth $W < B_c$ are said to be narrowband and undergo flat fading, while transmissions with bandwidth $W > B_c$ are considered wideband (and undergo frequency-selective fading). One definition of the coherence bandwidth is the frequency spacing for which the frequency correlation function drops to 1/2 of the value for $\Delta f = 0$:

$$\left| \rho_f(\Delta f) \right|_{\Delta f = B_c} = \frac{1}{2} |\rho_f(0)|. \qquad (2.78)$$

It is worth emphasizing that this choice of 50% correlation value is not universal; as with definitions of bandwidth, different values may be used. One common alternative is a value of 90% correlation. An easier way to calculate the coherence bandwidth to bound the coherence bandwidth between 50% and 90% of correlation [6]:

$$\frac{1}{50 \tau_{RMS}} < B_c \leq \frac{1}{5 \tau_{RMS}}. \qquad (2.79)$$

2.5 Channel Impulse Response, Multipath, and Small-Scale Fading

The upper and lower limits above are only meant for approximate calculation. Alternatively, the "uncertainty relation" derived in [11]

$$B_c \geq \frac{\cos^{-1}(c)}{2\pi \, \tau_{RMS}} \qquad (2.80)$$

can be used, where $c \in [0, 1]$ is the coherence level interval.

2.5.7 Coherence Time

Analogous to the coherence bandwidth is the concept of coherence time. As mentioned previously, the channel changes at the rate of the Doppler shift due to motion in the channel. The Doppler shift is given by $f_D = \pm v/\lambda_c = \pm f_c v/c$ and has a "-" sign if the relative motion if away from the receiver and a "+" sign if toward the receiver. Here, v represents the *relative* speed to the propagation direction.

Example 2.5 Determine the maximum Doppler frequency at highway speeds using 5G cellular frequencies at 2.18 GHz.

Solution Let the highway speed be $v = 110$ km/h or $v = 110000/3600 = 30.55$ m/s. For the Doppler shift, we have the expression $f_d = \pm f_c v/c$. Thus, the maximum Doppler shifts for these two scenarios are

$$f_D = \pm 2.18 \times 10^9 \times 30.55/3 \times 10^8 = \pm 222 \text{ Hz}.$$

The *coherence time* T_c is defined as the time range over which the channel does not change significantly; over this time period, the channel values are highly correlated. It is related to the Doppler frequency shift as [6]

$$\frac{9}{16\pi f_D} < T_c \leq \frac{0.423}{f_D} \qquad (2.81)$$

The value $T_c = 0.423/f_D$ is typically used in design calculations. For example, if the Doppler rate is $f_D = 213.85$ Hz, the coherence time is in the 2 ms range, whereas 4G and 5G systems operating below 3 GHz have transmission/processing units of 1 ms; thus, even at highway speeds, we see a nearly constant channel over the time duration of an entire unit. Higher Doppler rates result in a lower coherence time; in this case, the power level changes rapidly and cannot be properly tracked at the receiver, significantly degrading performance.

Problems

1. Calculate the free space loss at the distance $d = 100$ meters for signals transmitted at 900 MHz, 2600 MHz, and 26 GHz. Determine the dB scale value for these pathlosses and comment on the differences between the values.
2. If the transmit power is $P_T = 53$ dBm over 100 MHz, determine the following:

   a. Transmit power over 5 MHz assuming a constant power spectral density over the entire 100 MHz transmission range.
   b. Power flux density (PFD) over 5 MHz at the distance of $d = 1$ km assuming $G_T = G_R = 0$ dBi and free space transmission.
   c. Effective antenna area if the transmission center frequency is 3550 MHz.
   d. Received power over 5 MHz for the calculated PFD at 1 km of distance and effective antenna area from (c).
3. The dB scale free space pathloss can be expressed as

$$PL(d)_{|dB} = 10\log_{10}\left(\frac{4\pi d}{\lambda}\right)^2$$

   Express this pathloss in terms of distance d and carrier frequency f_c, and write simplified formulas the the following input options:

   a. d is in kilometers and f_c is in MHz.
   b. d is in kilometers and f_c is in GHz.
4. Assuming a transmitter with a transmit power of $P_T = 43$ dBm over a channel bandwidth of 10 MHz, determine the PFD value of the received signal over 1 MHz of measurement assuming an antenna gain at the transmitter of $G_T = 17$ dBi, carrier frequency of $f_c = 3550$ MHz, the distance between the transmitter and receiver is $d = 10$ km, and the pathloss exponent is $n = 2.5$. Assume constant power spectral density of the transmitted/received signal.
5. A cellular radio operating in the frequency range from 2180 to 2200 MHz emits a certain level of out of band emission in the frequency range above 2200 MHz. If the out of band emission level is -50 dBm/MHz, and a ground satellite receiver (earth station or ES) operating at 2234 MHz cannot tolerate more than -224 dBW/Hz of the interference power spectral density, calculate the minimal distance between the base station and an ES for urban and suburban morphology using COST Hata model. Assume transmit antenna gain of $G_T = 16$ dBi, received antenna gain of $G_R = 20$ dBi, base station antenna height $h_{BS} = 35$ m for urban, $h_{BS} = 70$ m for suburban scenario, and an ES antenna height $h_{ES} = 8$ m.
6. A simple way to illustrate the impact of Doppler frequency shifts relate to the rate change of the channel is to consider a scenario where the transmitted signal is $s(t) = \cos(2\pi f_c t)$, and at the receiver, we receive two multipath components

2.5 Channel Impulse Response, Multipath, and Small-Scale Fading

with different Doppler shifts $f_{D1} = f_1 - f_c$, where $f_1 < f_c$, and $f_{D2} = f_2 - f_c$, where $f_2 > f_c$. Determine the expression for the combined received signal and explain the rate of received signal amplitude change.

7. Determine the maximum Doppler frequency at pedestrian speeds for 5.2 GHz WLAN and at highway speeds cellular signal at 2.1 GHz assuming a highway speed of $v = 110$ km/h. How does the Doppler rate change if the cellular frequency is changed from 2.1 to 900 MHz?

8. The signal fading due to channel multipath happens over very small scales and is known as a *small-scale fading*. Let $s_T(t) = e^{j\omega_0 t}$ be the transmitted complex exponential signal, where $\omega_0 = 2\pi f_0$ is the angular frequency. If the signal arrives via two paths, with two different delays τ_1 and τ_2, as shown in the figure below, determine the expression for the received signal and the magnitude of the amplitude distortion.

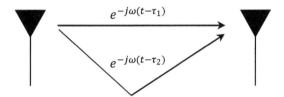

9. Let the power delay profile be $P(\tau) = \frac{1}{c} e^{-\tau/c}$ for $\tau \geq 0$ and $P(\tau) = 0$ for $\tau < 0$. Determine τ_{RMS} and the coherence bandwidth B_c if $c = 1\,\mu s$.

10. For the ITU-R vehicular channel B model given below, calculate the RMS delay spread of the power delay profile.

τ_i (μs)	0	0.3	8.9	12.9	17.1	20
P_i (dB)	−2.5	0	−12.8	−10	−25.2	−16

11. A discrete multipath power profile is shown in the figure below.

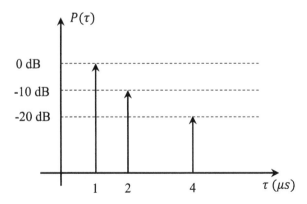

Determine the following:

a. Mean excess delay $\bar{\tau}$ and the RMS delay τ_{RMS}
b. Frequency correlation function $\rho_f(\Delta f)$
c. Ratio $\rho_f(0)/|\rho_f(\Delta f)|$ for $\Delta f = 250\,\text{kHz}$. Is $\Delta f > B_c$, where B_c is the coherence bandwidth of the channel?

12. Rayleigh distributed fast fading amplitude distribution is

$$p(\alpha) = \frac{\alpha}{\sigma^2} e^{-\frac{\alpha^2}{2\sigma^2}} \tag{2.82}$$

derive the expression for the power distribution.

13. To meet a certain outage level target under a Rayleigh fading scenario, we change the operating received power level to a higher level:

$$\bar{y}_{|dB} = M_{|dB} + y_{OUTAGE|dB} \tag{2.83}$$

The amount of the power increase is called the Rayleigh fading margin M and in linear scale it can be represented as

$$M = \frac{\bar{y}}{y_{OUTAGE}}. \tag{2.84}$$

Derive the expression for the Rayleigh margin as a function of outage probability.

14. A wireless communication system, subjected to Rayleigh fading, is designed with the outage target of 5%. Determine the required margin M.

Appendix: Selected 3GPP Defined Pathloss Models

The expanded pathloss formulas for rural macro deployment are as follows (Table 2.3):

$$PL_1 = 20\log_{10}(40\pi d_{3D} f_c/3) + \min\left(0.03h^{1.72}, 10\right)\log_{10}(d_{3D}) \tag{2.85}$$

$$-\min\left(0.044h^{1.72}, 14.77\right) + 0.002\log_{10}(h)d_{3D} \tag{2.86}$$

$$PL_2 = PL_1(d_{BP}) + 40\log_{10}(d_{3D}/d_{BP}) \tag{2.87}$$

$$PL'_{\text{RMa-NLOS}} = 161.04 - 7.1\log_{10}(W) + 7.5\log_{10}(h) \tag{2.88}$$

$$-(24.37 - 3.7(h/h_{BS})^2)\log_{10}(h_{BS}) \tag{2.89}$$

$$+(43.42 - 3.1\log_{10}(h_{BS}))(\log_{10}(d_{3D}) - 3) \tag{2.90}$$

$$+20\log_{10}(f_c) - (3.2(\log_{10}(11.75h_{UT}))^2 - 4.97 \tag{2.91}$$

2.5 Channel Impulse Response, Multipath, and Small-Scale Fading

Table 2.3 3GPP specified selected pathloss models in dB scale, f_c is in GHz and d is in meters

Scenario	LOS/NLOS	Pathloss	Applicability range
Rural	LOS	$PL_{\text{RMa-LOS}} = \begin{cases} PL_1 & , 10\text{m} \leq d_{2D} \leq d_{BP} \\ PL_2 & , d_{BP} \leq d_{2D} \leq 10\text{km} \end{cases}$	$10 \leq h_{BS} \leq 150$ m, $1 \leq h_{UT} \leq 10$ m, $5\text{ m} \leq W \leq 50$ m
	NLOS	$PL_{\text{RMa-NLOS}} = \max(PL_{\text{RMa-LOS}}, PL'_{\text{RMa-NLOS}})$	$5\text{ m} \leq h \leq 50$ m,
Urban	LOS	$PL_{\text{UMa-LOS}} = \begin{cases} PL_1 & , 10\text{m} \leq d_{2D} \leq d'_{BP} \\ PL_2 & , d'_{BP} \leq d_{2D} \leq 5\text{ km} \end{cases}$	$h_{BS} = 25$ m, $1.5 \leq h_{UT} \leq 22.5$ m,
	NLOS	$PL_{\text{UMa-NLOS}} = \max(PL_{\text{UMa-LOS}}, PL'_{\text{UMa-NLOS}})$	

and break point distance $d_{BP} = 2\pi h_{BS} h_{UT} f_c/c$, where f_c is the center frequency in Hz and $c = 3.0 \times 10^8$ m/s is the propagation velocity in free space.

Similarly, the expanded pathloss formulas for urban macro deployment are

$$PL_1 = 28 + 22 \log_{10}(d_{3D}) + 20 \log_{10}(f_c) \tag{2.92}$$

$$PL_2 = 28 + 40 \log_{10}(d_{3D}) + 20 \log_{10}(f_c) \tag{2.93}$$

$$-9 \log_{10}((d'_{BP})^2 + (h_{BS} - h_{UT})^2) \tag{2.94}$$

$$PL'_{\text{UMa-NLOS}} = 13.54 + 39.08 \log_{10}(d_{3D}) \tag{2.95}$$

$$+20 \log_{10}(f_c) - 0.6(h_{UT} - 1.5) \tag{2.96}$$

and the breakpoint distance $d'_{BP} = 4(h_{BS} - 1)(h_{UT} - 1) f_c/c$. It is worth noting that the pathloss model specified in TR 38.901 is limited to a certain range of parameters, especially for urban micro model, and also has probabilistic parameters intended for simulations. For a wider range of applicability, the TR 36.873 [12] specified channel model can be used.

References

1. Hata, M., Empirical formula for propagation loss in land mobile radio services. IEEE Trans. Veh. Technol. **VT-29**(3), 317–325 (1980)
2. Sharing Arrangement Between the Department of Industry of Canada and the Federal Communications Commission of the United States of America Concerning the Use of the Frequency Bands 806–824 MHz and 851–869 MHz by the Land Mobile Service Along the Canada-United States Border. https://ised-isde.canada.ca/site/spectrum-management-telecommunications/en/official-publications/legislation-regulations-and-treaties/terrestrial-radiocom-agreements-and-arrangements-traa/arrangement-f. Cited Oct 21, 2023
3. Y. Okumura et al., Field strength and its variability in UHF and VHF land-mobile radio service. Rev. Elec. Commun. Lab **16**, 825–873 (1968)
4. Final report of the COST Action 231. http://www.lx.it.pt/cost231/final_report.htm. Cited Oct 21, 2023
5. 3rd Generation Partnership Project, Technical Specification Group Radio Access Network, in *Radio Network Planning Aspects*. https://www.3gpp.org/ftp/Specs/archive/43_series/43.030/43030-i00.zip. Cited May 2, 2024
6. T.S. Rappaport, *Wireless Communications* (Prentice Hall, New York, 1999)
7. Study on channel model for frequencies from 0.5 to 100 GHz. https://www.3gpp.org/ftp/Specs/archive/38_series/38.901/38901-h00.zip. Cited Oct 21, 2023
8. Influence of tarrain irregularitis and vegetation on troposheric propagation. https://search.itu.int/history/HistoryDigitalCollectionDocLibrary/4.282.43.en.1006.pdf. Cited Oct 21, 2023
9. A.F. Molisch, *Wireless Communications* (Wiley, New York, 2011)
10. Digital land mobile radio communications, COST 207 report. https://op.europa.eu/en/publication-detail/-/publication/61fc77e7-bca2-4229-8eb4-77741f0d2ab2/language-en/format-PDF/source-search. Cited Oct 21, 2023
11. B. H. Fleury, An uncertaintly relation for WSS processes and its application to WSSUS systems. IEEE Trans. Commun. **44**(12), 1632–1634 (1996)
12. Study on 3D channel model for LTE. https://www.3gpp.org/ftp//Specs/archive/36_series/36.873/36873-c70.zip. Cited Oct 21, 2023

Chapter 3
Noise and Interference

3.1 Introduction

In the previous chapters, we developed the basics of communications and wireless channels. This chapter focuses on phenomena that limit our ability to communicate; specifically, we develop models for noise (which is always present in any practical receiver) and interference (signals received due to unwanted transmissions). We explore the basics of noise analysis including the use of a noise factor, an equivalent noise figure, receiver sensitivity, and noise rise. Distinguishing noise from interference, we introduce the key notions of signal-to-noise ratio (SNR) and signal-to-interference-plus-noise ratio (SINR). These ratios largely determine the achievable communication rates. We also introduce a step-by-step approach to link budget calculations; link budgets are crucial in system design. Finally, we examine how component nonlinearities generate intermodulation products, leading to additional interference and adding to the complexity of analyzing system performance.

3.2 Noise and Interference Sources

Unwanted signals at the wireless radio receiver can be categorized as following Fig. 3.1:

1. Internal noise, in the form of thermal noise generated by receiver's components

---

Supplementary Information The online version contains supplementary material available at (https://doi.org/10.1007/978-3-031-76455-4_3).

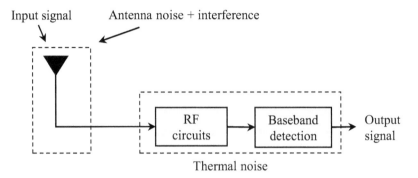

Fig. 3.1 Block diagram of a wireless receiver with sources of noise and interference

2. External noise, comprising antenna noise generated by surroundings and interference from other sources (e.g., other communication and/or radar systems)

An alternate categorization is that noise is caused by natural phenomena, while interference is due to man-made sources of radiation (Fig. 3.1).

3.2.1 Thermal Noise

Thermal noise is the fluctuation of voltage caused by the random motion of electrons within the electrical components at the receiver. This random motion is caused by the non-zero (in Kelvin) circuit temperature (and hence the term "thermal"). The overall measurable impact is due to the superposition of the random motion of the many electrons within the components; by the central limit theorem, any time sample of the overall noise can be treated as Gaussian and zero mean since the electrons do not have a preferred direction of motion. Furthermore, the rapid electron motion implies that two time samples, even if closely spaced, are uncorrelated. Consequently, the autocorrelation function can be modeled as a δ-function, i.e., the power spectral density of thermal noise is essentially constant (at least over the range of cellular frequencies). An alternative interpretation is that, on average, all frequencies contribute the same noise power. As seen in Fig. 3.2, the PSD can be represented as two sided (when modelling both positive and negative frequencies) or single sided (when only considering positive frequencies).

Because of the flat power spectral density, every frequency is affected the same way, so the thermal noise is also known as a white noise. Since thermal white noise adds to the receive signal, it is referred to as additive white Gaussian noise (AWGN)

For the two-sided case, the noise power spectral density is $S_n(f) = N_0/2$ W/Hz. The linear scale value of N_0 is given by

3.2 Noise and Interference Sources

Fig. 3.2 Two-sided and single-sided noise power spectral densities. Note the doubling of the noise PSD when considering positive frequencies only. (**a**) two-sided noise PSD. (**b**) single-sided noise PSD

Fig. 3.3 Bandlimited two-sided power spectral density of the thermal noise

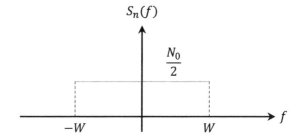

$$N_0 = k_B T_0 = 4 \times 10^{-21} \text{ W/Hz} \quad (3.1)$$

where $k_B = 1.38064852 \times 10^{-23}$ J/K is Boltzmann's constant and T_0 is temperature. For cellular systems, one usually considers a reference temperature of $T_0 = 290$ K. Note that $1 \text{ J} = 1 \text{ W} \times \text{s} = 1$ W/Hz. In the dB scale, the noise power spectral density is

$$N_{0|\text{dBW}} = 10 \log_{10}(k_B T_0) = -203.98 \approx -204 \text{ dBW/Hz} \quad \text{or} \quad -174 \text{ dBm/Hz} \quad (3.2)$$

As defined, mathematically, thermal noise has infinite power (due to the infinite integration over the constant value of PSD).[1] However, in practical systems, the receiver's processing chain includes filtering of the received signal, thus limiting the frequency range within the signal bandwidth W. This filtering operation limits the noise bandwidth and hence the effective noise power.

Figure 3.3 plots the noise PSD after ideal low-pass filtering. The noise power, also called the noise floor, over this limited frequency range is given by

---

[1] The flat—or white—PSD is valid up to frequencies beyond the THz range. Clearly, no physical signal can have infinite power.

$$P_N = \int_{-\infty}^{\infty} S_n(f)df = \int_{-W}^{W} \frac{N_0}{2} df = N_0 W. \tag{3.3}$$

In the dBm scale, the total power of the bandlimited noise is

$$P_{N|\text{dBm}} = N_{0|\text{dBm}} + 10\log_{10}(W) = -174 + 10\log_{10}(W) \tag{3.4}$$

Example 3.1 Determine the power of the thermal noise for a communication system with a bandwidth of $W = 10\,\text{MHz}$.

Solution Using (3.4), the thermal noise power is

$$P_{N|\text{dBm}} = -174 + 10\log_{10}\left(10 \times 10^6\right) = -174 + 70 = -104 \text{ dBm.} \tag{3.5}$$

Since thermal noise is white and zero-mean Gaussian, time samples are characterized by their variance, generally denoted as σ^2. Importantly, since we model the noise process as wide-sense stationary, the average power in a single time sample is the same as its time-averaged power, given by P_N, i.e., $\sigma^2 = P_N = N_0 W$. ∎

3.2.1.1 Noise Factor

A receiver, at a high level, includes the antenna and the RF chain used to retrieve the message signal. The RF chain, itself, consists of a series of processing steps such as filtering, down-conversion, and sampling. While the antenna picks up noise from the environment, each processing stage also adds thermal noise. A crucial measure of signal quality is the signal-to-noise ratio (SNR), the ratio of the power in the useful signal to the average power in the noise. Since each stage adds noise, the SNR changes from the input to the final output. This change in SNR is characterized by the *noise factor* of each stage.

To illustrate the use of a noise factor, consider an active component with gain G as shown in Fig. 3.4, the output noise power is

$$N_{OUT} = N_{IN} \times G + N_a, \quad Na = k_B T_e G W \tag{3.6}$$

where T_e is the equivalent noise temperature of the active component and W is the bandwidth of the receiver. This noise temperature is not the physical device temperature but a temperature equivalent to the temperature of a resistor that would generate thermal noise of power $P_N = k_B T_e W$. The output signal power, after the

3.2 Noise and Interference Sources

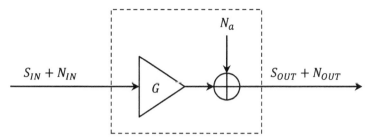

Fig. 3.4 Block diagram of an electric component that adds N_a noise

applied gain, is

$$S_{OUT} = S_{IN} \times G$$

as illustrated in Fig. 3.4.

By examining the expressions above, we observe that both the input signal and the input noise are amplified by the component gain G, but the noise power is further increased by the amount N_a. The output SNR is now:

$$SNR_{OUT} = \frac{S_{OUT}}{N_{OUT}} = SNR_{IN} \frac{G}{G + \frac{N_a}{N_{IN}}} \quad (3.7)$$

Evidently, due to the noise added by the component, the SNR at the output of an active component is worse (lower) than the SNR at the input. The SNR degradation is measured through the noise factor F, which is defined as the ratio of the input and outputs SNRs:

$$F = \frac{SNR_{IN}}{SNR_{OUT}} = \frac{G + \frac{N_a}{N_{IN}}}{G} = 1 + \frac{N_a}{G N_{IN}} > 1. \quad (3.8)$$

Importantly, since output SNR is always lower than the input SNR, the noise factor is always greater than unity. The noise factor value F is specified for the input temperature $T_{IN} = T_0$, as defined by the Institute of Radio Engineers (IRE), a predecessor of IEEE, in [1].

Effectively, the input signal and noise are subjected to different gains. While the signal is subjected to gain G, the output noise has a gain that is a product of G and the noise factor F:

$$N_{OUT} = N_{IN} \times G + N_a = N_{IN} \times G \left(1 + \frac{N_a}{G N_{IN}}\right) = N_{IN} \times G \times F. \quad (3.9)$$

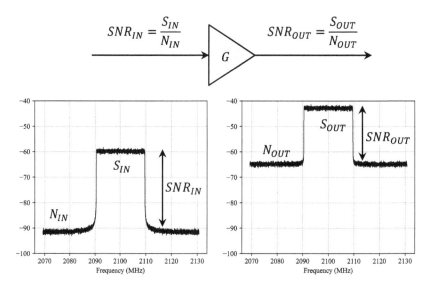

Fig. 3.5 Input power level (dBm) on the left and output power level (dBm) on the right. The output SNR is lower than the input SNR

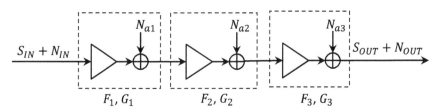

Fig. 3.6 Three components in series

The change in SNR is illustrated in Fig. 3.5. As mentioned before, both the input signal S_{IN} and the input noise N_{IN} are amplified, but the output S_{OUT}/N_{OUT} ratio has been reduced by about 10 dB, which corresponds to a noise factor $F = 10$.

It is also convenient to express the noise factor in terms of component and input temperatures:

$$F = 1 + \frac{N_a}{GN_{IN}} = 1 + \frac{k_B T_e W G}{k_B T_0 W G}$$

$$= 1 + \frac{T_e}{T_0}. \tag{3.10}$$

Equivalently, $T_e = T_0(F - 1)$ and $N_{OUT} = N_{IN} GF = k_B W (T_0 + T_e) G$.

As mentioned, the receiver comprises multiple components cascaded in series, resulting in composite noise factor. Figure 3.6 illustrates a three-stage processing chain.

3.2 Noise and Interference Sources

The output noise power after three stages is

$$N_{OUT} = N_{IN}G_1G_2G_3 + N_{a1}G_2G_3 + N_{a2}G_3 + N_{a3}$$
$$= N_{IN}G_1G_2G_3 \left(F_1 + \frac{F_2 - 1}{G_1} + \frac{F_3 - 1}{G_1G_2} \right) \quad (3.11)$$

where F_i are noise factor values for individual components

$$F_i = 1 + \frac{N_{ai}}{G_i N_{IN}}, \quad i = 1, 2, 3 \quad (3.12)$$

and the output signal is

$$S_{OUT} = G_1 G_2 G_3 S_{IN}. \quad (3.13)$$

We are now able to express an equivalent noise factor given by

$$F_{eq} = F_1 + \frac{F_2 - 1}{G_1} + \frac{F_3 - 1}{G_1 G_2}, \quad (3.14)$$

the output noise power is expressed as

$$N_{OUT} = N_{IN} G_1 G_2 G_3 F_{eq} \quad (3.15)$$

corresponding to an output SNR of

$$\frac{S_{OUT}}{N_{OUT}} = \frac{S_{IN} G_1 G_2 G_3}{N_{IN} G_1 G_2 G_3 F_{eq}} = \frac{S_{IN}}{N_{IN}} \frac{1}{F_{eq}} \quad (3.16)$$

which is the same expression as for a single component. Therefore, for the cascade of three components, the equivalent noise factor is calculated by combining individual noise factors and gains, as shown above. Similarly, the equivalent temperature of the three components in a series can be calculated as

$$T_{eq} = T_0 (F_{eq} - 1)$$
$$= T_0 \left(1 + \frac{T_1}{T_0} - 1 + \frac{T_2}{T_0 G_1} + \frac{T_3}{T_0 G_1 G_2} \right)$$
$$= T_1 + \frac{T_2}{G_1} + \frac{T_3}{G_1 G_2} \quad (3.17)$$

where $F_i - 1 = T_i / T_0$, $i = 1, 2, 3$. The output noise power, expressed in terms of temperatures, can be written as

$$N_{OUT} = N_{IN} G_1 G_2 G_3 F_{eq}$$
$$= k_B T_0 W G_1 G_2 G_3 \left(1 + \frac{T_{eq}}{T_0}\right)$$
$$= k_B (T_0 + T_{eq}) W G_1 G_2 G_3 \quad (3.18)$$

In general, the noise factor and temperature expressions, as well as the output signal and noise levels, for a series of n components are

$$F_{eq} = F_1 + \frac{F_2 - 1}{G_1} + \frac{F_3 - 1}{G_1 G_2} + \cdots + \frac{F_n - 1}{G_1 G_2 G_3 \cdots G_{n-1}}$$

$$T_{eq} = T_1 + \frac{T_2}{G_1} + \frac{T_3}{G_1 G_2} + \cdots + \frac{T_n}{G_1 G_2 G_3 \cdots G_{n-1}} \quad (3.19)$$

$$N_{OUT} = N_{IN} F_{eq} \prod_{i=1}^{n} G_i = k_B (T_0 + T_{eq}) W \prod_{i=1}^{n} G_i$$

$$S_{OUT} = S_{IN} \prod_{i}^{n=1} G_i \quad (3.20)$$

These expressions lead to an interesting conclusion: since $G_i > 1$, the equivalent noise temperature in (3.19) is dominated by the first stage. In a processing chain with multiple stages, it is best to make the first stage contribute as little noise as possible. In an RF receiver chain, this is usually done using a low-noise amplifier (LNA).

3.2.1.2 Passive Devices

Passive devices, operating at a temperature T, usually maintain a temperature equilibrium, and therefore the noise power remains unchanged between input and output. For a passive device gain of $G < 1$, we define a loss as $L = 1/G$. The output noise is

$$N_{OUT} = N_{IN} G + N_a - N_{IN} \quad (3.21)$$

where N_a is the added noise deficit that maintains the temperature equilibrium. From (3.21), the added noise N_a is

$$N_a = N_{IN}(1 - G) = k_B T W (1 - G) = k_B T_e W G \quad (3.22)$$

which gives us the equivalent noise temperature of the passive device operating at room temperature

3.2 Noise and Interference Sources

$$T_e = T\left(\frac{1}{G} - 1\right) = T(L-1) \tag{3.23}$$

and the noise figure

$$F = 1 + \frac{T_e}{T_0} = 1 + (L-1)\frac{T}{T_0}. \tag{3.24}$$

If the passive device operates at the room temperature T_0, the noise figure is

$$F = L = \frac{1}{G}. \tag{3.25}$$

In cellular communication systems, passive components are considered to operate at room temperature.

3.2.1.3 Noise Figure

Since the noise factor is related to power, it is usually mentioned in terms of the dB scale, called the noise figure:

$$NF = 10\log_{10}(F). \tag{3.26}$$

Typical noise figure values for cellular base stations are between 2 and 3 dB. For user devices, the noise figure values are between 6 and 10 dB.

To analyze cellular system receivers, we usually calculate the equipment noise floor as the product of an input noise level and the equivalent noise factor (noise factor from all the components in the receiver). Gains apply equally to both signal and noise and are usually ignored for the SNR considerations. With this in mind, given our discussion on the thermal noise PSD, the equipment noise floor in the dB scale is calculated as

$$P_{N|\text{dBm}} = -174 + 10\log_{10}(W) + NF. \tag{3.27}$$

Example 3.2 Calculate the user equipment noise floor if the operating bandwidth is 9 MHz and the noise figure is $NF = 7\,\text{dB}$.

Solution For the given set of parameters, the thermal noise floor is

$$-174 + 10\log_{10}\left(9 \times 10^6\right) = -104.46\ \text{dBm}, \tag{3.28}$$

and the equipment noise floor is

$$P_{N|dBm} = -174 + 10\log_{10}\left(9 \times 10^6\right) + NF = -104.46 + 7 = -97.46 \text{ dBm}. \tag{3.29}$$

This increase in noise floor due to noise figure has an important implication on the minimum required signal level at the receiver; this issue will be explored further in Sect. 3.2.3.2.

■

3.2.2 Antenna Noise and System Temperature

As the name suggests, antenna noise arises from radiation picked up by the receive antenna from its surroundings. In cellular systems, base station antennas usually point toward the ground, and ground noise forms the dominant noise source. For such antennas, sky noise is essentially negligible. Ground noise power is largely independent of frequency and can be modelled in a manner similar to thermal noise. For terrestrial systems, it is reasonable to assume that the noise temperature of the antenna is the room temperature, denoted as T_0. However, in general, the antenna temperature may be different, especially when satellite communications are considered. If T_A is the antenna noise temperature,[2] we are often interested in the total system temperature:

$$T_{system} = T_A + T_R \tag{3.30}$$

where $T_R = T_0(F-1)$ is the receiver's noise temperature and F is receiver's noise figure. If $T_A = T_0$, the total or system temperature becomes $T_{system} = T_0 + T_0(F-1) = FT_0$. However, for lower antenna temperatures (e.g., 50K–70K is typically used for earth stations that receive signals from satellites), the equivalent system temperature in dB scale is

$$\begin{aligned} 10\log_{10}\left(T_{system}\right) &= 10\log_{10}\left(T_A + T_R\right) \\ &= 10\log_{10}\left(T_A + \left(10^{0.1NF} - 1\right)T_0\right) \\ &= 10\log_{10}\left[10^{0.1NF}\left(T_A 10^{-0.1NF} + T_0 - T_0 10^{-0.1NF}\right)\right] \\ &= NF + 10\log_{10}\left(T_0 + (T_A - T_0)10^{-0.1NF}\right) \end{aligned} \tag{3.31}$$

which was, for example, used by 3GPP in its study of 5G networks to support non-terrestrial networks [2].

---

[2] The T_A is mostly the external noise; the internally added noise by the antenna as a passive device is not considered here and is usually negligible.

3.2.3 Interference and Signal-to-Interference-Plus-Noise Ratio in AWGN Channels

The key benefit of dividing a large geographical region into cells is that the same operating frequencies can be reused in different cells. However, when multiple regions operate on the same frequency, the receiver sees *interference*—undesired signals from other cells. Only one of the transmissions is desired (the signal), while the remaining interfere with the signal reception. In early cellular systems, neighboring regions used different sub-bands of the available spectrum, thereby reducing this *intercell interference*. However, this approach also limits the frequencies available within each cell to serve users. Modern systems have chosen to use all the available bandwidth in every cell, called frequency reuse one; the tradeoff is that modern systems deal with high levels of intercell interference. A key issue is the resulting impact on so-called cell-edge users, users near the boundary of two cells who seem approximately equal signal and interference powers.

In theory, with frequency reuse one, all cells interfere with all other cells. However, due to propagation losses (pathloss) the strongest interference encountered is from a neighboring cell. An illustration of two cells facing each other, at the intersite distance $2D$ (D is the radius of a single cell), is shown in Fig. 3.7.

Ignoring shadowing effects, as we saw in Chap. 2, the received signal power at some distance $d < D$ away from the transmitter, in dB scale, is

$$P(d)_{R|\text{dBm}} = P_{T|\text{dBm}} + G_{T|\text{dBi}} + G_{R|\text{dBi}} - PL(d)_{|\text{dB}}$$
$$= EIRP_{|\text{dBm}} + G_{R|\text{dBi}} - PL(d)_{|\text{dB}} \qquad (3.32)$$

where the mean pathloss, defined in Chap. 2, is

$$PL(d)_{|\text{dB}} = b + 10n \log_{10}(d) = b + 10 \log_{10}(d^n). \qquad (3.33)$$

Using these two expressions, in linear scale, we have

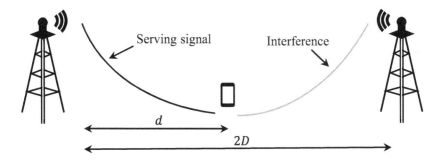

Fig. 3.7 Serving and interfering cells at a cell-edge receiver located almost equidistant from two cellular sites

$$P_R(d) = \frac{P_T G_T G_R}{10^{\frac{b}{10}} d^n} \propto P_T d^{-n} \qquad (3.34)$$

At the receiver, the SNR is given by

$$SNR = \frac{P_R(d)}{P_N} \propto \frac{P_T d^{-n}}{N_0 W}, \qquad (3.35)$$

where W is the signal bandwidth. As one would expect, the SNR increases with increasing transmit power but decreases (rapidly for large n) with distance. Interestingly, the noise power increases with increasing signal bandwidth. This expression, however, ignores the impact of the second base station in Fig. 3.7.

The transmission from the second base station, within the same band, acts as a source of interference that increases the effective noise power level. The power received from the interfering base station is

$$P_I(d) = \frac{P_T G_T G_R}{10^{\frac{b}{10}} (2D-d)^n} \propto P_T (2D-d)^{-n}, \qquad (3.36)$$

where for simplicity we assumed that the interfering base station uses the same transmit power P_T and the same transmit antenna gain G_T and that the receiver's antenna gain is G_R.

While SNR is the key metric when interference is absent ($P_I(d) = 0$ or a *noise-limited system*), the important metric, now, is the signal-to-interference-plus-noise ratio (SINR) given by

$$SINR = \frac{P_R(d)}{P_N + P_I(d)} = \frac{C P_T d^{-n}}{N_0 W + C P_T (2D-d)^{-n}}. \qquad (3.37)$$

In many practical cases, we have $P_I(d) \gg P_N$ which leads to *interference-limited system*. In this case, ignoring the noise term in the denominator, we have the simpler expression:

$$SINR = \frac{C P_T d^{-n}}{C P_T (2D-d)^{-n}} = \left(\frac{2D}{d} - 1\right)^n. \qquad (3.38)$$

Because the SINR is a function of distance, the SINR is also called a geometry factor or G-factor [3]. Crucially, in an interference-limited system with equal power transmissions, the interference-limited SINR is *independent* of the transmit power!

Since cell-edge users are, literally, the users near the edge of the cell being served by a base station, these users are at a distance $d \simeq D$ from their serving *and interfering* base stations. Using (3.38), this results in an SINR of unity (0 dB), i.e., the signal and interference are of the same level. In fact, with shadowing, the signal power could be lower than the interference! One possible solution is fractional

3.2 Noise and Interference Sources

frequency reuse, wherein frequencies used in all cells are reserved for users near the base stations and frequencies used for the cell edge are not shared with neighbors.

3.2.3.1 SINR and Cellular Data Rates

Using an information theoretic analysis due to Shannon [4], one can show that the maximum achievable reliable communications data rate

$$R_{\max} = W \log_2 (1 + SINR),$$

measured in bits per second (bps); here W is the transmission bandwidth used. This expression is closely tied to the maximum achievable *spectral efficiency*, measured in bit per second per Hz (bps/Hz):

$$\text{Spectral efficiency} = \frac{\text{data rate}}{\text{bandwidth used}} = \frac{R}{W} = \log_2 (1 + SINR).$$

In the noise-limited case, the SINR is replaced with SNR.

Unfortunately, achieving this theoretical upper bound requires infinitely long error control codes and Gaussian symbols. In practice, there is a penalty for finite-length codes and discrete modulations, and the practically achievable spectral efficiency is lower than the above expression. Importantly, the use of a finite set of modulations and coding settings also limits the true data rate to some finite choices. These choices are usually listed in a so-called *Modulation Coding Scheme* (MCS) table [5, 6].[3] However, for evaluation purposes, we can convert the SNR (or the SINR value) into a realistic cellular data rate. For modern systems, radio infrastructure vendors usually provide specific achievable rates as a function of SINR. However, for evaluation purposes, following the 3GPP technical reports TR 36.942 [7] (for 4G) and TR 38.803 [8] (for 5G), we can use

$$R_{\text{data}} = \alpha W \log_2 (1 + SINR) \tag{3.39}$$

where the SINR value is in linear scale and the value for α is

$$\alpha = \begin{cases} 0.6, & \text{downlink} \\ 0.4 & \text{uplink.} \end{cases} \tag{3.40}$$

In the industry, this is known as a truncated Shannon's formula because it applies to a limited range of SINR values, or alpha-Shannon formula due to scaling,

---

[3] In practice, the modulation order and coding rate are chosen based on the measured SINR, thereby adjusting the data rate; this is referred to as *adaptive modulation and coding*.

and provides the baseline performance for 4G and 5G systems. Actual network performance is usually better than this baseline.

Example 3.3 Using the alpha-Shannon formula, determine the SINR values for a data rate of $R_{UL} = 50$ kbps when 180 kHz and 360 kHz are used.

Solution

$$R_{UL} = 0.4W \log_2(1 + SINR) = 50 \text{ kbps}$$
$$\Rightarrow SINR = 2^{\frac{R_{UL}}{0.4W}} - 1$$
$$\Rightarrow SINR = [0.6183, 0.2721]$$
$$\Rightarrow SINR_{|dB} = 10\log_{10}(SINR) = [-2.1, -5.65] \text{ (dB)} \quad (3.41)$$

where $W = [180, 360]$ kHz. ∎

3.2.3.2 Receiver Sensitivity

Cellular systems based on 4G and 5G technologies can achieve data rates that can go as high as a few hundred Mbps for a single channel. On the other hand, due to adaptive modulation and coding, very low data rates are also possible. For operators and equipment manufacturers, it is beneficial to be able to compare actual receivers operating at low signal levels to a known reference. For example, for base stations, 3GPP has adopted a method where the SNR level that defines receiver's sensitivity for different reference channels achieves 95% of the throughput when QPSK modulation is used and the target code rate is 1/3 [9, 10]. A similar definition is also used for user equipment [11].

This reference minimum SNR corresponds to the ratio of the sensitivity level power and noise is

$$SNR_{min} = \frac{P_{REFSENS}}{P_N} \Rightarrow SNR_{min|dB} = P_{REFSENS|dBm} - P_{N|dBm}$$
$$\Rightarrow P_{REFSENS|dBm} = P_{N|dBm} + SNR_{min|dB}$$
$$= -174 + 10\log_{10}(W) + NF + SNR_{min|dB} \quad (3.42)$$

Example 3.4 Determine the receiver sensitivity at the user terminal (UT) for $W = 9$ MHz of occupied bandwidth. The noise figure at the UT is $NF = 7$ dB and the minimum required $SNR_{min} = -4$ dB.

3.2 Noise and Interference Sources

Solution The sensitivity level is

$$P_{\text{REFSENS}|\text{dBm}} = -174 + 10\log_{10}(W) + NF + SNR_{\min} \quad (3.43)$$

$$= -174 + 10\log_{10}\left(9 \times 10^6\right) + 7 - 4 = -101.46 \text{ dBm}. \quad (3.44)$$

∎

3.2.3.3 Noise Rise and Interference Margin

While technically the receiver sensitivity is the minimum usable signal level in AWGN conditions (purely for testing and declaring a reference performance of a receiver), sometimes the reference sensitivity includes the interference in the noise power calculations. However, more often, the impact of interference is accounted for via the interference margin, which along with the other margins, is added to the receiver sensitivity to determine the minimum operating signal level.

Noise and interference, added together, can be viewed as a scaled version of the noise:

$$P_N + P_I = P_N\left(1 + \frac{P_I}{P_N}\right) = P_N \times n_r \Rightarrow (P_N + P_I)_{|\text{dBm}} = P_{N|\text{dBm}} + n_{r|\text{dB}} \quad (3.45)$$

where $n_r = 1 + P_I/P_N$ is the linear scale noise rise.

For example, if the interference level is 6 dB lower than the equipment noise floor P_N, or $P_{I|\text{dBm}} = P_{N|\text{dBm}} - 6\text{ dB}$, the linear scale ratio is

$$\frac{P_I}{P_N} = 10^{-6/10} = 0.25, \quad (3.46)$$

and the noise rise value is

$$n_{r|\text{dB}} = 10\log_{10}(1.25) = 0.973 \text{ dB}. \quad (3.47)$$

In cellular systems, a noise rise of up to 1 dB is considered tolerable in scenarios where other services are adjacent (in spectrum) to the signal. This is why the interference to noise power ratio of $-6\,\text{dB}$ is an important number to keep in mind. We note that the noise rise caused by the base stations of the same operator (due to neighboring cells and frequency reuse) is usually much higher than 1 dB. However, in satellite communications, it is not uncommon to see the $(P_I/P_N)_{|\text{dB}} \equiv (I/N)_{|\text{dB}} = -12\,\text{dB}$.

In link budget calculations, the topic of the next section, we normally use an interference margin that corresponds to noise rise due to the presence of other sites/cells:

$$IM = \frac{P_I^{\text{allowed}} + P_N}{P_N} = 1 + \frac{P_I^{\text{allowed}}}{P_N} = n_r^{\text{allowed}}. \tag{3.48}$$

Typical values for the interference margin are between 4 and 13 dB, depending on the network morphology and antenna technology (using a large antenna array allows for narrower beams and reduced intercell interference).

3.3 Link Budget

The link budget is an important tool in cellular network design. It is used to determine the cell radius or, in other words, the cellular coverage area. By knowing the cell radius, we know how closely spaced cellular sites need to be. For instance, if we have a 10 km × 10 km area, and if we determine that sites need to be located every 1 km (individual cell radius is 500 m), we need approximately 100 sites to cover the entire area. If, on the other hand, sites need to be located every 250 m, we need approximately $40 \times 40 = 1600$ sites, which implies a much higher cost for network deployment.

All the necessary elements of the link budget have been covered in previous sections, and in this section, we will see how it all fits together. We start by defining the link budget as a measure of Maximum Allowable Pathloss (MAPL or simply MPL) between the transmitter and the receiver both in the downlink and uplink direction. From the MAPL, and by making use of one of the previously introduced pathloss models, we can determine the cell radius.

A link budget diagram is illustrated in Fig. 3.8. The *up* and *down* directions of vertical arrows do not necessarily represent gains and losses (respectively), but

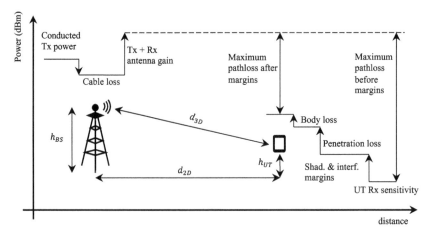

Fig. 3.8 Link budget illustration

rather their direction indicates whether they increase or decrease the maximum pathloss (and, consequently, cell radius). Parameters such as the transmit power, base station antenna height h_{BS}, and user equipment heights h_{UT} are usually provided as input parameters, although they may also be part of the link budget evaluation. The transmit EIRP value and the UT receiver sensitivity are generally known parameters, while losses and gains are determined as part of the link budget calculation. The MAPL value (before margins, in dB scale) is simply the difference between the EIRP level and the receiver sensitivity:

$$MAPL = EIRP + G_{UT} - P_{UT\ sensitivity} \tag{3.49}$$

and if we include the clutter (shadowing margin), interference margin, and body loss, the MAPL value is reduced, resulting in shorter cell radius, as illustrated in the figure. It should be emphasized that the target data rate does not necessarily correspond to the data rate associated with the sensitivity level. For example, our target rate may be higher than our allowed minimum rate. In this case, the sensitivity power in (3.49) is replaced by the minimum power required for the target data rate.

Example 3.5 (5G Link Budget) For the 5G link budget calculations, we will assume the following:

- Carrier frequency: $f_c = 2150$ MHz.
- Channel bandwidth: $BW = 40$ MHz. For signal and noise bandwidths, assume a slightly narrower bandwidth of $BW_N = 38.88$ MHz (for reasons that will be explained in Chap. 8]).
- Conducted transmit power: $P_T = 200$ watts or 53 dBm (see Fig. 3.9).
- Cable loss: $L_{cable} = 0.5$ dB.
- Tx antenna gain (base station side): $G_T = 18$ dBi.
- Rx antenna gain (at the UT): $G_R = -1$ dBi.
- UT noise figure: $NF = 7$ dB.
- Alpha-Shannon scaling factor: $\alpha = 0.6$.
- Control plane overhead: $OH_{CP} = 0.3$, that is, 30%.
- Pathloss model: Urban macro, non-line-of-sight (NLOS), following 3GPP TS 38.901. This model assumes $h_{BS} = 25$ m and $h_{UT} = 1.5$ m. (see Chap. 2).
- Shadowing standard deviation = $\sigma_L = 6$ dB.
- Interference margin: $IM = 6$ dB.
- Body loss: $L_{body} = 0$ dB (for voice, we typically assume 3 dB).
- Building penetration loss: $BPL = 18$ dB.
- Cell edge reliability: 85% ($P_{OUTAGE} = 0.15$).
- Target cell edge data rate: $R_{data} = 20$ Mbps.

Determine the maximum cell radius allowed.

Fig. 3.9 The EIRP value at the output of the antenna consists of the total conducted power of all output radio ports and antenna gain, reduced by the cable loss

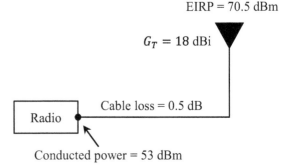

Solution Each of the steps is outlined below.

1. The EIRP value at the transmitter, in the dBm scale, is

$$EIRP_{|\text{dBm}} = P_{T|\text{dBm}} - L_{cable} + G_T = 53 - 0.5 + 18 = 70.5 \text{ dBm}. \quad (3.50)$$

2. The minimum signal power at the UE, for the target data rate of 20 Mbps, is calculated through the following steps:
   The receiver noise floor is

$$P_{N|\text{dBm}} = -174 + 10 \log_{10}(BW_N) + NF = -91.1 \text{ dBm} \quad (3.51)$$

The required rate is impacted by the overhead, i.e., achievable rate is

$$R_{data} = \alpha W (1 - OH_{CP}) \log_2(1 + SNR)$$
$$\Rightarrow 20 \times 10^6 = 0.6 \times 38.88 \times (1 - 0.3) \times 10^6 \times \log_2(1 + SNR)$$
$$\Rightarrow \text{Required SNR} = 2^{1.22} - 1 = 1.34 \; (\equiv 1.26 \text{ dB}) \quad (3.52)$$

The minimum received signal power is therefore

$$P_{\min|\text{dBm}} = P_{N|\text{dBm}} + SNR_{|\text{dB}} = -91.1 + 1.26 = -89.84 \text{ dBm}. \quad (3.53)$$

After adjusting for the receiver's antenna gain and interference margin, the minimum signal level required to arrive at the user is

$$P_{UT|\text{dBm}} = P_{\min|\text{dBm}} - G_R + IM = -82.84 \text{ dBm}. \quad (3.54)$$

3. We can now calculate the maximum allowable pathloss, before the other margins are applied, as

3.3 Link Budget

$$MAPL_{No\ Margins} = EIRP_{|dBm} - P_{UT|dBm} = 70.5 - (-82.84) = 153.54 \text{ dB}. \tag{3.55}$$

4. This maximum allowed path loss is reduced by the margins due to shadowing, the penetration loss into the building, and the loss in the user body. Based on the shadowing standard deviation $\sigma_L = 6$ dB and the target of 85% cell edge reliability, the shadowing margin is calculated as

$$M = \sqrt{2}\sigma_L \text{ erfc}^{-1}(2P_{OUTAGE}) = 6.22 \text{ dB} \tag{3.56}$$

The building penetration loss is specified as 18 dB (for carrier frequencies near 2 GHz, which corresponds to mid-band in terms of frequency ranges used, wall penetration is typically between 17 and 20 dB).
The specified body loss of 0 dB is also typical at these operating frequencies for data transmission.
All the margins taken together add up to

$$Margins_{total} = M + BPL + L_{body} = 24.22 \text{ dB}, \tag{3.57}$$

where $M = 6.22$ dB is the previously calculated shadowing margin.

5. Maximum pathloss, after the margins are applied, is

$$MAPL = MAPL_{No\ Margins} - Margins_{total} = 153.54 - 24.22 = 129.3 \text{ dB}. \tag{3.58}$$

6. The cell radius is determined by inverting the urban macro formula for NLOS condition from the TS 38.901:

$$PL_{NLOS}(d_{3D}) = 13.54 + 39.08 \log_{10}(d_{3D}) + 20 \log_{10}(f_c)$$
$$-0.6(h_{UT} - 1.5)$$
$$\Rightarrow 129.3 = 13.54 + 20 \log_{10}(2.15) + 39.08 \log_{10}(d_{3D})$$
$$\Rightarrow 39.08 \log_{10}(d_{3D}) = 109.1 \Rightarrow d_{3D} = 10^{\frac{109.1}{39.08}} = 619\text{m}$$
$$d_{2D} = Cell\ Radius = \sqrt{d_{3D}^2 - (h_{BS} - h_{UT})^2} \simeq 618 \text{ m}. \tag{3.59}$$

∎

It is worth noting that the UL radius determined by MAPL is typically lower than that of the DL and typically thus determines the overall cell radius.

Example 3.6 The UL MAPL after all the margins are applied is 135 dB. Determine the expected UL data rate at the base station if the base station is equipped with a 17 dBi antenna gain and the base station noise figure is 2.5 dB. Assume that the UT transmits 23 dBm over a bandwidth of $W = 720$ kHz, the UL Alpha-Shannon parameter is $\alpha = 0.4$, and the control plane overhead in the UL is $OH_{CP} = 0.1$.

Solution The UL signal at the base station receiver is

$$P_{R|\text{dBm}} = 23 - 135 + 17 = -95 \, (\text{dBm}). \tag{3.60}$$

The noise at the receiver is

$$P_{N|\text{dBm}} = -174 + 10 \log_{10}\left(720 \times 10^3\right) + 2.5 = -112.93 \, (\text{dBm}) \tag{3.61}$$

which results in the SNR value of

$$SNR_{|\text{dB}} = -95 - (-112.93) = 17.93 \, (\text{dB}). \tag{3.62}$$

After applying the modified Shannon formula, we estimate that the UL data rate is

$$\begin{aligned} R_{data}^{UL} &= \alpha W \left(1 - OH_{CP}\right) \log_2 \left(1 + SNR\right) \\ &= 0.4 \times 0.72 \times 0.9 \log_2 \left(1 + 10^{1.793}\right) \\ &= 1.55 \, \text{Mbps}. \end{aligned} \tag{3.63}$$

∎

For the purpose of satellite link budget calculation, the figure of merit used is the ratio between the antenna gain and the system temperature, as is the case in TR 38.811 [2], where the G/T in dB scale is given as

$$(G/T)_{|\text{dB}} = G_A - NF - 10 \log_{10}\left(T_0 + (T_A - T_0)10^{-0.1 NF}\right) \tag{3.64}$$

where G_A is the receive antenna gain in dB scale.

Example 3.7 A user terminal on the ground has a received antenna gain of $G_A = -3$ dBi and a figure of merit $(G/T)_{|\text{dB}} = -36.7$ dB/K. Calculate the system temperature and the noise figure if $T_A = T_0$.

3.3 Link Budget

Solution Given the $(G/T)_{|dB} = -36.7\,\text{dB/K}$, for $G_A = -3\,\text{dBi}$, we have $T = 33.7\,\text{dBK}$ or, in linear scale, $T = 10^{33.7/10} = 2344\,\text{K}$. The noise factor is

$$F = \frac{T}{T_0} = 8 \qquad (3.65)$$

and the noise figure is $NF = 9\,\text{dB}$.

■

3.3.1 Coverage Difference

Different bands have different propagation properties, because different signal wavelengths interact differently with the surrounding clutter. Low bands have much better propagation properties, whereas high bands attenuate much more as they go through different materials. For instance, if we use a concrete wall as an example, a 650 MHz signal is attenuated by only 7.6 dB (approximate value, depends on the wall thickness), a 3500 MHz signal is attenuated by 19 dB, whereas a 28 GHz signal is attenuated by as much as 117 dB (practically, is blocked by the wall), as illustrated below (Fig. 3.10).

Additionally, antenna gains are different for different frequency bands and so are the channel bandwidths. For instance, base station antenna gains for bands operating below 1 GHz typically are around 14–17 dBi, and these bands have small channel bandwidths, whereas bands between 1 and 3 GHz have a gain about 17–18 dBi

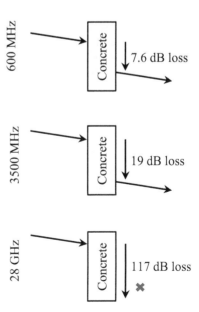

Fig. 3.10 Impact of different penetration losses due to different operating frequencies on received power

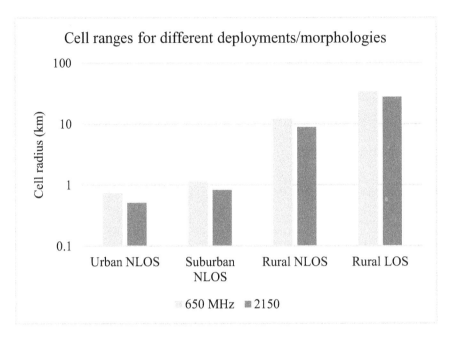

Fig. 3.11 Differences in coverage between different bands, for different deployment frequencies/morphologies

and generally have more bandwidth available. Consequently, the cell radius for the same level of service is very different for different bands. Furthermore, high target data rates (e.g., 100 Mbps) are not even possible with lower bands because they have lower bandwidths (typically no more than 10–20 MHz per operator for the 600 MHz operating band which has 35 MHz of total bandwidth available) despite their superior penetration properties.

In addition to different bands having different propagation properties within the same scenario (also called morphologies, e.g., urban morphology), there are significant coverage differences between different morphologies (i.e., urban, suburban, and rural). In the example below, we see differences between two 3GPP bands, designated n71 (600 MHz) and n66 (2150 MHz). The 3GPP band notation, band numbers, and corresponding frequency ranges will be covered in Chap. 4. As shown below, typical cell coverage in urban areas is less than 1 km, whereas for rural areas, it can be 10s of kilometers (Fig. 3.11).

3.4 Passive and Active Intermodulation

In general, system design assumes that signals that do not overlap in frequency do not interact. However, some nonideal nonlinearities always exist in implemented

3.4 Passive and Active Intermodulation

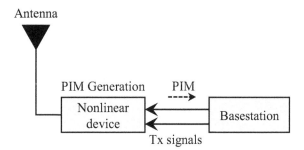

Fig. 3.12 Illustration of passive intermodulation generation at the transmitter at the base station

systems. These nonlinearities lead to *intermodulation products* in both passive (combiners, antennas, connectors) and active components (e.g., power amplifier). The resulting interaction manifests as interference and is a common problem in cellular communication systems (Fig. 3.12).

If the source of intermodulation is a passive device, the interfering signal is referred to as Passive InterModulation (PIM). This problem often happens at the base station, where two signals are combined to be transmitted by a common antenna. However, in newer systems where the UT can transmit on multiple carriers, the same problem is becoming common at the UT, as well.

3.4.1 Narrowband Intermodulation

The standard approach to analyze the PIM is via a Taylor series, where for an input signal $x(t)$, the output of a nonlinear device is

$$y(t) = \underbrace{a_1 x(t)}_{\text{Desired output}} + \underbrace{a_2 x^2(t) + a_3 x^3(t) + a_4 x^4(t) + a_5 x^5(t) + a_6 x^6(t) + \cdots}_{\text{Distortion (unwanted)}}$$
(3.66)

If two narrowband or single tone signals operating at frequencies f_1 and f_2 are input to a nonlinear device, the output of the nonlinear device comprises frequencies f_1 and f_2 (which is a desired outcome) but also will have unwanted frequency components that are located at $\pm m f_1 + \pm n f_2$, where m and n are integers. The sum of $n + m$ is the product order: if $m + n = 3$, the PIM is referred to as a third-order product, if $m + n = 5$ the PIM is referred to as a fifth-order product and so on (Fig. 3.13).

At the base station receiver, the even-order intermodulation distortion (IMD) terms usually do not fall in the vicinity of the received signal, so traditionally they were not considered in the intermodulation analysis. However, in 5G systems where several downlink carriers may be received by the UT simultaneously via carrier aggregation (discussed in Chap. 12), the even-order terms of the UL transmission may overlap with some of the received carriers. Consequently, even-order terms are also considered in the IMD analysis. For the base station, the third-order IMD term

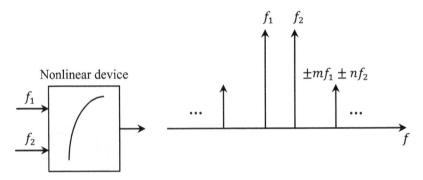

Fig. 3.13 Narrowband intermodulation due to nonlinearity of the device

is typically the strongest and as such the third-order intermodulation product is of most interest for base station design. Given the very low receive powers at the UT, the fourth and fifth order are also important.

3.4.1.1 Second-Order Intermodulation and a Second-Order Harmonics

Let $x_1(t) = s_1(t)\cos(2\pi f_1 t)$ and $x_2 = s_2(t)\cos(2\pi f_2 t)$ be two narrowband passband signals. The second-order IMD term is derived from the square term $x^2(t)$, where $x(t) = x_1(t) + x_2(t)$. Using basic trigonometric identities, we can write

$$x^2(t) = [s_1(t)\cos(2\pi f_1 t) + s_2(t)\cos(2\pi f_2 t)]^2 \tag{3.67}$$

$$= s_1^2(t)\cos^2(2\pi f_1 t) + 2s_1(t)s_2(t)\cos(2\pi f_1 t)\cos(2\pi f_2 t) + s_2^2(t)\cos^2(2\pi f_2 t)$$

$$= \frac{1}{2}s_1^2(t) + \frac{1}{2}s_2^2(t) + \quad \text{(DC terms)}$$

$$\frac{1}{2}s_1^2(t)\cos(4\pi f_1 t) + \frac{1}{2}s_2^2(t)\cos(4\pi f_2 t) + \quad (2^{nd}\text{ harmonics at } 2f_1 \text{ and } 2f_2)$$

$$s_1(t)s_2(t)\cos(2\pi(f_2 - f_1)t) + s_1(t)s_2(t)\cos(2\pi(f_1 + f_2)t)(2^{nd} \text{ order IMD})$$

The terms in the expression above that do not involve a product of two signals are second harmonics and are the result of the same nonlinearity process. The last line that involves a product between two input signals $s_1(t)$ and $s_2(t)$ is the second-order IMD. In the illustration below, the second-order IMD terms and harmonics are shown, while the DC terms are omitted (Fig. 3.14).

3.4 Passive and Active Intermodulation

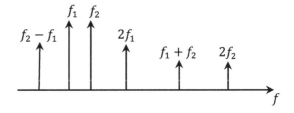

Fig. 3.14 Second-order intermodulation and harmonic terms

3.4.1.2 Third-Order Intermodulation (IM3)

A similar derivation process applies to third-order intermodulation. For shorter notation, $\omega_i = 2\pi f_i$, where $i = 1, 2$ is used below.

$$x^3(t) = [s_1(t)\cos(\omega_1 t) + s_2(t)\cos(\omega_2 t)]^3 \qquad (3.68)$$

$$= \frac{3}{4}s_1^3(t)\cos(\omega_1 t) + \frac{3}{2}s_1(t)s_2^2(t)\cos(\omega_1 t) + \quad \text{(spectrum widening around } f_1\text{)}$$

$$\frac{3}{4}s_2^3(t)\cos(\omega_2 t) + \frac{3}{2}s_1^2(t)s_2(t)\cos(\omega_2 t) + \quad \text{(spectrum widening around } f_2\text{)}$$

$$\frac{1}{4}s_1^3(t)\cos(3\omega_1 t) + \frac{1}{4}s_2^3(t)\cos(3\omega_2 t) + \quad \text{(third harmonics)}$$

$$\frac{3}{4}s_1^2(t)s_2(t)\cos((2\omega_1 - \omega_2)t) + \quad \text{(IM3-A)}$$

$$\frac{3}{4}s_1(t)s_2^2(t)\cos((2\omega_2 - \omega_1)t) + \quad \text{(IM3-B)}$$

$$\frac{3}{4}s_1^2(t)s_2(t)\cos((2\omega_1 + \omega_2)t) + \quad \text{(IM3-C)}$$

$$\frac{3}{4}s_1(t)s_2^2(t)\cos((2\omega_2 + \omega_1)t) \quad \text{(IM3-D)}$$

We will refer to the first two IM3 terms, IM3-A and IM3-B, in subsequent sections. Using the same approach, it is possible to derive higher-order IM terms.

3.4.1.3 Harmonics and the dBc Scale

Relative dB-scale power levels of the harmonics, when compared to the carrier power (which is in the intermodulation analysis called the fundamental), are referred to as dB relative to the carrier (dBc) (Fig. 3.15).
Using the same power level values and the same illustrative example as in [12]:

- Fundamental power is $P_{\text{fundamental}} = 33$ dBm.
- Second harmonic power is $P_{H2} = -27$ dBm, which corresponds to $X_{\text{dBc}} = -60$ dBc.
- Third harmonic power is $P_{H3} = -47$ dBm, which corresponds to $Y_{\text{dBc}} = -80$ dBc.

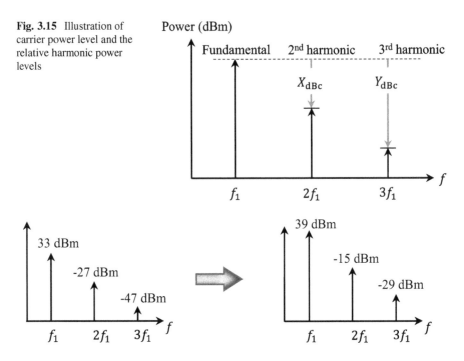

Fig. 3.15 Illustration of carrier power level and the relative harmonic power levels

Fig. 3.16 Illustration of impact of carrier power increase on the harmonic power levels

We can determine impact of power change of the fundamental power on the harmonic values.

The nonlinearities play an interesting role in the choice of carrier power. We recall that for a signal with an amplitude $x(t)$, the second-order harmonic has the amplitude $a_2 x^2(t)$ and the third-order harmonic has an amplitude $a_3 x^3(t)$, where a_1 and a_2 are device-dependent nonlinear expansion coefficients. If we double the amplitude of $x(t)$ (new amplitude is $2x(t)$), the fundamental power increases by a factor of 4 (6 dB). Given that the second harmonic has an amplitude proportional to $x^2(t)$, the amplitude of the second harmonic is proportional to $4x^2(t)$, that is, the second harmonic has 16 times higher power if the fundamental power is increased four times. This corresponds to 12 dB. Similarly, the third-order harmonic has an amplitude proportional to $8x^3(t)$ or is $8^2 = 64$ times higher (18 dB). This effect is illustrated in Fig. 3.16. Theoretically, a 1 dB change in carrier power corresponds to a 2 dB change for the second harmonic power, a 3 dB change for the third harmonic power, etc. These are known as the 1:2 and 1:3 rule.

3.4 Passive and Active Intermodulation

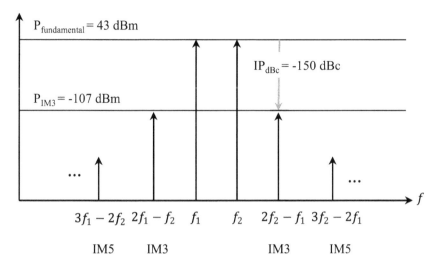

Fig. 3.17 Illustration of product intermodulation and dBc scale

3.4.1.4 Product Intermodulation and dBc

The IP_{dBc} is the level (in dBc) of the intermodulation product relative to the fundamental (i.e., carrier power). For example, if two carriers have equal fundamental powers of $P_{fundamental} = 43$ dBm and $P_{IM3} = -107$ dBm, the $IP_{dBc} = -150$ dBc.

The PIM dBc value is usually stated for antennas and other passive components and assumes two continuous wave input signals at a stated input level (Fig. 3.17).

The same 1:3 dB rule of power change applies for IM3 as for the third harmonic, at least in theory. In practice, for every 1 dB of the total two carriers powers (where carrier powers are equal), the change in IM3 power is around 2.4 dB to 2.8 dB for input power levels above 40 dBm [13].

If we change only one signal at a time, the changes will depend on which IM3 term is being considered. For instance, the IM3-B term $s_1(t)s_2^2(t)\cos((2\omega_2 - \omega_1)t)$ will change its power as below:

- Increases by 2 dB for every 1 dB increase in s_2 power (1:2 rule)
- Increases by 1 dB for every 1 dB increase in s_1 power (1:1 rule)

These rules can be derived if we look at what happens at the amplitudes and power of the components involved in the product of signals.

3.4.1.5 Third-Order Intercept

In the theoretical model, the third-order IM power increases following the 1:3 rule (Fig. 3.18), that is, for every 1 dB of the combined signals power, the IM3 power increases by 3 dB. Let us consider a power amplifier that adds gain to the input

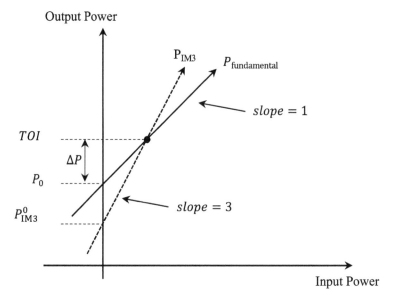

Fig. 3.18 Third-order intercept illustration, where the third-order intermodulation power grows at 3× the slope of the fundamental power

power. If we plot the input and output powers and extrapolate the IM3 growth over a wide range of input powers, there will be a point where the IM3 power is equal to the carrier (also referred to as a fundamental) power. This point is called the third-order intercept (TOI) point.

For some nominal power level P_0, the IM3 power is determined by using the IP_{dBc} rating:

$$P_{IM3}^0 = P_0 - |IP_{dBc}| \tag{3.69}$$

The output fundamental and a third-order PIM powers can be expressed as

$$P_{\text{fundamental}} = P_0 + \Delta P \tag{3.70}$$
$$P_{IM3} = P_{IM3}^0 + 3\Delta P$$

The intercept point is the point where these two powers are equal:

$$P_{\text{fundamental}} = P_{IM3} \tag{3.71}$$
$$P_0 + \Delta P = P_0 - |IP_{dBc}| + 3\Delta P$$
$$\Rightarrow \Delta P = \frac{|IP_{dBc}|}{2}$$

3.4 Passive and Active Intermodulation

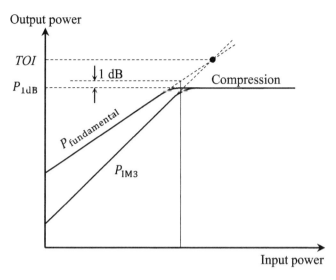

Fig. 3.19 Illustration of third-order intercept point and its position relative to output compression

$$\Rightarrow TOI = P_0 + \frac{|IP_{dBc}|}{2}$$

For example, if $IP_{dBc} = -150$ dBc and $P_0 = 43$ dBm, the third-order intercept is

$$TOI = P_0 + \frac{|IP_{dBc}|}{2} \quad (3.72)$$
$$= 118 \text{ dBm}.$$

As per this calculation, if the fundamental power is 118 dBm, the third-order PIM has the same power as the carrier. In practice, 118 dBm is well outside the operating range of the power amplifier. Namely, power amplifiers operate in a linear region up to a certain output level, after which the output power begins to saturate or compress. The point inside the nonlinear region where the output power level decreases by 1 dB from the expected linear gain is called the 1 dB compression point, as illustrated in Fig. 3.19. We normally consider that a power amplifier operates in a linear region up to the 1 dB compression point.

Therefore, the TOI point is a linearly extrapolated point above the compression point $P_{1 \text{ dB}}$ but is still a useful parameter. By knowing the TOI point, we can determine the power of third-order intermodulation product for any output (or, similarly, for any input power level). For example, if the nominal output power is $P_{nominal} = 33$ dBm, and the $TOI = 118$ dBm, the dBc scale offset is $|IP_{dBc}| = 2(TOI - P_{nominal}) = 2(118 - 33) = 170$ dBc, and the third-order IM power is $P_{IM3} = 33 - 170 = -137$ dBm. Clearly, the higher the TOI value, the lower the IM3 power.

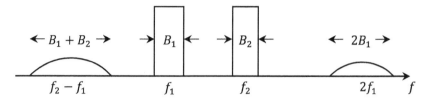

Fig. 3.20 Second-order harmonic and intermodulation term indicating their respective center frequencies and bandwidths

3.4.2 Wideband Intermodulation

All the concepts described for narrowband intermodulation largely apply to wideband signals, with some important differences. While for a single tone, continuous waveform (CW), signals the intermodulation products are CW signals, for wideband signals the intermodulation products are wider than the fundamental signals. To understand this process, we need to determine bandwidth of a product of two signals as explained next.

3.4.2.1 Frequency Domain Analysis of PIM Bandwidth

Let $s_1(t)$ and $s_2(t)$ be wideband signals (e.g., 4G or 5G signals). The Fourier transform of the product of these two signals is [4]

$$\mathcal{FT}[s_1(t)s_2(t)] = \int_{-\infty}^{\infty} S_1(\lambda) S_2(f - \lambda) d\lambda \tag{3.73}$$

which is a definition of convolution in the frequency domain. Therefore, if the bandwidth of $s_1(t)$ is B_1 and the bandwidth of $s_2(t)$ is B_2, the bandwidth of their product is equal to the sum of their bandwidths, $B_1 + B_2$. That is because convolution "stretches" the signals so that their bandwidths add up.

The implication is that the second-order harmonic at $2f_1$ and IM2 at $f_2 - f_1$ will have bandwidths as shown in the next figure (Fig. 3.20).

Similarly, 3rd harmonics will have bandwidth three times as wide as the fundamental and so on.

Example 3.8 Determine the lower and upper frequencies for the IM3-A term as a function of f_{1L}, f_{1H}, f_{2L}, and f_{2H} based on the diagram shown below.

3.4 Passive and Active Intermodulation

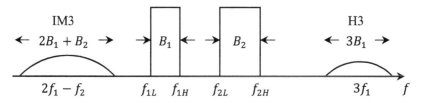

Fig. 3.21 Two carrier signals with channel edges f_{1L}, f_{1H}, f_{2L}, and f_{2H} along with the third-order intermodulation and harmonic

Solution The IM3-A term, if we ignore the scaling factor, is proportional to $s_1^2(t)s_2(t)\cos(2\pi(2f_1 - f_2)t)$, and the IM3 and H3 bandwidths are as shown in Fig. 3.21 ($2B_1 + B2$ and $3B_1$, respectively). Given that

$$f_1 = \frac{f_{1L} + f_{1H}}{2} \tag{3.74}$$

$$f_2 = \frac{f_{2L} + f_{2H}}{2} \tag{3.75}$$

we can determine high and low frequencies for the IM3-A term as following:

$$f_L^{IM3-A} = 2f_1 - f_2 - \frac{2B_1 + B_2}{2} \tag{3.76}$$

$$= \frac{2(f_{1L} + f_{1H}) - (f_{2L}) + f_{2H}) - 2(f_{1H} - f_{1L}) - f_{2H} + f_{2L}}{2} \tag{3.77}$$

$$= 2f_{1L} - f_{2H} \tag{3.78}$$

$$f_H^{IM3-A} = 2f_1 - f_2 + \frac{2B_1 + B_2}{2} \tag{3.79}$$

$$= \frac{2(f_{1L} + f_{1H}) - (f_{2L}) + f_{2H}) - 2(f_{1H} - f_{1L}) + f_{2H} - f_{2L}}{2} \tag{3.80}$$

$$= 2f_{1H} - f_{2L}. \tag{3.81}$$

∎

This approach of expressing intermodulation terms in terms of high and low frequencies for individual signals has been adopted by 3GPP. An example of second- and third-order intermodulation terms, from [14], is shown below (Table 3.1).

A convenient alternative approach is to indicate center frequencies and bandwidths of the IM terms, similar to what is shown in Fig. 3.20. A summary table of all third-order intermodulation and harmonic terms, along with their center frequencies and IM bandwidths, based on (3.68), is provided in Table 3.2.

The last few years have seen significant progress in the area of digital cancellation of IM3 products, so that many of the associated problems are reduced but not completely eliminated. For example, the PIM impact increases with time, as the product ages, and even if today's equipment does not experience interference caused

Table 3.1 Intermodulation products due to second- and third-order nonlinearity using the 3GPP approach

IM term	f_{1L}	f_{1H}	f_{2L}	f_{2H}
IMD2	$\|f_{2L} - f_{1H}\|$	$\|f_{2H} - f_{1L}\|$	$f_{2L} + f_{1L}$	$f_{2H} + f_{1H}$
IMD3	$\|2f_{1L} - f_{2H}\|$	$\|2f_{1H} - f_{2L}\|$	$\|2f_{2L} - f_{1H}\|$	$\|2f_{2H} - f_{1L}\|$
IMD3	$2f_{1L} + f_{2L}$	$2f_{1H} + f_{2H}$	$2f_{2L} + f_{1L}$	$2f_{2H} + f_{1H}$
IMD3	$f_{1L} - B_2$	$f_{1H} + B_2$	$f_{2L} - B_1$	$f_{2H} + B_1$

Table 3.2 Intermodulation products and harmonics due to third-order nonlinearity

IM no.	Center frequency	IM bandwidth	IM term
1	f_1	$\max(3B_1, B1 + 2B2)$	widening around f_1
2	f_2	$\max(3B_2, 2B1 + B2)$	widening around f_2
3	$\|2f_1 - f_2\|$	$2B_1 + B_2$	IM3-A
4	$\|2f_2 - f_1\|$	$B_1 + 2B_2$	IM3-B
5	$2f_1 + f_2$	$2B_1 + B_2$	IM3-C
6	$f_1 + 2f_2$	$B_1 + 2B_2$	IM3-D
7	$3f_1$	$3B_1$	third harmonic
8	$3f_2$	$3B_2$	third harmonic

by PIM, there is no guarantee that within a few years the problem will not (re)appear. Therefore, a careful system design is always required.

Problems

1. Determine the interference power value if the SINR at the receiver is SINR = 0.1 (linear). Assume that the received signal power is $P_{S|dBm} = -97\,dBm$, the receiver operates with a $W = 10\,MHz$ channel with noise power spectral density of $N_{0|dBm} = -174\,dBm/Hz$ and has a noise figure of $NF = 7\,dB$.
2. If the signal level in the Example 6 drops by 6 dB due to fading, what is the new expected UL data rate? How would the data rate change if the signal level was increased by 6 dB.
3. Three active components are placed in a cascade. Their respective gains and noise figures are $G_1 = 1\,dB$, $NF_1 = 6\,dB$, $G_2 = 10\,dB$, $NF_2 = 3\,dB$, and $G_3 = 20\,dB$, $NF_3 = 2\,dB$.

    a. Calculate the equivalent noise figure for the given cascade.
    b. Reverse the component order and calculate the equivalent noise figure. Comment on the result.

4. The stated noise figure for devices, NF, is measured at the reference temperature $T_0 = 290\,K$. If the input temperature is $T_1 \neq T_0$, the SNR degradation at the output can be higher or lower than the expected noise factor F. Show that this is case by considering a single active device.

3.4 Passive and Active Intermodulation

5. An RF filter that acts as a passive device at the front end and is used to eliminate interference from adjacent services has an insertion loss of 3 dB. If the receiver consists of an LNA with $G_2 = 20$ dB and $NF_2 = 2$ dB, followed by another amplifier with $G_3 = 10$ dB and $NF_3 = 2$ dB, determine the overall noise figure. How does the noise figure change if the front end filter is replaced with a filter with the insertion loss of 1 dB?

6. A mobile system using a bandwidth of $W = 10$ MHz has a cable loss of $L_{cable} = 0.9$ dB, the antenna temperature is equal to the reference temperature $T_0 = 290$K, the gain of the amplifier following the cable is $G_2 = 20$ dB, the equivalent noise temperature of the amplifier is $T_2 = 800$K, the gain of the third component in the receive chain is $G_3 = 0$ dB, while the noise figure is $NF_3 = 2$ dB. Determine the following:

   a. Values for: T_1, F_1, G_1, F_2, F_3, T_3.
   b. Value of the noise output of the cable as a function of input noise N_{IN}.
   c. Total equivalent noise figure and total equivalent noise temperature (for all three components).
   d. Noise power at the input, N_{IN}, and at the output, N_{OUT}, for the given bandwidth.
   e. Signal power at the output S_{OUT} if the input signal power is $S_{IN} = -122$ dBW. Compare input and output SINR values.

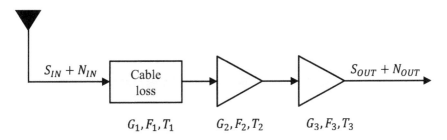

7. An Rx chain consists of an antenna, with an input temperature T_A, low noise amplifier (LNA) with the equivalent internal noise temperature T_{e1} and gain G_1, a cable with cable loss L, and a receiver with a known noise figure F. Assuming the cable and the receiver operate at the room temperature T_0, determine the overall system noise.

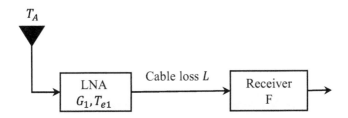

8. Determine the geometry (SINR) values within the serving cell at 200 m, 350 m, and at the cell edge. Assume that there are two neighboring cells that are facing each other with the inter-site distance (ISD) of 1 km. Use $PL(d) = 128.1 + 37.6 \log_{10}(d)$ formula for the pathloss, where the distance d is in kilometers. The EIRP at the transmitter is 60 dBm (same for both base stations), and the occupied bandwidth is 9 MHz.

9. Consider two base stations facing each other with the intercell distance of $2D = 1$ km. Within a serving cell, there are seven users:

    - One near user with the $SINR = 20$ dB
    - Two middle users with the $SINR = 10$ dB each
    - Four cell-edge users with the $SINR = 0$ dB each

    The pathloss exponent is $n = 3.5$. If the scheduler at the base station allocates the frequency domain resource blocks (RBs), where each RB occupies 180 kHz, in the following way:

    a. 2 RBs for near, 4 RBs for middle, 14 RBs for cell-edge users
    b. 4 RBs for near, 6 RBs for middle, 8 RBs for cell-edge users

    then using the alpha-Shannon formula, determine user data rates, the average cell data rate, and the required bandwidth for (a) and (b) schemes. Also, determine the distance of individual users from the base station (assuming the SINR is dominated by the intercell interference).

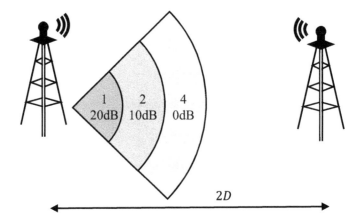

10. Two cellular base stations, shown below, are configured to operate with a 9 MHz channel and are separated by the distance of D meters. If only the first base station is turned on, a UT at a distance d from the first base station measures the SNR value of $SNR_{|dB} = 10$ dB. Assuming a noise figure at the UE receiver of $NF = 7$ dB, determine the received signal value at the UT and the distance from the first base station for a given pathloss of $PL(d) = 128.1 + 37.6 \log_{10}(d)$, where d is in km, if the EIRP of the first base station is $EIRP_1 = 50$ dBm.

3.4 Passive and Active Intermodulation 119

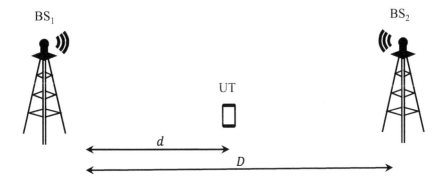

After the second base station is turned on, the UT attached to the first base station now measures $SINR_{|dB} = 3$. Determine the interference level at the UT and the distance D if the EIRP level of the second base station is $EIRP_2 = 46$ dBm.

11. A heterogeneous network consists of one macro base station (BS) and two small cells. The macro BS transmits at 43 dBm, while the small cells (SC) transmit at 23 dBm each. Three UTs, indicated by squares in the figure below, are being served by this network, and the distances indicated in the figure are $d_1 = 230$ m, $d_2 = 70$ m, and $d = 298$ m. Assuming the same pathloss model for macro and small cells, determine the following:

    a. Serving cell for each of the UEs (maximal received power is the criteria for the serving cell choice), assuming the pathloss exponent value of $n = 3.5$
    b. Geometry/SINR for each UE, ignoring the thermal noise
    c. Maximal spectral efficiency (b/s/Hz, based on Shannon's capacity formula) for each UE based on the SINRs calculated in (b)

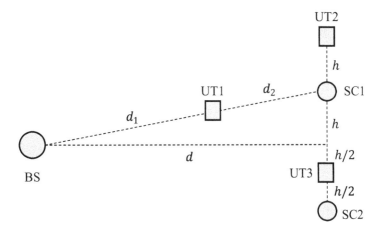

12. Five hexagonal sites cover a certain geographic area as shown in the figure below. The cell sites are located at the center of each hexagon (small circles),

while a single UT is located at the cell edge of the serving cell 1 (black rectangle). The pathloss exponent is $n = 3.9$. If the thermal noise has a negligible effect, determine the geometry (SINR) at UT's receiver. Each cell site transmits with equal power.

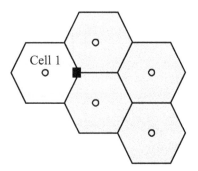

13. Calculate the UT receiver sensitivity if the channel bandwidth is $W = 98$ MHz, the noise figure is $NR = 7$ dB, and the required SNR for a data service is 0.8 in linear scale.
14. The coexistence requirements between the fixed satellite systems and 5G cellular systems are typically expressed in terms of maximally allowed power flux density (PFD) value that satellite earth station can generate at certain distance. Calculate the missing values in the table below. Make sure to use the proper units.

Parameter	Value	Comment
I/N in dB	-6	Target interf. to noise power ratio
Frequency (GHz)	28	Operating frequency
Wavelength (m)		λ based on the carrier frequency
Rx antenna gain (dBi)	16	Assumed 5G Rx antenna gain
Implementation loss (db)	3	Add to noise figure
Rx noise figure (dB)	5	
Rx noise temperature T_0 (K)	290	Room temperature
Boltzmann constant (W/K/Hz)	1.38×10^{-23}	Constant
Thermal noise PSD (dBW/Hz)		
Rx noise floor over 1 MHz (dBm/MHz)		Accounts for NF and impl. loss
Interf. power over 1 MHz (dBm/MHz)		
PFD (dBm/m²/MHz)		Calculated coexistence requirement

15. A Low Earth Orbit (LEO) satellite orbits the earth with an altitude of 550 km and has the parameters in the table below (next page). Add the missing values in the table. Using the alpha-Shannon formula, determine the peak downlink data rate if the value for alpha is $\alpha = 0.65$.

3.4 Passive and Active Intermodulation 121

Parameter	Value	Unit
Bandwidth	4500	kHz
Frequency	960	MHz
Transmit power	50	W
Transmit feeder loss	1	dB
Transmit antenna gain	22.4	dBi
Transmit EIRP		dBW
LEO altitude	550	km
Free space pathloss		dB
Atmospheric loss	0.5	dB
Polarization loss	3	dB
Receive antenna gain	−3	dBi
Received signal power		dBW
Receiver system temperature	2344	K
Boltzmann constant	1.38×10^{-23}	W/Hz/K
Receiver's noise power		dBW
Signal-to-noise ratio		dB

16. The 3500 MHz band spans the range from 3450 to 3650 MHz and is divided into 20 equal bandwidth blocks of 10 MHz, as illustrated below (figure next page). An operator operates the following UL blocks: 1740–1755 MHz and 1770–1780 MHz. A second harmonic from the UL band generates an interfering signal that affects the reception of the 3500 MHz band signal inside the phone. Determine all the 3500 MHz blocks that are affected by this interference. Explain the calculation.

A B C D E F G H J K L M N P Q R S T U V

3450   3650

17. A power amplifier (PA) operating in a small signal region (well below the compression point) has a third-order intercept (TOI) point of $TOI = 66$ dBm. If two continuous wave (CW) signals operating at f_1 and f_2 carrier frequencies are outputs of the same PA with $P_{O1} = 43$ dBm and $P_{O2} = 33$ dBm, respectively, determine the P_{IM3} levels of the intermodulation products (generated by the two CWs) found at $2f_1 - f_2$ and $2f_2 - f_1$ frequencies.
18. Operator A holds 600 MHz band licenses for blocks A and B, whereas Operator B holds block E license, as shown below.

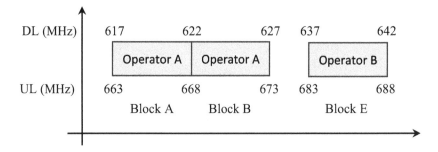

If operators A and B were to share the 600 MHz spectrum, do the following steps:

a. Determine the third-order PIM (IM3) when two downlink signals are combined at the radio (same power amplifier).
b. Plot the third-order PIM and the affected UL block(s) if any.
c. What percentage of IM3 product falls within the affected block(s)?
d. Why is an IM3 problematic?

19. The PIM caused interference can happen on the device side via a slightly different mechanism. Namely, the local oscillator (LO) is not an ideal sinusoidal wave, and as such a receive LO generates harmonics that can mix with the transmit signal harmonics, which is then down-mixed to baseband, thus causing intermodulation distortion. An example is a local oscillator operating at $f_{LO} = 2625$ MHz for the receive signal block in the [2620, 2630] MHz range. If the user terminal transmitter operates in the uplink frequency range from 3930 to 3940 MHz, show that a product of the third-order harmonic of the receive LO, mixed with the second harmonic of the transmit signal (the so called 3DL/2UL MSD scenario in 3GPP terminology), can cause interference in the baseband domain.

References

1. R. Adler at al., *Description of the Noise Performance of Amplifiers and Receiving Systems*, in Sponsored by IRE subcommittee 7.9 on Noise, Proceedings of the IEEE (1963), pp. 436–442
2. Study on New Radio (NR) to support non-terrestrial networks. https://www.3gpp.org/ftp//Specs/archive/38_series/38.811/38811-f40.zip. Cited Oct 23, 2023
3. H. Holma, A. Toskala, *LTE for UMTS Evolution to LTE-Advanced*, 2nd edn. (Wiley, New York,, 2011)
4. S. Haykin, M. Moher, *Communication Systems* (Wileay, New York, 2009)
5. 3GPP TS 36.213, Physical layer procedures, Rel 14. https://www.3gpp.org/ftp/Specs/archive/36_series/36.213/36213-eh0.zip. Cited 4 Nov 2023
6. 3GPP TS 38.214, NR; Physical layer procedures for data. https://www.3gpp.org/ftp/Specs/archive/38_series/38.214/38214-i00.zip. Cited 4 Nov 2023
7. Evolved Universal Terrestrial Radio Access (E-UTRA). Radio Frequency (RF) system scenarios. https://www.3gpp.org/ftp/Specs/archive/36_series/36.942/36942-h00.zip. Cited Oct 22, 2023

References

8. Study on new radio access technology: Radio Frequency (RF) and co-existence aspects. https://www.3gpp.org/ftp/Specs/archive/38_series/38.803/38803-e30.zip. Cited Oct 22, 2023
9. Simulaton results FRC for REFSENS—FR1. https://www.3gpp.org/ftp/tsg_ran/WG4_Radio/TSGR4_86/Docs/R4-1802310.zip. Cited Oct 22, 2023
10. General aspects for Base Station (BS) Radio Frequency (RF) for NR. https://www.3gpp.org/ftp/Specs/archive/38_series/38.817-02/38817-02-fb0.zip. Cited Oct 23, 2023
11. User Equipment (UE) radio transmission and reception; Part 1: Range 1 Standalone. https://www.3gpp.org/ftp/Specs/archive/38_series/38.101-1/38101-1-i30.zip. Cited Oct 23, 2023
12. Intermodulation Distortion (IMD) Measurements, Anritsu. https://reld.phys.strath.ac.uk/local/manuals/Anritsu37xxxVNA-intermod.pdf. Cited Oct 23, 2023
13. Passive Intermodulation (PIM) handling for Base Stations (BS). https://www.3gpp.org/ftp/Specs/archive/37_series/37.808/37808-c00.zip. Cited Oct 23, 2023
14. TR 38.716-02-00, NR inter-band Carrier Aggregation (CA)/Dual Connectivity (DC) Rel-16 for 2 bands Down Link (DL)/x bands Up Link (UL) (x = 1, 2). https://www.3gpp.org/ftp//Specs/archive/38_series/38.716-02-00/38716-02-00-g00.zip Cited 5 Nov 2023

Chapter 4
Spectrum and RF Characteristics of Cellular Systems

4.1 Introduction

The radio spectrum is the range of radiofrequencies that enable wireless communication systems. As discussed earlier, electrical signals in these systems have a specific bandwidth, requiring a designated range within the spectrum. Multiple signal bandwidths, or channels, are typically grouped into larger segments called bands. As an analogy, imagine the radio spectrum as a network of highways. Each highway represents a band, with traffic lanes within it corresponding to channels. These "highways" can vary in width and may have different regulations, just like real highways in different locations. Rules and regulations that govern radio spectrum are regulated by governments in their respective countries. For example, *Innovation, Science and Economic Development Canada* (ISED) regulates and manages Canadian spectrum, the *Federal Communications Commission* (FCC) regulates most of the US spectrum,[1] etc.

Different countries may impose slightly different regulations on the same bands, but they generally share a high degree of commonality to ensure seamless operation of devices across borders. These commonalities are achieved through the existence of global standards. This chapter explores the role of the International Telecommunication Union (ITU) and 3GPP in defining these standards.

---

[1] The FCC regulates commercial spectrum use, while the National Telecommunications and Information Administration (NTIA) regulates spectrum for federal (US government).

Supplementary Information The online version contains supplementary material available at (https://doi.org/10.1007/978-3-031-76455-4_4).

While current spectrum allocation spans from 3 kHz to 300 GHz, here we focus on the RF ranges from 450 MHz to 71 GHz.

4.2 Spectrum Licenses

In Canada[2] there are two basic license types for terrestrial services [1, 2]: radio licenses and spectrum licenses. Fixed and mobile radio licenses are issued for each radio apparatus based on the technical details of land-based communications systems such that the use of the spectrum does not interfere with other users. Radio licenses are used by taxis, mining, public safety, etc. With an ever-growing demand for spectrum due to the proliferation of wireless devices and services, spectrum licenses were introduced in Canada in 1996 to provide access to spectrum with minimal administrative burden. Spectrum licenses are issued for a given geographic area and block of frequency with the licensee being responsible for ensuring their network is properly planned and coordinated in compliance with the technical and regulatory requirements set for the band. This includes adhering to power limits, avoiding interference and fulfilling coverage obligations among others.

In general, globally, there are three types of spectrum licensing processes currently in use:

1. Licensed spectrum is allocated to operators on an *exclusive use* basis for the most part (i.e., there could be exceptions, where overlay licenses are issued), which guarantees interference-free operation in their respective allocated channels. To date, over 40 cellular bands have been allocated for license use globally. This type of allocation remains the preferred paradigm for operators as it provides access (and therefore coverage and performance) guarantees. On the other hand, it requires large financial investments by the operators since allocations are typically made via spectrum auctions.
2. License exempt (unlicensed spectrum) is allocated for shared use among different operators or users. The best-known examples are WiFi operating in 2400–2483.5 MHz and 5150–5895 MHz. New allocations in the 6 GHz range (5925–7125 MHz) and in the mmWave range (57–61 GHz) have opened up large amounts of unlicensed spectrum for everyday use. For operators, this type of spectrum is quite important to realize traffic offloading. For users, it is usually provided as a complimentary service with other types of connectivity (e.g., fiber or cable connection). While there are mechanisms to control the interference between the users who are sharing the same band, this interference can in some scenarios be significant (e.g., airports), thus limiting the quality of experience.
3. Shared spectrum involves spectrum allocation to different operators on the co-primary basis but may also include other forms of sharing of spectrum that may

---

[2] With regard to spectrum licensing, Canada is representative of a broad swath of countries. Most countries follow similar licensing regimes.

involve dynamic spectrum access (e.g., TV white spaces). The first commercial example of the shared spectrum has been introduced by the FCC for the 3500 MHz band (3550–3700 MHz) [3], which is known as the *Citizens Broadband Radio Service* (CBRS) band. This shared spectrum scheme uses a spectrum manager to allocate access to spectrum either at a relatively low cost or free of cost and yet provides the benefits on having an interference-free operation when compared to unlicensed spectrum.

In regulating the use of this band, the FCC created three tiers of users:

- Tier 1: Incumbent Access—Primarily for federal use (military radar), but also legacy Fixed Satellite Services (FSS).
- Tier 2: Priority Access—Priority Access Licenses (PALs) are auctioned, nonrenewable, three-year licenses for 10 MHz channels in the 3550–3650 MHz range. Individual operators are allowed up to four channels out of seven channels available for PAL. Geographic resolution of licenses is at the county level.
- Tier 3: General Authorized Access (GAA)—Access to any unallocated portion of the band, and opportunistic use of priority access channels on a no- interference/no-protection basis.

The CBRS band is facilitated by a *Spectrum Access System* (SAS) administered by third-party coordinator, which grants use of spectrum to GAA users opportunistically while managing interference and protection for higher-tier users.

ISED has also introduced the Non-Competitive Local Licensing (NCL) framework to support private wireless networks and enterprises that is localized and based on first come, first served basis [4]. Other shared spectrum frameworks are under development (e.g., UK/Ofcom with spectrum in 1.8 GHz, 2.3 GHz, 3.8–4.2 GHz, and 24.25–27.5 GHz [5]).

4.3 Role of ITU-R and 3GPP in Definition of Spectrum Bands and Standards

The International Telecommunication Union (ITU) is the UN's specialized agency which traces its origins back to 1865, when the International Telegraph Convention was signed by 20 founding members in Paris [6]. The next major milestone in the history of ITU occurred in 1932, in Madrid, when the International Telecommunication Union name has been adopted. Finally, in 1947, the ITU became an agency for global telecommunications after an agreement with the United Nations, which became effective in 1949. ITU is now based in Geneva, Switzerland, and includes 193 countries and almost 800 private-sector companies and academic institutions as members.

Within the ITU, the ITU Radiocommunications sector (ITU-R) is tasked to allocate radio spectrum in the form of Radio Regulations. ITU-R also creates

Table 4.1 ITU-R defined radio bands

Band no.	Band name	Range of frequencies	Wavelengths
4	Very low frequency (VLF)	3–30 kHz	100–10 km
5	Low frequency (LF)	30–300 kHz	10–1 km
6	Medium frequency (MF)	300–3000 kHz	1000–100 m
7	High frequency (HF)	3–30 MHz	100–10 m
8	Very high frequency (VHF)	30–300 MHz	10–1 m
9	Ultra high frequency (UHF)	300–3000 MHz	100–10 cm
10	Super high frequency (SHF)	3–30 GHz	10–1 cm
11	Extremely high frequency (EHF)	30–30 GHz	10–1 mm
12		300–3000 GHz	1–0.1 mm

Note: Band no. N extends from 0.3×10^N to 3×10^N Hz

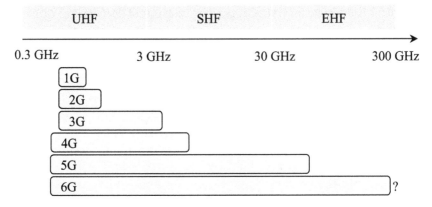

Fig. 4.1 Cellular technologies from 1G to 5G have continuously expanded the frequency range

standards for International Mobile Telecommunications (IMT) and develops reports and handbooks that provide best practices in implementing these standards.

4.3.1 ITU Band Nomenclature

The ITU-R's Radio Regulation (RR) issued in 2020 (there may be a 2023 version by the time this is published), available at [7], lists nine bands reproduced in Table 4.1.

Different portions of these frequency bands are used for different RF systems, which is covered in more detail in [8]. Commercial cellular systems operate in a subset of these bands because of favorable propagation properties and antenna size constraints imposed by the signal wavelength (antenna sizes are in the $\lambda/2$ range and need to fit inside a handheld phone). The evolution of frequency ranges with cellular technologies is shown in Fig. 4.1.

Table 4.2 lists the frequency bands for IMT as identified by ITU-R.

4.3 Role of ITU-R and 3GPP in Definition of Spectrum Bands and Standards

Table 4.2 ITU-R identified frequency bands that might be used for IMT systems

Frequency bands identified for IMT (MHz)	Available bandwidth (MHz)
450–470	20
470–698	228
694/698–960	262
1427–1518	91
1710–2025	315
2110–2200	90
2300–2400	100
2500–2690	190
3300–3400	100
3400–3600	200
3600–3700	100
4800–4990	190
24250–27500*	3250
37000–43500*	6500
45500–47000*	1500
47200–48200*	1000
66000–71000*	5000

*Bands identified for IMT after the RR was published in 2020

4.3.2 ITU-R Defined Geographic Regions

While the goal of ITU-R is to define globally harmonized spectrum bands and standards, there are variations from country to country. To help group these variations, as shown in Fig. 4.2, three geographic regions have been defined:

- **Region 1** includes Europe and Africa and parts of Asia (the whole of the territory of Armenia, Azerbaijan, the Russian Federation, Georgia, Kazakhstan, Mongolia, Uzbekistan, Kyrgyzstan, Tajikistan, Turkmenistan, Turkey, and Ukraine and the area to the north of Russian Federation which lies between lines A and C).
  Region 1 is organized in four regional telecommunication organizations (RTOs): the Arab Spectrum Management Group (ASMG), the African Telecommunications Union (ATU), the European Conference of Postal and Telecommunications Administrations (CEPT), and the Regional Commonwealth in the Field of Communications (RCC).
- **Region 2** includes North and South Americas and is between lines B and C. The Inter-American Telecommunication Commission (CITEL) is the RTO for this region.
- **Region 3** includes Australia, New Zealand, and parts of Asia not included in Region 1. The RTO for Region 3 is the Asia-Pacific Telecommunity (APT).

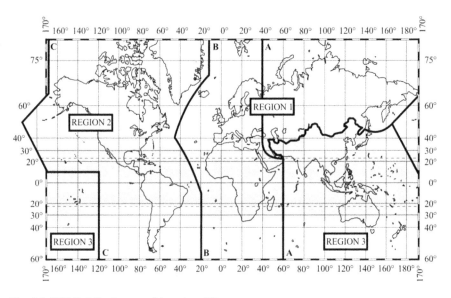

Fig. 4.2 ITU-R defined geographic regions [7]

4.3.3 WRC Conferences

The harmonized use of spectrum, identified in ITU-R's Radio Regulation, is updated every 4 years following the ITU World Radiocommunication Conference (WRC). The WRC activities go beyond harmonization: this is where the Radio Regulations are updated, including the outcomes of sharing/compatibility studies and the new allocations, adoption of resolutions to protect existing services, etc. The latest WRC event, WRC-23, took place in November to December 2023, in Dubai.

Key outcomes of the WRC-23 include:

- Harmonization of mobile allocation and IMT identification across Region 2 in 3300–3400 MHz, 3600–3700 MHz, and most of Region 2 in 3700–3800 MHz
- Global identification of 694–960 MHz, 1710–1885 MHz, and 2500–2690 MHz mobile bands for the use of High-Altitude Platform Stations (HAPS) as IMT Base Stations (HIBS)
- Enablement of the use of earth stations in motion (ESIMs) communicating with non-geostationary satellites in the 17.7–18.6 GHz, 18.8 19.3 GHz, 19.7–20.2 GHz (space-to-Earth), 27.5–29.1 GHz, and 29.5–30 GHz (Earth-to-space) frequency bands
- Establishment of agenda for WRC-2027 (triggering/initiating new studies in the ITU Working Groups) and preliminary agenda for WRC-2031

4.3 Role of ITU-R and 3GPP in Definition of Spectrum Bands and Standards

4.3.4 IMT Standardization

In addition to harmonizing and identifying available radio spectrum, one of the key roles of ITU-R is to define standards (service requirements) for IMT systems. So far, standards for three cellular generations have been defined by ITU; the framework for IMT-2030 was defined in December 2023.

- **IMT-2000** Standard is ITU-R's 3G mobile communication standard, which targeted data rates of up to 2 Mbps, and had requirements set in Recommendation M.816 [9] in 1997. The Recommendation ITU-R M.1457 [10], first approved in 2000, contains detailed specifications of the terrestrial radio interfaces of IMT-2000.
- **IMT-Advanced** Standard, marketed as 4G, comprises many recommendation documents. Requirements related to the technical performance for the IMT-Advanced radio interface, issued in 2008, are found in Report ITU-R M.2134 [11], where one of the key targets was a peak data rate of 1 Gbps. The ITU-R Recommendation M.2012 [12], approved originally in 2012 and subsequently revised, identifies the terrestrial radio interface technologies of IMT-Advanced and provides detailed radio interface specifications.
- **IMT-2020** Standard, better known as 5G. Targets and requirements were first identified in 2015 in Recommendation ITU-R M.2083 [13] that set the IMT-2020 vision. Further, a set of minimum requirements was defined in 2017, in Report ITU-R M.2410 [14]. Finally, detailed specifications of the radio interfaces for the terrestrial component of IMT-2020 were defined in 2021, in ITU-R M.2150 [15].
- **IMT-2030** Report ITU-R M.2516 [16], published in 2022, identifies future technology trends of terrestrial IMT systems toward 2030 and beyond. The IMT Framework for IMT-2030, which sets the overall objectives, has been issued as Recommendation ITU-R M.2160-0 [17] in November 2023. The definition of IMT-2030 requirements and evaluation criteria are left for 2024–2027 time frame.

These specifications are developed in collaboration with the radio interface technology proponent organizations, global partnership projects, and regional standards developing organizations (SDOs). However, these standards only represent a set of key performance indicators (KPIs) and use cases but do not prescribe how to implement the IMT systems. This is where the third-generation partnership project (3GPP) and other organizations come in to play.

4.3.5 The Role of 3GPP

3GPP defines technical specifications that represent the blueprint for radio equipment implementation and associated protocols and are designed to meet the IMT targets. Within the Project Coordination Group (PCG), which is the decision-making

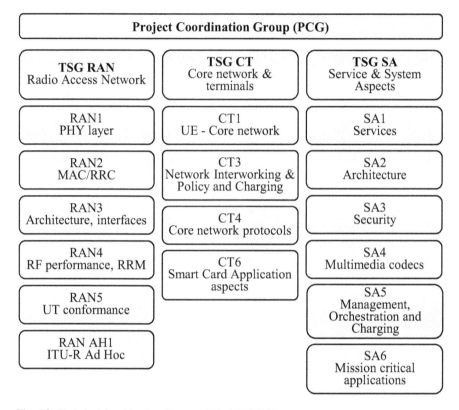

Fig. 4.3 Technical Specification Groups within 3GPP [18]

umbrella for Technical Specification Groups (TSG), there are three TSGs: Radio Access Network (TSG RAN), Core Network and Terminals (TSG CT), and Service and System Aspects (TSG SA), as illustrated in Fig. 4.3. The high level structure with three TSGs has been stable from the start of 3GPP work, but the work groups have evolved over time (e.g., RAN6 working group, responsible for the GSM/EDGE RAN and UTRAN radio and protocol work, was closed in 2020).

The main focus of this textbook is the work carried out within the TSG RAN group. In addition to preparing physical layer, protocol, and architecture specifications, the TSG RAN group also defines the bands that are to be used by the radio equipment. The 3GPP band definition is aligned with the ITU-R frequency allocations, as well as with the national regulatory bodies band definitions. These band definitions are based on requests made by operators or infrastructure manufacturers and are related to their specific market needs. Given that there is no perfect alignment between different geographic regions, or countries, 3GPP has defined many bands. Within the 5th generation of wireless networks, the

4.3 Role of ITU-R and 3GPP in Definition of Spectrum Bands and Standards

Table 4.3 Selected list of NR bands defined by 3GPP in the FR1 range, applicable to the United States and Canada

NR operating band	Uplink frequency range	Downlink frequency range	Duplex mode
n2	1850 MHz–1910 MHz	1930 MHz–1990 MHz	FDD
n5	824 MHz–849 MHz	869 MHz–894 MHz	FDD
n7	2500 MHz–2570 MHz	2620 MHz–2690 MHz	FDD
n12	699 MHz–716 MHz	729 MHz–746 MHz	FDD
n13	777 MHz–787 MHz	746 MHz–756 MHz	FDD
n14	788 MHz–798 MHz	758 MHz–768 MHz	FDD
n25	1850 MHz–1915 MHz	1930 MHz–1995 MHz	FDD
n26	814 MHz–849 MHz	859 MHz–894 MHz	FDD
n29	N/A	717 MHz –728 MHz	SDL
n30	2305 MHz–2315 MHz	2350 MHz–2360 MHz	FDD
n38	2570 MHz–2620 MHz	2570 MHz–2620 MHz	TDD
n41	2496 MHz–2690 MHz	2496 MHz–2690 MHz	TDD
n46	5150 MHz–5925 MHz	5150 MHz–5925 MHz	TDD
n54	1670 MHz–1675 MHz	1670 MHz–1675 MHz	TDD
n66	1710 MHz–1780 MHz	2110 MHz–2200 MHz	FDD
n71	663 MHz–698 MHz	617 MHz–652 MHz	FDD
n77	3300 MHz–4200 MHz	3300 MHz–4200 MHz	TDD
n78	3300 MHz–3800 MHz	3300 MHz–3800 MHz	TDD
n80	1710 MHz–1785 MHz	N/A	SUL
n85	698 MHz–716 MHz	728 MHz–746 MHz	FDD
n86	1710 MHz–1780 MHz	N/A	SUL
n89	824 MHz–849 MHz	N/A	SUL
n96	5925 MHz–7125 MHz	5925 MHz–7125 MHz	TDD

3GPP bands are currently divided in two ranges [19, 20] within terrestrial networks framework:[3]

- The frequency range 1 (FR1) covers 410–7125 MHz.
- The frequency range 2 (FR2) covers 24250 MHz –71000 MHz and is subdivided in FR2-1 (24250 MHz–52600 MHz) and FR2-2 (52600 MHz–71000 MHz).

Study item on channel models for the FR3 range spanning 7–24 GHz range has been initiated in 3GPP Rel-19. A subset of 3GPP bands within the FR1 range is shown in Table 4.3.

---

[3] The non-terrestrial network (NTN) component has similar division into NTN-FR1 (410–7125 MHz) and NTN-FR2 (17.3–30 GHz) frequency ranges.

Example 4.1 Determine the amount of spectrum below 1 GHz in Table 4.3.

Solution By examining Table 4.3, we can see that the following bands have the frequency ranges below 1 GHz (low bands): n5, n12, n13, n14, n26, n29, and n71. We also observe that n89 and n5 bands are subsets of n26 and n12 is a subset of band n85, so to avoid double counting, these two bands will not be included. The total amount of low bands spectrum is 35×2 (n26) $+ 18 \times 2$ (n85) $+ 10 \times 2$ (n13) $+ 10 \times 2$ (n14) $+ 11$ (n29) $+ 35 \times 2$ (n71) $= 227$ MHz. ∎

4.3.6 Frequency Bands and Characteristics of Each Frequency Range (sub-7GHz, mmWave, THz)

As we have seen, the propagation channel between the transmitter and receiver determines the performance of the communication system. These propagation characteristics, especially the penetration loss as a wave passes through obstacles, are heavily dependent on the frequency band used. There are three[4] broad frequency ranges in consideration for wireless communications: the FR1 (microwave or sub-7GHz) and FR2 (or mmWave) ranges and, speculatively, even higher in frequency in the THz range, as illustrated in Fig. 4.4.

Broadly, as the operating frequency increases, it is possible to transmit with larger bandwidths (keeping *fractional* bandwidth[5] constant) and achieve narrower

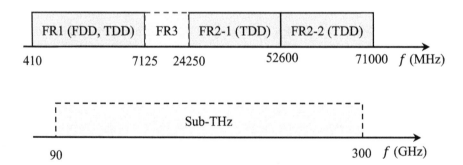

Fig. 4.4 Frequency ranges in consideration for wireless communication

---

[4] Four, if we include the FR3 range mentioned in the 3GPP section.

[5] Fractional bandwidth is defined as $FBW = 200 \frac{f_{high} - f_{low}}{f_{low} + f_{high}} \%$, where f_{low} and f_{high} and the low and high edges of the channel frequency range.

transmit/receive beams, but pathloss increases dramatically. We briefly review the characteristics of these three frequency ranges.

4.3.6.1 FR1: Sub-7 GHz or Microwave Range

The discussion on channel models in Chap. 2 largely assumes this frequency range. This remains the most important and widely used frequency range. Specifically, this band provides an ideal balance between propagation characteristics and system design. Within this band, a popular empirical model for urban environments, the COST-231 extension to the Hata model discussed in Chap. 2, shows that the dominant frequency-dependent pathloss (in dB) is $33.8 \log(f_c)$, where f_c is the operating frequency, i.e., in the linear scale, the loss is proportional to $f_c^{3.38}$. Indeed, because of the lower penetration loss, the sub-1GHz is especially well suited for communicating with transceivers within buildings. However, since antenna is most efficient if its size is at least comparable to wavelength (0.33m at 900 MHz), systems in this frequency range must use physically large antennas or suffer from reduced antenna efficiency. On the other hand, since pathloss is substantially higher at the higher frequencies, the sub-1GHz band may be the only viable choice if the signal must penetrate multiple and/or thick concrete walls.

In most scenarios, the 2–6 GHz band provides an ideal balance between antenna size and propagation characteristics. However, this is true for many applications, not just wireless communications. The colorful Canadian Table of Frequency Allocations chart [21] lists the many uses for this spectral region; wireless communication must coexist with these other applications in a non-interfering manner (more below). In the recent years, as the discussion on auctions below shows, a larger fraction of this prime spectral "real estate" has been repurposed for cellular communications. This repurposing is driven by the growing importance of communications in the broad economy.

4.3.6.2 FR2: The mmWave Range

The mmWave frequency range has received significant attention recently, including in auctions [22, 23]. The main attraction is the vastly increased bandwidths available at these higher operating frequencies. For example, the FCC *Auction 103* offered a total bandwidth of 3.4 GHz, vastly larger than anything possible in the FR1 range. The proposed usage scenarios for this band, e.g., in [23], list high data-rate applications such as augmented/virtual reality, short-range applications like industrial automation, and low-latency applications such as connected vehicles.

The main issue limiting the widespread use of mmWave frequencies is the associated pathloss and penetration loss. A mmWave signal, essentially, does not penetrate through blockages. For example, a human hand, covering a receiver antenna, would block the mmWave signal. Researchers have considered many potential workarounds, especially the use of multiple antennas and transmission paths (we will discuss multiple antennas in a later chapter). Even with these enhancements, providing a stable signal remains a challenge.

A second issue with mmWave propagation is the remarkably narrow beamwidths. For a given antenna size, beamwidths are inversely proportional to frequency; indeed, to overcome the pathloss, focusing the mmWave energy into a narrow beamwidth helps. However, the drawback is the attendant requirement of a precise alignment between transmitter and receiver. This has been implemented in fixed wireless broadband systems but would be challenging in a mobile and dynamic environment. To address this issue, researchers have suggested channel tracking methods, e.g., [24].

It is worth emphasizing that the large pathloss necessitates a new analysis of mmWave channels. As far as possible, an implementation would use line-of-sight channels (since reflection losses are also high). Similarly, very few paths between the transmitter and receiver would be strong enough to effectively contribute—a Rayleigh multipath channel model may not be appropriate in a mmWave system.

4.3.6.3 THz Transmissions

If mmWave frequency ranges provide substantial bandwidth, the THz range provides another order of magnitude more. Furthermore, given the miniature wavelength, THz communications could enable device-to-device communication on a nanoscale [25]. While early proposals depended on noncoherent communications, i.e., did not measure the phase of the received signal, more recent implementations overcome this limitation [26]. At THz frequencies, every aspect of the communication systems requires reformulation. Antenna technologies at these frequencies are not as mature as at lower frequencies; applications require line of sight with the extremely high pathlosses making for extremely limited transmission ranges. However, THz systems do open up new application domains such as high-speed UAV [27] and intra-body communications [28]. At this time, communications in the THz region is yet in its infancy.

4.4 Mobile Spectrum Allocation, Tier Organization, and Spectrum Auctions

ITU-R typically identifies frequency bands for IMT before these bands are allocated for mobile. When a frequency band is allocated to more than one service, the table of allocations establishes categories of services by identifying primary and secondary services the latter cannot cause of claim harmful interference to or from a primary service.

4.4.1 The Canadian Example

Canadian spectrum allocation for radio services is closely aligned with ITU-R frequency band allocations, but domestic regulatory policy occasionally deviates when necessary to meet the specific needs of Canadian spectrum requirements. Taking into account the results of WRC-19 and domestic requirements, the latest Canadian frequency allocation table [29] was updated in 2022 (note that the allocation table may be updated to reflect WRC-23 outcomes around the time of publication of this textbook) and covers a wide range, from 8.3 kHz to 275 GHz. A subset of frequency ranges within this allocation is dedicated for mobile communication services.

ISED revises the Canadian Table periodically, normally following a WRC by adopting changes to the frequency allocations for amateur, fixed, mobile, radiolocation, navigation, science, mobile-satellite, and fixed-satellite services that reflect WRC decision as well as Canadian specific policy decisions and footnotes.

4.4.1.1 Canadian Tier Organization

Mobile radio spectrum is typically licensed differently in different service (geographic) areas and not uniformly across the entire nation. In Canada, five tiers of service areas have been established by ISED [30]:

- **Tier 1** is a single national service area covering the entire territory of Canada.
- **Tier 2** consists of 16 provincial and large regional service areas covering the entire territory of Canada. Of these, 14 Tier 2 areas (2-001 to 2-014) cover the entire Canada.
- **Tier 3** consists of 59 smaller regional service areas covering the entire territory of Canada.
- **Tier 4** includes 172 localized service areas covering the entire territory of Canada.
- **Tier 5** includes smaller service areas, designated as metropolitan, urban (medium and large population centers), rural, and remote categories.

Tiers 1–4 were first introduced by ISED in 1998. Later, in 2019, ISED introduced Tier 5 service areas which are nested within Tier 4 service areas. An example of a Tier 4 area and the smaller Tier 5 areas contained within is illustrated in Fig. 4.5.

In general, tiers of the nth order are nested within tiers with order of $n-1$, where $n = 2, 3, 4, 5$.

Additionally, there are 66 legacy local telephone (TEL) areas whose borders correspond to the historical wireline service area boundaries of local telephone companies, but these legacy TEL areas are slowly being retired.

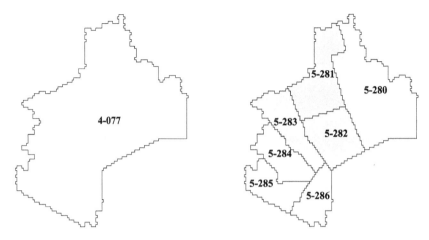

Fig. 4.5 Illustration of nested Tier 5 service areas within a Tier 4 service area 4-077 (Toronto)

Table 4.4 Tier 5 population numbers for the Toronto market

Service area	Service area name	2016 population
5-280	Durham	643,578.03
5-281	York	1,129,350.69
5-282	Toronto	2,723,714
5-283	Peel	1,379,668.97
5-284	Halton	554,214.77
5-285	Hamilton	533,046.66
5-286	Niagara	67,176.56
Total population		7,030,749.68

Example 4.2 Tier 4 service area 4-077, representing Toronto, consists of seven Tier 5 service areas: 5-280, 5-281, 5-282, 5-283, 5-284, 5-285, and 5-286. Verify that the sum of populations of these Tier 5 areas adds up to the population of 4-077.

Solution Based on the information provided in [30], we find the population values (based on 2016 census data) as listed in Table 4.4.

If we compare the total with a 4-077 population of 7,030,750, we see that the numbers are in agreement. The reason why fractional numbers appear in population count is a consequence of the total Canadian geography being divided into small rectangular grid areas with a basic building block being 1 minute by 1 minute and represents the product of local population density and the area of a grid area.

4.4 Mobile Spectrum Allocation, Tier Organization, and Spectrum Auctions

4.4.1.2 Canadian Auctions

Canadian spectrum was originally awarded to various regional operators using a process referred to as comparative selection (similar to an RFP process, sometimes called a "beauty contest"). First, the cellular (850 MHz) spectrum band was awarded in 1985. The cellular spectrum band was initially split into sub-band A and sub-band B. Sub-band A was awarded to CANTEL (Rogers Wireless Inc.), while sub-band B was allocated to the local telephone companies according to their respective operating territories (TEL areas) where they provided public switched telephone service. Since spectrum area licenses were not yet introduced in Canada, the first cellular stations were managed via radio licenses and later converted to spectrum licenses. Following that, parts of the Personal Communications Services (PCS) band (1850 MHz to 1990 MHz) were awarded in 1995, again to different regional operators. In the subsequent years, a series of mergers and acquisitions occurred, and three strong national operators emerged. Since then, all the subsequent spectrum allocations have been based on auctions. The first auction happened in 2001, and it included 40 MHz of PCS spectrum that had been held in reserve in 1995, and an additional spectrum band returned to ISED by TELUS in 2000. A timeline of Canadian auctions up to 2023 year is shown in Fig. 4.6.

Residual auctions in (2015, 2018) and BRS auction (2015) are not shown in the timeline. The combined auction proceeds from these auctions have exceeded $28.5B, as shown in Table 4.5.

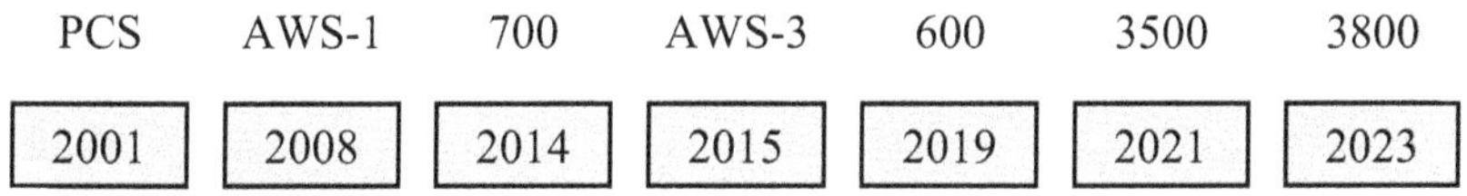

PCS	AWS-1	700	AWS-3	600	3500	3800
2001	2008	2014	2015	2019	2021	2023

Fig. 4.6 Canadian auction history

Table 4.5 Summary of auction costs in Canada in the period from 2001 to 2023

Auction	Total
PCS (2021)	$1,481,920,000
AWS-1 (2008)	$4,254,710,327
700 MHz (2014)	$5,270,636,002
AWS-3 (2015)	$2,109,147,421
Residual (2015)	$58,509,286
BRS (2015)	$755,371,001
Residual (2018)	$43,436,322
600 MHz (2019)	$3,470,328,000
3500 MHz (2021)	$8,911,563,543
3800 MHz (2023)	$2,157,696,532
Total	$28,513,318,434

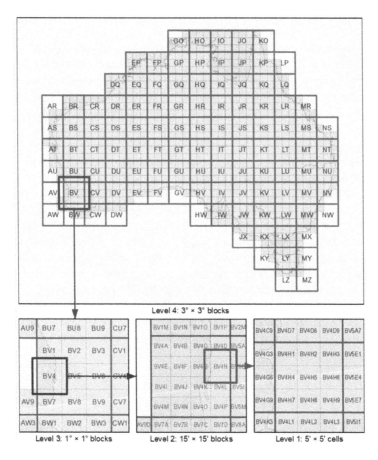

Fig. 4.7 Illustrating the hierarchical Australian spectrum map grid [31] ©Commonwealth of Australia (Australian Communications and Media Authority) 2020

4.4.2 The Australian Example

Allocations and auctions in Australia follow a pattern similar to the Canadian experience. The Australian spectrum map grid was revised in 2023 [31]. The process consolidates multiple versions of datum: the process of mapping coordinates to locations on the Earth surface. The country is divided into 163 cells with defined boundaries. A hierarchical cell identification scheme (HCIS) provides a unique identifier to each region covering 5 minutes of arc. In all there are four levels in the HCIS. Figure 4.7 illustrates the hierarchical nomenclature used to designate areas [31]. It is worth noting the similarity of this figure to the Canadian example of nested tiers.

In Australia, access to the radio spectrum is provided by the Australian Communications and Media Authority (ACMA) via a number of different legislative

4.4 Mobile Spectrum Allocation, Tier Organization, and Spectrum Auctions

tools. The Australians use three forms of license: apparatus, class, and spectrum. As the name suggests, the apparatus license allows for the operation of transmitters (and receivers). This type of licensing is often used outside of areas that have been allocated using spectrum licensing where companies like mobile network (cellular) operators still wish to install small numbers of transmitters providing cellular network coverage to serve remote communities. A class license sets out usable equipment including the frequency ranges of operation and rules, such as EIRP, of use. The spectrum license specifies the geographical area and frequency band of operation. Like Canada, Australia is a signatory to the ITU convention, and its spectrum regulation regime is consistent with ITU regulations [32].

The most common method of providing spectrum access at least for the mobile (cellular) telecommunications industry is via "Spectrum Licensing." These licenses are sold using auction-based mechanisms and hold a similar standing to that of leasing land. They are tradeable and leasable on the secondary spectrum market. Spectrum licenses also confer all the responsibilities or the maintenance of the spectrum onto the licensee, similar to the Canadian example.

As with other countries, Australia publishes a five-year spectrum outlook[6] on a sliding window basis. Among other specific matters, a key goal is to keep the approach to licensing relevant to the ever-evolving needs of the country. The published review framework, for example, considers licensed spectrum, below 6 GHz, and how to regulate spectrum in the context of new technologies [34]. However, it is worth noting that, as in other countries, the spectrum plan covers a wide range of frequencies, not just those used for cellular communications. For example, the Australian plan covers frequencies from 4.3 kHz to 420 THz!

More recently, the Australian regulator has introduced a variation to the apparatus licence, known as an apparatus "Area Wide Licence" (AWL). The difference that the AWL brings is that rather than the applicant seeking permission to operate a single transmitter that provides service to a notional area, instead the applicant applies for a geographic area that encompasses the region that they wish to provide services. In this case, multiple apparatus transmitting base stations can be registered by the license holder within the geographic area they applied for on their license, and all they need to do is to register the transmitters with the ACMA, after first confirming that each transmitter in and of itself would not radiate signals outside of the license area owned by the licensee. In many ways, this is like a spectrum license, except that the applicant applies for the area they wish to operate in, as opposed to the ACMA defining the area that all licensees holding spectrum in a given band can operate

---

[6] At the time of writing, the latest outlook covers 2023–2028.

within. The areas covered by an AWL are typically much smaller than a spectrum license service area too, although that is not always the case.

4.4.2.1 Spectrum Auctions

In the previous section, we discussed the history of spectrum auctions in Canada and the associated costs. A similar process has played out (and is playing out) in Australia starting with PCS allocations in the 800 MHz and 1800 MHz bands in 1998 [35]. The most recent auction covered the 850/900 MHz band (2021) and the 3.4/3.7 GHz bands (2023). It is worth noting the similarities between the auction bands in Canada and in Australia. Indeed, a search for auctions conducted, and spectrum allocated, around the globe is likely to see similar trends. Reliable, high data-rate, communications is now considered essential for any economy. Worldwide the 3.4–3.7 GHz band is attractive for the bandwidth available for communications, while the sub-1 GHz range is attractive for its propagation characteristics, especially the ability to communicate from/to within buildings. As in Canada, and other parts of the globe, substantial monies are raised through these auctions. While the 1998 auction raised $380 million, the 2021 auction in the 850/900 MHz band raised more than $2 billion.

4.4.3 The US Example

Two agencies are responsible for spectrum management in the United States. The National Telecommunications and Information Administration (NTIA) manages the use of spectrum by the federal government. Additionally, the NTIA identifies spectrum for potential commercial use.

The second agency is the Federal Communications Commission (FCC), which manages spectrum for commercial use. The FCC is responsible for auctioning spectrum and issuing spectrum licenses on behalf of the US government, a role it was authorized to perform by Congress in 1993. The FCC regulates various communication services in the United States, including wireless, wireline, satellite, and broadcast services. The rules and regulations established by the FCC are published in Title 47 of the Code of Federal Regulations (47 CFR, [36]), which consists of several parts that cover different aspects of telecommunications. For instance, Part 15 covers unlicensed RF devices, Part 24 covers personal communication services, and Part 25 covers satellite communications.

As part of ITU-R Region 2, the US spectrum, as in cases of Canadian and Australian examples, is closely aligned with the ITU's table of frequency allocations. The online FCC table of frequency allocation [37] provides a good side-by-side comparison with the ITU-R table of frequency allocation.

The FCC spectrum licenses can be assigned either on geographic or location (site) specific basis [38]. Similar to spectrum licenses in Canada, geographic

4.4 Mobile Spectrum Allocation, Tier Organization, and Spectrum Auctions 143

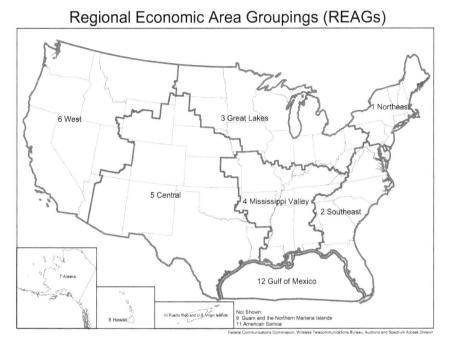

Fig. 4.8 FCC's Regional Economic Area Groupings (REAG) illustration [40], reproduced with permission from the FCC, Wireless Telecommunications Bureau, Auctions and Spectrum Access Division

licenses in the United States allow for use of spectrum anywhere within a given geographic area, whereas location-specific licenses are tied to specific location and specific frequency. The FCC issues spectrum licenses across various geographical levels [39], the first of which is the Cellular Market Area (CMA) first defined in 1981. Since then, several other areas were used by the FCC to assign spectrum licenses, including Economic Areas (EA), Partial Economic Areas (PEA), Major Economic Areas (MEA), Regional Economic Areas (REA), and Economic Area Groupings (EAG) which is also known as Regional Economic Area Groupings (REAG) (Fig. 4.8). Basic Trading Areas (BTA) and Major Trading Areas (MTA), where BTAs are subsets of MTAs, were also used. Recent FCC spectrum auctions tended to use PEA (C-band and 3.45 GHz) or county-based areas (CBRS and 28 GHz). The choice of the type of area balances several factors, such as the type of the service, technical performance, policy objectives, and market demand.

4.4.3.1 Spectrum Auctions

The first US cellular licenses were the 850 MHz licenses, which date back to 1981, when the FCC sets aside 40 MHz of spectrum, divided into two blocks of 20 MHz and awarded it initially via comparative hearings and later through lotteries [41]. After that, the FCC conducted its first auction for spectrum licenses in the Personal Communications Services (PCS) band in 1994. Since that time, it held over 100 auctions for various spectrum licenses spanning a wide swath of frequencies ranging from 220 MHz to millimeter wave band (e.g., 47 GHz) and for various applications, such as personal mobile services, location and monitoring services, and broadcast and paging services. Selected auctions, with a total of $224.7B gross proceeds, are illustrated in Table 4.6.

For a complete list of FCC auctions, a reader is referred to the FCC's online auction summary [42].

4.5 Spectrum Cost Comparison

Important metrics in spectrum cost comparisons are the product of bandwidth in MHz and the population (MHz-PoP) and the average cost per MHz-PoP.

The MHz-PoP, a measure used to compare spectrum holdings among operators, is defined as sum over tiers of the products of allocated spectrum (in MHz) per given area and population count for the same area:

$$\text{MHz-PoP} = \sum_{Tier} BW_{Tier} \times Population_{Tier} \qquad (4.1)$$

The total cost of spectrum divided by the total MHz-PoP gives average cost per MHz-PoP. For example, if we look at the 600 MHz band auction from 2019, the total cost to all the operators who participated in the auction was approximately $3.5B. Using the Canadian census information from 2016, the total Canadian population was approximately 35 million, and the total auctioned spectrum bandwidth of 70 MHz, we get the average cost per MHz-PoP of $1.41.

It is instructive to look at the cost per MHz-PoP for individual operators because dividing (normalizing) the cost by MHz allows us to compare different quantities of spectrum via a consistent unit price. Similarly, dividing by pop supports an *apples-to-apples* comparison of spectrum costs in different geographies.

Example 4.3 Xplornet participated in the 600 MHz auction and won four licenses in four Tier 4 areas: 2-001, 2-002, 2-003, and 2-010. In each of these areas, their total amount of spectrum is 10 MHz. Given their final auction price of $35, 755, 000 [43], determine their average cost per MHz-PoP.

Solution Using the population numbers from the ISED's service area website, we calculate MHz-PoP values for individual tiers and the total MHz-PoP value, which is summarized in Table 4.7.

4.5 Spectrum Cost Comparison

Table 4.6 Summary of selected US auctions

Auction #	Year completed	Band name	Band (MHz)	License area	Gross proceeds
110	2022	3.45 GHz	3450–3550	PEA	$22,513,631,811
107	2021	3700 MHz/C-band	3700–3980	PEA	$81,168,677,645
105	2020	3500 MHz/CBRS	3550–3650	County	$4,858,663,345
103	2020	37/39/47 GHz	Upper 37 GHz, 37,600–38,600 39 GHz, 38,600–40,000 47 GHz, 47,200–48,200	PEA	$7,569,983,122
102	2019	24 GHz	24,250–24,450 and 24,750–25,250	PEA	$2,024,268,941
101	2019	28 GHz	27,500–28,350	County	$702,572,410
1002	2017	600 MHz	617-652/663-698	PEA	$19,768,437,378
97	2015	AWS-3	A1, 1695–1700 B1, 1700–1710 G, 1755–1760/2155–2160 H, 1760–1765/2160–2165 I, 1765–1770/2165–2170 J, 1770–1780/2170–2180	A1, B1, H, I, and J, EA G, CMA	$44,899,451,600
96	2014	H-block	1915–1920/1995–2000	EA	$1,564,000,000
73	2008	700 MHz	A, 698-704/728-734 B, 704-710/734-740 E, 722-728 C, 746-757/776-787 D, 758-763 / 788-793	A, EA B, CMA E, EA C, REAG D, nationwide license	$19,120,378,000

Table 4.7 Summary of 600 MHz auction costs for Xplornet

Service area	Service area name	Bandwidth (MHz)	Population	MHz-PoP
2-001	Newfoundland and Labrador	10	520,176	5,201,760
2-002	Nova Scotia and PEI	10	1,066,470	10,664,700
2-003	New Brunswick	10	745,596	7,455,960
2-010	Manitoba	10	1,278,016	12,780,160
Total MHz-PoP				36,102,580

Given the total price of \$35,755,000, the average cost per MHz-PoP was \$0.99. ∎

Different bands have been auctioned on a different tier basis (e.g., 600 and 700 MHz bands were auctioned on Tier 2 basis, BRS band on a Tier 3 basis, 3500 MHz on a Tier 4 basis, etc.).

4.6 Conducted and Radiated RF Requirements for 5G BS

We discussed how the spectrum is allocated and how the licensed spectrum becomes available to operators via spectrum auctions. Now, let us explore what limitations are imposed on its usage.

A crucial aspect of regulation is limiting the power transmitted over the radio waves. The limits set allow for safe radiation (we introduced the Safety Code 6 in Chap. 1) but also limit the (always present) out of band interference to acceptable levels. These limits are often aligned with 3GPP specifications, although deviations from 3GPP are possible for some bands. In this section, we outline the 3GPP limits and provide references to Canadian regulation.

For 5G networks, 3GPP specifies output powers per BS class (wide area, medium range, local area) and per BS type (BS type 1-C, BS type 1-H, BS type 1-O, and BS type 2-O) in TS 38.104 [20]. Similar requirements are defined for 4G LTE in TS 36.104 [44].

Conventional base stations used in 3G and 4G networks have accessible radio ports, which allow for various signal measurements directly at the port. Similar base stations, especially for lower frequency bands, are used in 5G, also called New Radio or NR. These types of base stations are classified as BS type 1-C, where the letter "C" indicates *conducted* power measurement (as opposed to radiated, over the air, power). As illustrated in Fig. 4.9, the measurement point is either Port A or, if external power amplifier (PA) and an external filter are available, the measurement is performed at Port B.

Power limits are listed in Table 4.8.

4.6 Conducted and Radiated RF Requirements for 5G BS

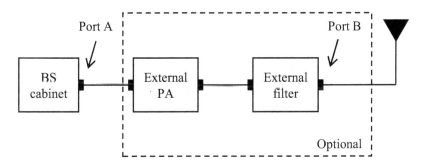

Fig. 4.9 Type 1-C transmitter interface, from 3GPP TS 38.104

Table 4.8 3GPP defined output power levels for BS type 1-C [20]

BS class	$P_{rated,c,AC}$
Wide area BS	No upper limit
Medium-range BS	$\leq$38 dBm
Local area BS	$\leq$24 dBm

Table 4.9 3GPP defined output power levels for BS type 1-H [20]

BS class	$P_{rated,c,sys}$	$P_{rated,c,TABC}$
Wide area BS	No upper limit	No upper limit
Medium-range BS	$\leq$38 dBm $+10\log(N_{TXU,counted})$	$\leq$38 dBm
Local area BS	$\leq$24 dBm $+10\log(N_{TXU,counted})$	$\leq$24 dBm

In systems with active antenna systems (AAS) operating in FR1 range, BS type 1-H hybrid requirements are defined. Transmit power requirements are defined for two points of reference:

1. Conducted characteristics are defined at individual or groups of Transceiver Array Boundary (TAB) connectors, as illustrated below.
2. Radiated characteristics are defined over the air (OTA), where the operating band specific radiated interface is referred to as the Radiated Interface Boundary (RIB).

If TAB connectors (TABC) are not accessible, OTA measurements are performed instead.

Power limits for BS type 1-H are listed in Table 4.9.

In these systems, $N_{TXU,active}$ is the actual number of active transmitter units and is implementation specific and is not used to define emission limits. Instead, 3GPP uses the number of *counted* active transmitter units, calculated as $N_{TXU,counted} = \min(N_{TXU,active}, 8 \times N_{cells})$ for the purposes of calculating the emission limit for AAS systems. For example, for single cells ($N_{cells} = 1$), value for is $N_{TXU,counted} = 9$ dB. The $N_{cells} > 1$ scenario may occur if the same transceiver is used to service multiple cells. These values are general, 3GPP guidelines. Regulation in each country or region may impose different power limits.

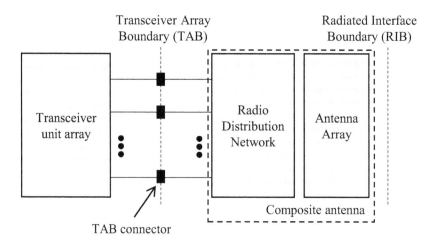

Fig. 4.10 Type 1-H transmitter interface, from 3GPP TS 38.104

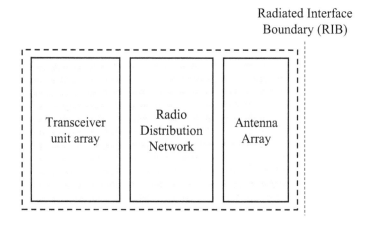

Fig. 4.11 Type 1-O/Type 2-O transmitter interface, from 3GPP TS 38.104

Table 4.10 3GPP defined output power levels for BS type 1-O

BS class	$P_{rated,c,TRP}$
Wide area BS	No upper limit
Medium-range BS	≤ 47 dBm
Local area BS	≤ 33 dBm

In AAS systems, where radio and antenna are integrated, and conducted measurements are not possible, can only be measured OTA, at the Radiated Interface Boundary (RIB) shown in Fig. 4.10. Type 1-O and Type 2-O systems apply to FR1 and FR2 frequency ranges, respectively (Fig. 4.11). Power limits are listed in Table 4.10.

The Total Radiated Power (TRP) is measured during the transmitter ON period, which means that in case of a TDD system (where the transmitter may be OFF while it receives) the OFF period is excluded from measurement. Importantly, this TRP is averaged over all the directions in 3D space and effectively is close to the total conducted power input into the antenna system (i.e., the TRP value of 47 dBm is the same as the conducted sum of all TAB connectors limit of 38 + 9dBm from the Type 1-H), excluding the antenna efficiency losses.

The upper limits for the wide area BS (macro base stations) as well as for the other types of base stations are imposed by national regulators (e.g., by FCC in the United States, by ISED in Canada, etc.). Each frequency band comes with its own regulation. Next example illustrates Canadian regulation for a band that overlaps with the 3GPP band n77.

> **Example 4.4** Canadian Radio Standards Specification RSS-192 [45] sets the regulatory framework for the band operating in frequency range from 3450 to 3900 MHz. It sets the maximal output EIRP for wide area non-AAS base station radios at 68 dBm/5MHz and maximal TRP power for AAS-based base stations at 47 dBm/5 MHz. Determine what is the maximal antenna gain for non-AAS system if its conducted power is equal to the AAS TRP power. Comment on additional operational requirements specified in Standard Radio System Plans document SRSP-520 [46].

Solution The TRP and conducted power are equal, so the maximal gain the non-AAS-based system is $G_T = 68 - 47 = 21$ dBi. The SRSP-520 imposes additional operation requirement that the total EIRP cannot exceed 68 dBm/5MHz regardless of whether AAS or non-AAS technology is deployed. In practical system, it is not necessary that 21 dBi gain antenna is deployed. It is possible to achieve the 47 dBm/5MHz and 68 dBm/5MHz limits by deploying, for example, radios that output 43 dBm/5MHz while utilizing 25 dBi antenna gains. ∎

4.7 Device Output Power and Dynamic Range

We now consider user devices (terminals). Band-specific power classes for terrestrial network devices are defined in TS 38.101-1 [19] and TS 38.101-2 [47]. For FDD bands, power class 3 (23 dBm, already mentioned in Chap. 1) is typical as it allows for maximal transmission power without exceeding safety limits for human electromagnetic field exposure. In TDD bands, due to DL/UL duty cycles, wherein devices transmit only a fraction of the time, higher-power classes have been defined, as summarized in the table below, which shows only a subset of bands in use (Table 4.11).

Note that 3GPP defined tables include more details such as implementation tolerance (e.g., ±2 dB) and notes explaining the use case (e.g., 31 dBm power level is not intended for smartphones but for public safety devices).

Table 4.11 3GPP defined output power levels for user terminals

NR band	Class 1 (dBm)	Class 1.5 (dBm)	Class 2 (dBm)	Class 3 (dBm)
n2				23
n5				23
n7				23
n12				23
n13				23
n14	31			23
n25				23
n26				23
n30				23
n38				23
n41		29	26	23
n66				23
n71	31			23
n77		29	26	23
n78		29	26	23

Obviously, these power levels, used for UL transmission, are much lower than the power levels on the base station side that carry DL transmission for multiple devices (e.g., 23 dBm vs. the EIRP of 68 dBm/5MHz). This results in noticeable differences in data rates, where DL data rates are sometimes an order of magnitude higher than the UL data rates (e.g., peak DL data rate of 2 Gbps vs. approximately 100 Mbps of the peak data rate in the UL). The question one could ask is if the transmit power of 23 dBm at the cell edge is sufficient for UL signal reception. The answer is provided in Example 3.6 in Chap. 3, where we have seen that a typical UL link budget allows for meaningful data rates in the UL even when a moderate amount of spectrum is used (e.g., 0.72 MHz).

Example 4.5 RSS-192 specifies the maximum device transmit power of 30 dBm per channel bandwidth for the band operating in frequency range from 3450 to 3900 MHz. Compare this value to the 3GPP table limits for the band n77 and comment on the supported power class in Canada.

Solution Given that the maximal transmit power allowed by the RSS-192 is 30 dBm per channel bandwidth, it allows for power class 1.5 operation. This provides additional 6 dB of the UL link budget improvement relative to power class 3, which is important given the low TDD duty cycle, which is normally <25% in the UL. ∎

As shown in Table 4.12, low-power wide area networks-based internet of things (LPWA IoT) devices, classified as Narrowband IoT (NB-IoT) and Category M (LTE-M) devices, have more power classes defined in TS 36.101.

4.8 Unwanted Emission Requirements

Table 4.12 3GPP defined output power levels for NB-IoT devices

EUTRA band	Class 3 (dBm)	Class 5 (dBm)	Class 6 (dBm)
2	23	20	14
4	23	20	14
5	23	20	14
7	23	20	14
12	23	20	14
13	23	20	14
14	23	20	14
17	23	20	14
25	23	20	14
26	23	20	14
66	23	20	14
71	23	20	14

4.8 Unwanted Emission Requirements

Operators, when deploying their networks, strictly follow the regulatory requirements related to output powers. Similarly, infrastructure vendors design their equipment such that it meets regional or national regulatory requirements. In addition to maximal output power within specified bands and occupied channels, both 3GPP and national regulators specify how much energy/power can leak outside of occupied channels. This energy leakage is known as unwanted emission, and keeping these unwanted emissions within a certain range ensures coexistence between operators operating in the same band or other services operating in adjacent bands.

4.8.1 Unwanted Emissions at the BS

Unwanted emissions are divided into two domains: unwanted out of band emission (OOBE) and spurious domain emission, as illustrated in Fig. 4.12. An operating channel (indicated as *CH*) is typically narrower than the operating band, and the operating band unwanted emission (OBUE) extends $\Delta f_{OBUE} = 10$ MHz or $\Delta f_{OBUE} = 40$ MHz outside of of operating band. Values for Δf_{OBUE} depend on channel bandwidth (e.g., if channel bandwidth ≥ 100 MHz, its value is 40 MHz) but also on the band (e.g., n96 band currently has $\Delta f_{OBUE} = 50$ MHz).

Operating band unwanted emission limits are not a rectangular function. Instead, it has a transition range before it levels off at a certain range (e.g., -13 dBm/MHz for Category A wide area base stations that operate above 1 GHz).

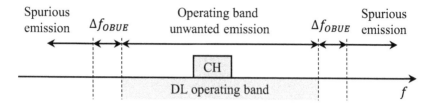

Fig. 4.12 Illustration of unwanted and spurious emission ranges at the base station side

Fig. 4.13 Illustration of unwanted emission levels at the base station side

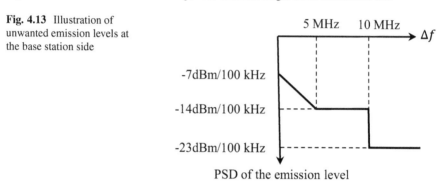

> **Example 4.6** Operating band unwanted emission from the channel edge of a 5G radio matches exactly the emission limit shown in Fig. 4.13. Calculate the total unwanted emission power in the range from 0 to $\Delta f = 10$ MHz.

Solution The unwanted emission power spectral density expression for the frequency range Δf from 0 to 5 MHz frequency offset is

$$S(\nu)|_{\text{dBm/100kHz}} = -7 - \frac{7}{50}\nu \quad (4.2)$$

where $\nu = \Delta f/100\text{kHz}$ is the normalized frequency where each frequency point covers 100 kHz.

This can be converted to linear scale PSD as

$$S(\nu) = 10^{\frac{S(\nu)[\text{dBm}]-30}{10}} = 10^{-3.7} \times 10^{-\frac{0.7}{50}\nu} \quad (4.3)$$

Linear power of the unwanted emission in the range from 0 to 5 MHz is

$$P_1 = \int_0^{50} S(\nu)\,d\nu \quad (4.4)$$

4.8 Unwanted Emission Requirements

$$= 10^{-3.7} \int_0^{50} 10^{-\frac{0.7}{50}v} \, dv$$

$$= 0.00495454 \text{ (watts)}.$$

Converted to dBm scale, the unwanted emission power is

$$P_{1|dBm} = 10 \log_{10}(P_1) + 30 = 6.94 \text{ (dBm)}. \tag{4.5}$$

In the second range, the PSD function is constant and the dBm scale power is

$$P_{2|dBm} = -64 + 10 \log_{10}\left((10-5) \times 10^6\right) = 3 \text{ (dBm)}, \tag{4.6}$$

where the PSD value $-14 - 10 \log_{10}(100 \times 10^3) = -64$ is calculated over 1 Hz. The total unwanted emission power over these two ranges is

$$P_{unwanted|dBm} = 10 \log_{10}\left(10^{\frac{6.94}{10}} + 10^{\frac{3}{10}}\right) = 8.42 \text{ (dBm)}. \tag{4.7}$$

■

The spurious domain range starts just outside the OBUE region and extends from 9 kHz to 12.75 GHz or even higher depending on where the fifth-order harmonics, introduced in Chap. 3, are for the FR1 bands. The 5th harmonic has been adopted by 3GPP because it is the frequency point where we may expect unwanted emission that impacts adjacent services/systems. The spurious domain emission level is aligned with the unwanted emission in the OBUE range (e.g., −13 dBm/MHz). A simplified table from the TS 38.104 3GPP specification is shown Table 4.13.

Table 4.13 General BS transmitter spurious emission limits in FR1, Category A

Spurious frequency range	Basic limit	Measurement bandwidth
9 kHz–150 kHz		1 kHz
150 kHz–30 MHz		10 kHz
30 MHz–1 GHz		100 kHz
1 GHz–12.75 GHz	−13 dBm	1 MHz
12.75 GHz–5th harmonic of the upper frequency edge of the DL operating band in GHz		1 MHz

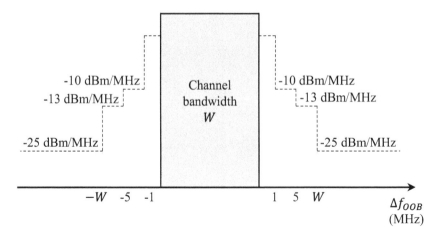

Fig. 4.14 Illustration of unwanted emission at the UT side

4.8.2 Unwanted Emissions at the UT

3GPP technical specifications for the user equipment also have the unwanted emissions divided into:

- OOBE caused by nonlinearity of the transmitter (e.g., spectrum regrowth immediately outside the occupied bandwidth) is constrained by a specific spectrum emission mask and adjacent channel leakage ratio (ACLR).
- Spurious emissions caused by harmonics and intermodulation products that go up to 5th harmonic or 26 GHz for FR1 range and up to 2nd harmonic for FR2 range.

The spectrum emission mask for the UT is specified for the out of band (OOB) frequency range Δf_{OOB} that is measured from the edges of the assigned channel bandwidth W. An example of FR1 range emission mask, specified in TS 38.101-1, for channel bandwidths $W = 10, 15, 20, 25, 30, 35, 40, 45$ MHz, is illustrated in Fig. 4.14. The first level, for the range $0 \geq \Delta f_{OOB} < 1$ MHz, has a limit of -13 dBm/1% of W.

An example of the same emission limit, imposed by ISED, for the 3500 MHz band (RSS-192), is shown in the Table 4.14. Spurious emissions start from $W + 5$ MHz and are defined in TS 38.101-1 (for FR1 range). Important limits are for the range $30 \leq f < 1000$, MHz, where spurious emission level cannot exceed -36 dBm/100 kHz, and for the range $1 \leq f < 12.75$ GHz, where the limit is -30 dBm/MHz.

4.9 Reference Sensitivity

Table 4.14 Unwanted emission limits for subscriber equipment (RSS-192). Offsets are from the edge of the frequency block group W, in MHz

Frequency block (W)	0–1 offset	1–5 offset	5–W	>W
10,20,30,40 MHz	−13 dBm/1% of W	−10 dBm/MHz	−13 dBm/MHz	−25 dBm/MHz
>40 MHz	−24 dBm/30 kHz	−10 dBm/MHz	−13 dBm/MHz	−25 dBm/MHz

Table 4.15 Selected 3GPP defined fixed reference channels

Reference channel	G-FR1-A1-1	G-FR1-A1-4
Subcarrier spacing (kHz)	15	15
Allocated resource blocks	25	106

4.9 Reference Sensitivity

We have already introduced the notion of a receiver sensitivity called REFSENS in 3GPP. We will revisit it here and define again the 3GPP specified reference values for different channel bandwidths. For base stations, the reference sensitivity $P_{REFSENS}$ is calculated as

$$P_{REFSENS} = -174 + 10\log_{10}(BW) + NF + IM + SNR \quad (4.8)$$

where the IM term stands for implementation loss margin and is the only new term not defined so far. Typically, its value is set to $IM = 2$ dB.

In the reference sensitivity calculation by the 3GPP RAN WG4, the noise figure is assumed to be $NF = 5$ dB for wide area BS, 10 dB for medium-range BS, and 13 dB for local area BS. The signal-to-noise ratio (SNR) value is bandwidth dependent. In the example below, we will use the value given in TR 38.817-02 [48], Annex B, where the $SNR = -1.2$ dB.

The bandwidth value BW depends on Fixed Reference Channels (FRC) specified in TS 38.104, Annex A.1. Selected FRCs are listed in Table 4.15.

Example 4.7 Determine the reference sensitivity level for 5, 10, or 15 MHz channels configured to use 15 kHz of subcarrier spacing; the reference measurement channel is G-FR1-A1-1. For the occupied channel bandwidth, use the following formula:

$$BW = 300 \times \text{Subcarrier spacing}. \quad (4.9)$$

The details of this channel bandwidth calculation will be introduced in Chap. 8.

Table 4.16 From TS 38.104, Table 7.2.2-1: NR wide area BS reference sensitivity levels

BS channel bandwidth (MHz)	Subcarrier spacing (kHz)	Reference measurement channel	Reference sensitivity power level, $P_{REFSENS}$ (dBm)
5, 10, 15	15	**G-FR1-A1-1**	**−101.7**
10, 15	30	G-FR1-A1-2	−101.8
10, 15	60	G-FR1-A1-3	−98.9
20, 25, 30, 40, 50	15	G-FR1-A1-4	−95.3
20, 25, 30, 40, 50, 60, 70, 80, 90, 100	30	G-FR1-A1-5	−95.6
20, 25, 30, 40, 50, 60, 70, 80, 90, 100	60	G-FR1-A1-6	−95.7

Solution Channel bandwidth is

$$BW = 300 \times 15 \text{ kHz} = 4.5 \text{ MHz}. \tag{4.10}$$

Using the parameters provided above, the reference sensitivity level is

$$P_{REFSENS} = -174 + 10\log_{10}(BW) + NF + IM + SNR \tag{4.11}$$

$$= -174 + +10\log_{10}\left(4.5 \times 10^6\right) + 5 + 2 + (-1.2) \tag{4.12}$$

$$= -101.66 \text{ dBm}. \tag{4.13}$$

■

This value calculated in Example 4.7 matches the value of -101.7 dBm provided in Table 7.2.2-1 in TS 38.104 [20], reproduced Table 4.16.

4.10 Adjacent Channel Selectivity, Adjacent Channel Leakage Ratio, and Adjacent Channel Interference Ratio

No cellular receiver can have an ideal, brick-wall, receive filter frequency response. Instead, it has a roll-off frequency range. As a consequence, a portion of the energy from adjacent channels (gray area in the figure below) is picked up by the receive filter, which acts as interference. The adjacent channel selectivity (ACS) is defined as a receiver's ability to suppress a signal on an adjacent channel, which means it measures the ratio of the receiver filter attenuation on the wanted signal to the receiver filter attenuation on the adjacent interfering signal channel (Fig. 4.15).

4.10 Adjacent Channel Selectivity, Adjacent Channel Leakage Ratio, and...

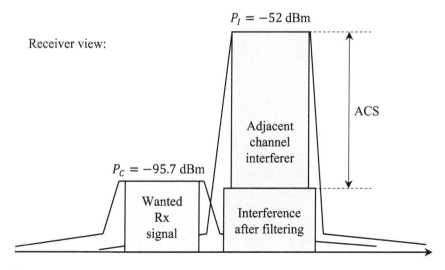

Fig. 4.15 Illustration of adjacent channel selectivity

Table 4.17 From TS 38.104, Table 7.4.1.2-1: Base station ACS requirement

BS channel bandwidth of the lowest/highest carrier received (MHz)	Wanted signal mean power (dBm)	Interfering signal mean power (dBm)
5, 10, 15, 20		Wide area: −52
25, 30, 40, 50, 60	$P_{REFSENS} + 6$ dB	Medium range: −47
70, 80, 90, 100		Local area: −44

The adjacent channel interference scenario may occur when a user receives a desired signal from a faraway serving cell while an adjacent channel (from a different operator) is in close proximity. 3GPP specifies requirements for this adjacent channel interference tolerance by specifying the minimum performance for 6 dB desensitization, as illustrated in Table 4.17. The 6 dB value has been chosen so that the thermal noise does not dominate the performance of the receiver.

Example 4.8 Calculate the base station ACS value for 5, 10, and 15 MHz channels using the requirements specified in Table 4.17.

Solution Calculations herein follow closely the steps outlined in [49], Chapter 14. In Example 7, we have calculated the reference sensitivity $P_{REFSENS} = -101.7$ dBm, and Table 4.17 specifies an interference level $P_I = -52$ dBm for wide area networks. The carrier signal level, in the presence of adjacent interference, is

$$P_{C|dBm} = P_{REFSENS} + 6 = -95.7 \text{ dBm}, \quad (4.14)$$

and the noise power is

$$P_{N|dBm} = -174 + 10\log_{10}(BW) + NF + IM \qquad (4.15)$$
$$= -174 + 10\log_{10}\left(4.5 \times 10^6\right) + 5 + 2 \qquad (4.16)$$
$$= -100.5 \text{ dBm}, \qquad (4.17)$$

which is, not surprisingly, exactly 1.2 dB above the reference sensitivity level (REFSENS was calculated for the $SNR_{|dB} = -1.2$ dB) as was the case in the Example 7. Due to the impact of adjacent channel interference, the same linear SNR value now applies to $P_C = 4 P_{\text{REFSENS}}$ operating level, so that in linear scale we have

$$\frac{P_C}{P_N + P_{I,\text{ Adjacent channel}}} = 10^{-\frac{1.2}{10}} = SNR \qquad (4.18)$$

where $P_{I,\text{ Adjacent channel}}$ is the adjacent channel interference area in the figure above (after filtering). Power of the filtered adjacent channel interference at the receiver can be calculated as

$$P_{I,\text{ Adjacent channel}} = \frac{P_C}{SNR} - P_N \qquad (4.19)$$

After calculating the linear value and converting it to dBm scale, the filtered adjacent channel interference level required to guarantee the same reference data rate as for the sensitivity test is

$$P_{I,\text{ Adjacent channel}}|dBm = -95.76 \text{ dBm}. \qquad (4.20)$$

The received filter suppressed the adjacent channel interference by the ACS amount to the allowed interference:

$$ACS = P_{I|dBm} - P_{I,\text{ Adjacent channel}|dBm} = -52 + 95.76 = 43.76 \text{ dB}. \qquad (4.21)$$

■

Adjacent Channel Leakage Ratio (ACLR) measures the amount of transmit power that leaks into the adjacent channel, as illustrated in Fig. 4.16.

The ACLR and ACS combined define the total leakage between transmissions on adjacent channels. That ratio is called the adjacent channel interference ratio (ACIR) and is defined as the ratio of the power transmitted on one channel to the total interference received by a receiver on the adjacent channel, due to both transmitter (ACLR) and receiver (ACS) imperfections.

Following the ITU-R methodology outlined in ETSI TR 102 742 document [50], the total interference is

4.10 Adjacent Channel Selectivity, Adjacent Channel Leakage Ratio, and...

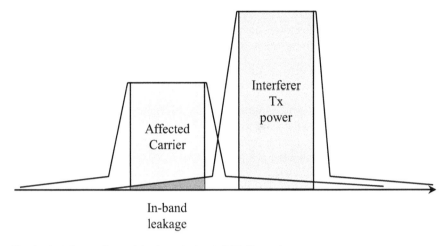

Fig. 4.16 Adjacent Channel Leakage Ratio (ACLR) illustration

$$P_{I,\text{TOTAL}} = \frac{P_T}{ACLR} + \frac{P_T}{ACS} \quad (4.22)$$

where the $P_{I,\text{In-band}} = P_T/ACLR$ component of interference corresponds to in-band (in-channel) interference portion and the $P_{I,\text{Adjacent channel}} = P_T/ACS$ part corresponds to out of band portion of the interference. The ACIR is then calculated as

$$ACIR = \frac{P_T}{P_{I,\text{TOTAL}}} = \frac{1}{\frac{1}{ACLR} + \frac{1}{ACS}}. \quad (4.23)$$

Problems

1. Populate the missing values in the table below. The uplink and downlink low and high frequencies are in MHz.

Operator	Area	Population	UL low	UL high	DL low	DL high	BW	MHz-PoP
ABC	T1	12,200	2540	2550	2660	2670		
DEF	T1	12,200	2520	2540	2640	2660		
GHI	T1	12,200	2600	2620	2600	2620		
GHI	T2	900,000	2540	2550	2660	2670		
DEF	T2	900,000	2520	2540	2640	2660		
ABC	T2	900,000	2600	2620	2600	2620		

Calculate the total MHz-PoP per company.
2. If an operator secured the total count of 255, 600, 000 MHz-PoP at the total price of $413,560,800, determine the following:

   a. The cost per MHz-PoP
   b. Normalized national level bandwidth if the total population is 18M people

3. A general 5G spectrum emission mask of the UT applies to frequencies Δf_{OOB} away from the edges of the assigned channel bandwidth. Given the partial table below, and the specified measurement and channel bandwidths, calculate the missing values in the table below.

Δf_{OOB}	5 MHz	10 MHz	20 MHz	30 MHz	40 MHz	50 MHz	60 MHz	Measurement BW
±0–1	−13	−13	−13	−13	−13			1% of CH BW
±0–1						−24	−24	30 kHz
±1–5	−10	−10	−10	−10	−10	−10	−10	1 MHz
±1–5								30 kHz

4. Two adjacent channel base stations are operating in the same geographic area and each one operates over a channel bandwidth of 10 MHz. The first base station transmits a power of $P_1 = 47$ dBm within its own channel and has an antenna gain of $G_T = 21$ dBi and the ACLR value of 45 dB. The second base station, operating in the adjacent channel, has a transmit power of $P_2 = 30$ dBm and an identical antenna gain of $G_T = 21$ dBm. The UT is attached to the second base station but is closer to the first base station, as shown below.

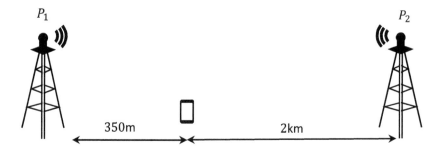

Given the pathloss expression $PL(d) = 128.1 + 37.6 \log_{10}(d)$, where d is expressed in km, calculate the expected SINR at the UE using the provided ACLR value.

5. A TV station operating at 600 MHz emits a strong signal with the *EIRP* = 1 MW over a 6 MHz bandwidth. A 4G system, designed to operate using a 5 MHz channel (4.5 MHz occupied bandwidth), has a downlink portion of the band that is 12 MHz away from the TV channel band edge and is subjected to strong out of band emission of −20 dBm over 6 MHz, as illustrated below.

4.10 Adjacent Channel Selectivity, Adjacent Channel Leakage Ratio, and...

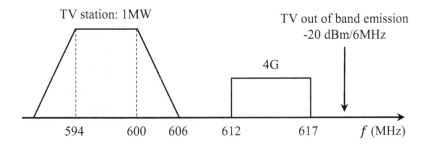

Determine the following:

a. Required TV filter attenuation to guarantee the out of band emission of -20 dBm over 6 MHz.
b. Interference level caused by the TV out of band emission at the UT receiver if the 4G UG is close to the TV station transmitter so that the pathloss between the TV transmitter and a 4G UT is only 50 dB. *Note*: make sure to normalize the interference power to the 4G occupied bandwidth.
c. Maximal interference level at the 4G UT receiver that ensures desensitization of $\leq$1dB. Assume the UT noise figure of $NF = 7$ dB.
d. Minimal distance $d_{\min}$ between the TV transmitter and 4G UT receiver that ensures that the maximal interference from the TV out of band emission corresponds to interference level calculated in c). The pathloss between the TV station and the 4G UT is determined by the free space pathloss.

6. An earth station (ES) operates at 3800 MHz and has a system temperature of $T = 142.8$K. If a 5G system operates at 3550 MHz and emits unwanted emission of -35 dBm/MHz in the frequency range where the ES operates, determine the minimal distance between the 5G base station and the ES using free space pathloss assuming the following: target $I/N = -6$ dB at the ES receiver, the ES antenna gain is 0 dBi (due to high elevation angle), and the 5G base station antenna gain in the direction of the ES is 10 dBi.
7. To protect an earth station (ES) that operates at 4100 MHz, mobile operator is required to coordinate with an ES operator to determine if a mobile service, operating at 3800 MHz with a conducted transmitted power of $P_T = 40$ dBm/MHz, can be enabled. The coordination trigger occurs when the mobile operator's in-band signal exceeds a PFD level of -90 dBW/m2/MHz at the ES location. Within the coordination process, the unwanted emission of -35 dBm/MHz in the frequency range where the ES operates is subjected to 3 dBi of transmit antenna gain and is not expected to exceed the PFD limit of -120 dBW/m2/MHz at 4100 MHz frequency. Determine which of the two conditions happens at the shorter distance.
8. If $ACS = 33$ dB and $ACLR = 45$ dB, determine the value for the ACIR. Comment on the result.

References

1. Radiocommunication Act (R.S.C., 1985, c. R-2). https://laws-lois.justice.gc.ca/eng/acts/R-2/page-1.html. Cited Jun 9, 2024
2. ISED, Licensing Procedure for Spectrum Licences for Terrestrial. Services, CPC-2-1-23 Issue 4, October 2015. https://www.ic.gc.ca/eic/site/smt-gst.nsf/vwapj/cpc-2-1-23-i4-2015-eng.pdf/FILE/cpc-2-1-23-i4-2015-eng.pdf. Cited Jun 9, 2024
3. FCC Authorizes Full Commercial Deployment In 3.5 GHz Band, Advancing American 5G Leadership, GN Docket No. 15-319. https://www.fcc.gov/document/fcc-authorizes-full-commercial-deployment-35-ghz-band. Cited Oct 29, 2023
4. Decision on a Non-Competitive Local Licensing Framework, Including Spectrum in the 3900-3980 MHz Band and Portions of the 26, 28 and 38 GHz Bands, ISED. https://ised-isde.canada.ca/site/spectrum-management-telecommunications/en/spectrum-allocation/decision-non-competitive-local-licensing-framework-including-spectrum-3900-3980-mhz-band-and. Cited Jun 8, 2024
5. OFCOM, Shared access licences. https://www.ofcom.org.uk/manage-your-licence/radiocommunication-licences/shared-access. Cited Jun 12, 2024
6. Overview of ITU's History. http://handle.itu.int/11.1004/020.2000/s.210. Cited Oct 29, 2023
7. Radio Regulation. http://handle.itu.int/11.1002/pub/814b0c44-en. Cited Oct 29, 2023
8. H. Takagi, B.H. Walke, *Spectrum Requirement Planning in Wireless Communications: Model and Methodology for IMT—Advanced* (Wiley Online Library, 2008)
9. Recommendation ITU-R M.816-1, Framework for services supported on International Mobile Telecommunications-2000 (IMT-2000). https://www.itu.int/rec/R-REC-M.816. Cited 29 Oct 2023
10. Recommendation ITU-R M.1457-15, Detailed specifications of the terrestrial radio interfaces of International Mobile Telecommunications-2000 (IMT-2000). https://www.itu.int/rec/R-REC-M.1457-15-202010-I/en. Cited 29 Oct 2023
11. Report ITU-R M.2134-0 (2008), Requirements related to technical performance for IMT-Advanced radio interface(s). https://www.itu.int/pub/R-REP-M.2134-2008. Cited 29 Oct 2023
12. Recommendation ITU-R M.2012, Detailed specifications of the terrestrial radio interfaces of International Mobile Telecommunications Advanced (IMT-Advanced). https://www.itu.int/rec/R-REC-M.2012. Cited 29 Oct 2023
13. Recommendation ITU-R M.2083, IMT Vision - Framework and overall objectives of the future development of IMT for 2020 and beyond. https://www.itu.int/rec/R-REC-M.2083. Cited 29 Oct 2023
14. Report ITU-R M.2410-0, Minimum requirements related to technical performance for IMT-2020 radio interface(s). https://www.itu.int/pub/R-REP-M.2410-2017. Cited 29 Oct 2023
15. Recommendation ITU-R M.2150, Detailed specifications of the terrestrial radio interfaces of International Mobile Telecommunications-2020 (IMT-2020). https://www.itu.int/rec/R-REC-M.2150/en. Cited 29 Oct 2023
16. Report ITU-R M.2516-0, Future technology trends of terrestrial International Mobile Telecommunications systems towards 2030 and beyond. https://www.itu.int/pub/R-REP-M.2516. Cited 29 Oct 2023
17. Recommendation ITU-R M.2160-0, Framework and overall objectives of the future development of IMT for 2030 and beyond. https://www.itu.int/rec/R-REC-M.2160-0-202311-I/en. Cited 5 Feb 2024
18. 3GPP Groups. https://www.3gpp.org/3gpp-groups. Cited 5 Feb 2024
19. User Equipment (UE) radio transmission and reception; Part 1: Range 1 Standalone. https://www.3gpp.org/ftp/Specs/archive/38_series/38.101-1/38101-1-i30.zip. Cited 29 Oct 2023
20. NR; Base Station (BS) radio transmission and reception. https://www.3gpp.org/ftp/Specs/archive/38_series/38.104/38104-i30.zip. Cited 29 Oct 2023

References

21. ISED, Canadian Table of Frequency Allocations. https://ised-isde.canada.ca/site/spectrum-management-telecommunications/sites/default/files/attachments/2022/2018_Canadian_Radio_Spectrum_Chart.PDF. Accessed 13 Feb 2024
22. Auction 103: Spectrum Frontiers - Upper 37 GHz, 39 GHz, and 47 GHz. https://www.fcc.gov/auction/103. Accessed 13 Feb 2024
23. The Impacts of mmWave 5G in India. https://www.gsma.com/spectrum/wp-content/uploads/2020/11/mmWave-5G-in-India.pdf. Accessed 13 Feb 2024
24. G. Liu, A. Liu, R. Zhang, M. Zhao, Angular-domain selective channel tracking and doppler compensation for high-mobility mmWave massive MIMO. IEEE Trans. Wirel. Commun. **20**(5), 2902–2916 (2021)
25. J.M. Jornet, I.F. Akyildiz, Channel modeling and capacity analysis for electromagnetic wireless nanonetworks in the terahertz band. IEEE Trans. Wirel. Commun. 10(10), 3211–3221 (2011)
26. C.T. Parisi, S. Badran, P. Sen, V. Petrov, J.M. Jornet, Modulations for terahertz band communications: joint analysis of phase noise impact and PAPR effects. IEEE Open J Commun. Soc. **5**, 412–429 (2024)
27. B. Chang, W. Tang, X. Yan, X. Tong, Z. Chen, Integrated scheduling of sensing, communication, and control for mmWave/THz communications in cellular connected UAV networks. IEEE J Sel Areas Commun. **40**(7), 2103–2113 (2022)
28. H. Elayan, C. Stefanini, R.M. Shubair, J.M. Jornet, End-to-end noise model for intra-body terahertz nanoscale communication. IEEE Trans. NanoBiosci. **17**(4), 464–473 (2018)
29. Canadian Table of Frequency Allocations (2022). https://ised-isde.canada.ca/site/spectrum-management-telecommunications/en/learn-more/key-documents/consultations/canadian-table-frequency-allocations-sf10759. Cited Nov 11 Nov 2023
30. Service areas for competitive licensing. https://ised-isde.canada.ca/site/spectrum-management-telecommunications/en/spectrum-allocation/service-areas-competitive-licensing. Cited 11 Nov 2023
31. Australian spectrum map grid. https://www.acma.gov.au/australian-spectrum-map-grid. Cited 28 July 2024
32. https://www.acma.gov.au/rules-manage-spectrum. Cited 28 July 2024
33. https://www.acma.gov.au/how-we-plan-and-manage-spectrum. Cited 28 July 2024
34. https://www.acma.gov.au/spectrum-licence-technical-framework-review. Cited 29 July 2024
35. https://www.acma.gov.au/spectrum-auctions. Cited 29 July 2024
36. Code of Federal Regulations. https://www.ecfr.gov/current/title-47. Cited Jul 31 2024
37. FCC Online Table of Frequency Allocations, 47 C.F.R. 2.106, Revised on July 1, 2022. https://transition.fcc.gov/oet/spectrum/table/fcctable.pdf. Cited Jul 31 2024
38. Review of Spectrum Management Practices, DOC-229047A1. https://docs.fcc.gov/public/attachments/DOC-229047A1.pdf. Cited Jul 31 2024
39. FCC, Auction Maps. https://www.fcc.gov/economics-analytics/auctions-division/auctions/auction-maps. Cited Jul 31 2024
40. FCC, Auction Maps, Regional Economic Area Groupings (REAGs). https://www.fcc.gov/sites/default/files/wireless/auctions/data/maps/REAG.pdf. Cited Jul 31 2024
41. FCC, 800 MHz Cellular Service. https://www.fcc.gov/wireless/bureau-divisions/mobility-division/800-mhz-cellular-service. Cited Aug 1 2024
42. FCC, Auctions Summary. https://www.fcc.gov/auctions-summary Cited Jul 31 2024
43. ISED, 600 MHz Auction—Final Results. https://ised-isde.canada.ca/site/spectrum-management-telecommunications/en/spectrum-allocation/auctions/auction-spectrum-licences-600-mhz-band/600-mhz-auction-final-results. Cited 11 Nov 2023
44. Evolved Universal Terrestrial Radio Access (E-UTRA); Base Station (BS) radio transmission and reception. https://www.3gpp.org/ftp/Specs/archive/36_series/36.104/36104-eb0.zip. Cited 29 Oct 2023
45. RSS-192—Flexible Use Broadband Equipment Operating in the Band 3450-3900 MHz, Issue 5, 2023. https://ised-isde.canada.ca/site/spectrum-management-telecommunications/en/devices-and-equipment/radio-equipment-standards/radio-standards-specifications-rss/rss-192-flexible-use-broadband-equipment-operating-band-3450-3900-mhz. Cited 11 Nov 2023

46. SRSP-520—Technical Requirements for Fixed and/or Mobile Systems, Including Flexible Use Broadband Systems, in the Band 3450-3900 MHz, Issue 3, 2023. https://ised-isde.canada.ca/site/spectrum-management-telecommunications/en/devices-and-equipment/standard-radio-system-plans/srsp-520-technical-requirements-fixed-andor-mobile-systems-including-flexible-use-broadband-systems. Cited 11 Nov 2023
47. NR; User Equipment (UE) radio transmission and reception; Part 2: Range 2 Standalone. https://www.3gpp.org/ftp/Specs/archive/38_series/38.101-2/38101-2-i30.zip. Cited 29 Oct 2023
48. General aspects for Base Station (BS) Radio Frequency (RF) for NR. https://www.3gpp.org/ftp/Specs/archive/38_series/38.817-02/38817-02-fb0.zip. Cited 29 Oct 2023
49. H. Holma, A. Toskala, *LTE for UMTS Evolution to LTE-Advanced*, 2nd edn. (John Wiley & Sons, 2011)
50. ETSI TR 102 742, Broadband Radio Access Networks (BRAN); Consideration of requirements for Mobile Terminal Station (TS) in Broadband Wireless Access Systems (BWA) in the 3 400 MHz to 3 800 MHz Frequency Band. https://www.etsi.org/deliver/etsi_tr/102700_102799/102742/01.01.01_60/tr_102742v010101p.pdf. Cited 29 Oct 2023

Chapter 5
Multiple-Input, Multiple-Output Systems

5.1 Introduction

In previous chapters, we have seen the possible gains in beamwidth and antenna gain by exploiting multiple antennas at the transmitter and/or the receiver. The analysis presented in these chapters drew directly from classical antenna array theory, assuming line-of-sight propagation. However, as we have also seen, wireless communication systems suffer from multipath propagation; as we will see, this significantly complicates any analysis, but also allows the use of multiple antennas in unique ways to obtain *better than expected* performance gains. The key is the statistical nature of wireless channels.

This chapter will focus on the use of multiple antennas at one or both ends of the wireless channel, developing concepts more-or-less as they appeared in the research literature. We will begin with a simple system with multiple antennas at the receiver; this will help illustrate the key role played by multipath propagation and tools used to measure the resulting performance gains. We will then consider multi-antenna transmitters and the need to incorporate the time dimension to achieve similar gains. While these early proposals focused on reliability, more modern systems focus their analysis on achievable rate. To do so, we will consider multiple antennas at both the basestation (BS) and the user (or multiple users)—multiple-input, multiple-output (MIMO) systems, illustrated in Fig. 5.1. We will conclude with modern concepts such as massive MIMO systems.

Supplementary Information The online version contains supplementary material available at (https://doi.org/10.1007/978-3-031-76455-4_5).

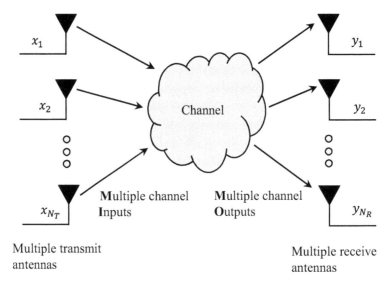

Fig. 5.1 Illustration of a MIMO system with N_T transmit and N_R receive antenna

In the most part, we will consider narrowband transmissions, i.e., frequency-flat fading. Essentially, we focus on a single subcarrier in a multicarrier system.

5.2 Receive Diversity

As mentioned in Chap. 3, the transmitted signal, after propagation to the receiver, is received in thermal noise. The quality of the signal is measured in terms of the signal-to-noise ratio (SNR), the ratio of the signal power to the average noise power. For a received SNR of $\gamma = P_S/P_N$, where P_S and $P_N = \sigma^2$ are average signal and noise powers, respectively, all modulations used in cellular systems have a bit error rate (BER) well approximated by BER $\simeq a Q(\sqrt{b\gamma})$ where the Q-function is closely related to the error function defined in Chap. 2 and is given by

$$Q(x) = \frac{1}{\sqrt{2\pi}} \int_x^\infty e^{-t^2/2} dt = \frac{1}{2}\mathrm{erfc}\left(\frac{x}{\sqrt{2}}\right) \qquad (5.1)$$

and a and b are constants related to the modulation scheme used.

In digital communication systems, the SNR ratio P_S/P_N can be related to the normalized SNR per bit $\gamma_B = E_b/N_0$, where E_b is the average energy per bit, as follows:

$$\gamma = \frac{P_S}{P_N} = \frac{E_s/T_s}{N_0 W} = \frac{E_b R_b}{N_0 W} = \gamma_B \frac{R_b}{W}. \qquad (5.2)$$

5.2 Receive Diversity

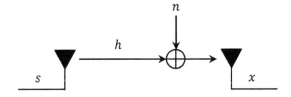

Fig. 5.2 Illustration of a single input, single output (SISO) system with *one* transmit and *one* receive antenna

For example, for BPSK modulation, the minimal baseband bandwidth is $W = R_b/2$, and the SNR is $\gamma = 2E_b/N_0 = 2\gamma_B$.[1] The bit error rate for BPSK and QPSK modulation schemes is BER $= Q(\sqrt{\gamma_B})$ [1]. In the remainder of this chapter, we will use the SNR value γ.

Importantly, due to fading in the propagation channel, the channel power is *random* and, consequently, so is the signal power and SNR. For the purposes of this chapter, we will consider Rayleigh fading, i.e., the channel is modeled as a complex Gaussian random variable with zero mean and an average channel power, determined by a combination of path loss and shadowing (the large-scale fading), denoted by σ_h^2:

$$h \sim \mathcal{CN}(0, \sigma_h^2) \quad \Rightarrow \quad p_h = |h|^2 \sim \frac{1}{\sigma_h^2} \exp\{-p_h/\sigma_h^2\}. \tag{5.3}$$

Here, p_h is the *instantaneous* channel power which follows an exponential distribution. The channel power is, therefore, often low, leading to a low SNR and a high error rate. The impact of this randomness can be seen when considering the *average* BER, averaged over the fading.

If the receiver were to use only a single receiver antenna, as shown in Fig. 5.2, the signal received from a single transmit antenna could be written as

$$x = \sqrt{P_T} h s + n, \tag{5.4}$$

where x denotes the receive signal, P_T the transmit power, s the transmitted symbol to be recovered, h the fading channel as described above, and n the noise. As developed in previous chapters, the noise can be modeled as $n \sim \mathcal{CN}(0, \sigma^2)$. For convenience, we set the average symbol power to unity, i.e., $\mathbb{E}[|s|^2] = 1$, resulting in the SNR $\gamma = P_T |h|^2/\sigma^2 = p_h/\sigma^2$.

Therefore, like the channel power, this SNR is an exponential random variable, i.e.,

$$\gamma \sim \frac{1}{\Gamma} \exp\{-\gamma/\Gamma\},$$

---

[1] This is also the output SNR for a matched-filter-based demodulator.

where $\Gamma = P_T \sigma_h^2/\sigma^2$ is the *average* SNR. The BER, averaged over channel realizations, is therefore

$$\overline{\text{BER}} = \int_0^\infty \text{BER}(\gamma) f(\gamma) d\gamma = \int_0^\infty a Q(\sqrt{b\gamma}) \frac{1}{\Gamma} \exp\{-\gamma/\Gamma\} d\gamma, \qquad (5.5)$$

where $f(\gamma)$ denotes the probability density function (PDF) of the received SNR. Using [2, Eqn. (3.61)], we have:

$$\overline{\text{BER}} = \frac{a}{\sqrt{2b}} \left(1 - \sqrt{\frac{\Gamma}{1+\Gamma}}\right). \qquad (5.6)$$

This is an important result, underlining the dramatic difference between *regular* digital communications over a wired channel and communications over a fading wireless channel. Without fading, the channel is deterministic, as is the received SNR—the error rate falls off *exponentially* with Γ. In a fading channel, on the other hand, the error rate falls off only *inversely* with SNR! To illustrate this, consider the asymptotic behavior as $\Gamma \to \infty$. We have:

$$\overline{\text{BER}} \to \frac{a}{2\sqrt{b}} \frac{1}{\Gamma}. \qquad (5.7)$$

> Another representation of (5.7) is to take a logarithm on both sides, i.e.,
>
> $$10 \log_{10}(\overline{\text{BER}}) = C - 10 \log_{10}(\Gamma),$$
>
> where C is some constant. Essentially, to reduce the average BER by a factor of 10, we need a 10 dB increase in the average SNR. At high SNR, therefore, a log-log plot of BER versus SNR (or semi-log plot of BER versus SNR in dB) is linear *with unit slope*.

This result, coupled with the fact that small-scale fading changes rapidly with distance, motivates the use of multiple antennas. Depending on the local propagation environment, even closely spaced antennas see very different small-scale fading terms (but, the same large-scale fading). Effectively, samples of the small-scale fading made at two points with some reasonable spacing have very low correlation; for the purposes of our initial exposition, we will model them as *statistically independent*.

5.2 Receive Diversity

> At a high level, consider the case where we need a receive SNR of Γ_0 to maintain a reliable communication link. The probability of a communication outage, or just outage, is the probability that the SNR falls below this threshold. To illustrate how multiple antennas help, let us assume that at a specific antenna, this outage probability is 0.1. If the system uses two antennas, outage happens if *both* antennas are in outage—and, if the fading is statistically independent, outage probability falls to $(0.1)^2 = 0.01$.s With M antennas, it would be $(0.1)^M$, i.e., we obtain exponential gains in reliability!

The simplest illustration of the use of multiple antennas is an array at the receiver—usually the basestation (BS). Extending the signal model of (5.4) to M receive antennas, we have a receive vector:

$$\mathbf{x} = \sqrt{P_T}\mathbf{h}s + \mathbf{n}, \tag{5.8}$$

where the transmitting user only uses a single antenna, transmitting symbol s. This is the so-called single-input, multiple-output (SIMO) case. Here $\mathbf{x} = [x_0, x_1, \ldots, x_{M-1}]^T$ is the receive vector, $\mathbf{h} = [h_0, h_1, \ldots, h_{M-1}]^T$ the channel vector and $\mathbf{n} = [n_0, n_1, \ldots, n_{M-1}]^T$ the noise (^{T} denotes transpose). All vectors are complex of length M.

Since the noise is statistically independent from one antenna to another, we can model the noise as $\mathbf{n} \sim \mathcal{CN}(\mathbf{0}, \sigma^2 \mathbf{I})$, i.e., the mean noise vector is the length-M vector of zeros and the covariance matrix is a scaled identity matrix. Importantly, since we model the channel coefficients at any two antennas as statistically independent, but with the same large-scale fading coefficients, we can similarly model the channel vector as $\mathbf{h} \sim \mathcal{CN}(\mathbf{0}, \sigma_h^2 \mathbf{I})$.

We have M samples to estimate the single symbol s; these samples must be combined in some way to recover the symbol. As shown in Fig. 5.3, the general approach is to use a linear combination based on a *weight vector* $\mathbf{w}$, i.e., for further processing we obtain:

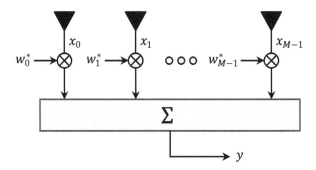

Fig. 5.3 The receiver in a diversity combining system

$$y = \sum_{m=0}^{M-1} w_m^* x_m = \mathbf{w}^H \mathbf{x}, \tag{5.9}$$

where * denotes the complex conjugate and ^{H} the Hermitian or conjugate transpose of a vector/matrix. There are three popular approaches for the choice of this weight vector.

5.2.1 Selection Combining

Selection combining is, in fact, a misnomer—instead of combining the M samples, this approach selects one antenna for further processing. The antenna selected is the one with the highest SNR, i.e.,

$$w_m = \begin{cases} 1 & \gamma_m = \max_k\{\gamma_k, k = 0, 1, \ldots, M-1\} \\ 0 & \text{otherwise} \end{cases} \tag{5.10}$$

and the output SNR of the combiner is $\gamma_{\text{out}} = \max_m\{\gamma_m, m = 0, 1, \ldots, M-1\}$. Note that since the channels are statistically independent across antennas, so are the SNRs.

While extremely simple, selection combining achieves most of the benefits of using multiple antennas. To illustrate this, let us set Γ_0 as the threshold required to achieve reliable communications. This threshold could be set by the maximum BER allowed for the chosen data rate and modulation. For any antenna, m, we have:

$$P[\gamma_m < \Gamma_0] = \int_0^{\Gamma_0} \frac{1}{\Gamma} e^{-\gamma_m/\Gamma} d\gamma_m = 1 - e^{-\Gamma_0/\Gamma}. \tag{5.11}$$

Since, in selection combining, we choose the antenna with the highest SNR, the outage probability is given by

$$P_{\text{out}} = P[\gamma_{\text{out}} < \Gamma_0] = P[\max_m \gamma_m < \Gamma_0] = \prod_{m=0}^{M-1} P[\gamma_m < \Gamma_0]$$

$$= \left(1 - e^{-\Gamma_0/\Gamma}\right)^M. \tag{5.12}$$

In (5.7), we saw that the BER, in the single-antenna case, is inversely proportional to average SNR. We have a similar result for the outage probability. Asymptotically, as Γ grows toward infinity, Γ_0/Γ becomes small and $e^{-\Gamma_0/\Gamma} \simeq 1 - \Gamma_0/\Gamma$, i.e., asymptotically,

5.2 Receive Diversity

$$P_{\text{out}} \to \left(\frac{\Gamma_0}{\Gamma}\right)^M. \tag{5.13}$$

Again, for $M = 1$, the outage probability is inversely proportional to average SNR—and, for $M > 1$, we get exponential gains in outage probability. As before, we have:

$$10\log_{10}(P_{\text{out}}) = C - 10M\log_{10}(\Gamma), \tag{5.14}$$

i.e., at high SNR, the semi-log plot of outage versus SNR (in dB) is linear with a slope of $-M$. This notion is captured in the notion of diversity order.

Diversity Order A system is said to have diversity order D if

$$D = -\lim_{\Gamma \to \infty} \frac{\log(P_{\text{out}})}{\log(\Gamma)} \tag{5.15}$$

Essentially, at large SNR, the outage $P_{\text{out}} \propto (\text{SNR})^{-D}$. From the relation in (5.13), the diversity order achieved by selection combining is M.

Intuitively, the diversity order captures the number of statistically independent measurements of the same symbol. Clearly, a system with M receive antennas can only have a maximum of M such measurements, Selection combining, therefore, achieves the maximum possible diversity order.

Figure 5.4 plots the outage probability as a function of the ratio between the threshold SNR, Γ_0, and the average receive SNR Γ. At the left (high Γ), we see the linear relationship between outage and SNR. We also see the slope increasing with increasing M, indicating increased diversity.

Gains Not from SNR A crucial aspect of using multiple antennas is that the gains in outage/BER do not arise from large gains in output SNR. To illustrate this, consider the average output SNR after selection combining. We recognize that the relation in (5.12) also represents the complementary distribution function (CDF) of the output SNR. The probability density function (PDF) is therefore given by

$$f(\gamma_{\text{out}}) = \frac{d}{d\gamma_{\text{out}}}\left[1 - e^{-\gamma_{\text{out}}/\Gamma}\right]^M = \frac{M}{\Gamma}e^{-\gamma_{\text{out}}/\Gamma}\left[1 - e^{-\gamma_{\text{out}}/\Gamma}\right]^{M-1}$$

$$\Rightarrow \mathbb{E}[\gamma_{\text{out}}] = \int_0^\infty \gamma_{\text{out}} f(\gamma_{\text{out}}) d\gamma_{\text{out}}$$

$$= \Gamma \sum_{m=1}^M \frac{1}{m} \simeq \Gamma \left(C + \ln M + \frac{1}{2M}\right), \tag{5.16}$$

where $C = 0.577216$ is Euler's constant and the approximation is valid for large M. Importantly, we see that the gain in average SNR is only on the order of $\ln(M)$, though we achieve exponential gains in outage (diversity order of M)!

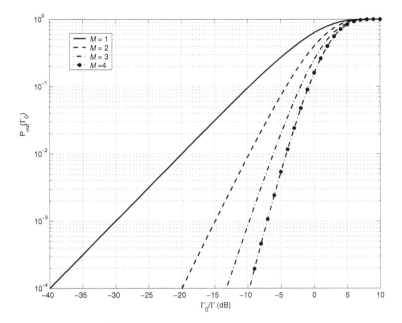

Fig. 5.4 Performance of a selection combining system

In the analysis above, we have assumed that the small-scale fading to any two receive antennas is statistically independent. Consider now the other extreme—where the fading is perfectly correlated across all the elements, i.e., at best there is a phase difference between the channels and, consequently, $\gamma_0 = \gamma_1 = \cdots = \gamma_{M-1}$. In this case, choosing the element with maximum SNR is the same as picking one at random and the outage is given by

$$P_{\text{out}} = P[\gamma_m < \Gamma_0] = \left[1 - e^{-\Gamma_0/\Gamma}\right].$$

From (5.11), this is the equivalent of having only one receive antenna, i.e., there are no gains using selection combining if all the fading terms are perfectly correlated. In this case, we achieve a diversity order of only one.

This result underlines the importance of the random nature of the wireless channel and the fact that the small-scale fading term changes quickly over short distances. The rapid changes in fading lead to our assumption of independent fading and a diversity order of M.

5.2.2 *Maximum Ratio Combining*

While selection combining achieves full diversity order, it is clearly not the best we can do—by picking only one antenna for further processing, we ignore the signals received at the other $(M-1)$ antennas. Maximum ratio combining (MRC) is optimal in the sense of maximizing the output SNR—when using the linear combination of received signals as in (5.9). We have, as the output

$$y = \mathbf{w}^H \mathbf{x} = \sqrt{P_T}\mathbf{w}^H \mathbf{h} s + \mathbf{n}.$$

Here, the signal power is $P_T|\mathbf{w}^H\mathbf{h}|^2$ (since, on average, the symbol s has unit power) and the noise power is given by

$$\mathbb{E}[|\mathbf{w}^H\mathbf{n}|^2] = \mathbb{E}[\mathbf{w}^H\mathbf{n}\mathbf{n}^H\mathbf{w}] = \sigma^2\mathbf{w}^H\mathbf{w} = \sigma^2||\mathbf{w}||^2,$$

since $\mathbb{E}[\mathbf{n}\mathbf{n}^H] = \sigma^2 \mathbf{I}$ is the definition of the covariance matrix of the noise term. The output SNR is, therefore, given by $\gamma_{out} = P_T|\mathbf{w}^H\mathbf{h}|^2/\sigma^2||\mathbf{w}||^2$.

Since multiplying by a constant does not change the output SNR, we can scale $\mathbf{w}$ such that $||\mathbf{w}||^2 = 1$, making the denominator in the SNR a constant σ^2. Our optimization problem to choose the weight vector $\mathbf{w}$ is

$$\max_{\mathbf{w}} |\mathbf{w}^H \mathbf{h}|^2 \text{ such that } ||\mathbf{w}|| = 1.$$

The Cauchy-Schwarz inequality immediately provides the answer: we need $\mathbf{w} \propto \mathbf{h}$. Again, since scale factors do not matter, we can choose:

$$\mathbf{w} = \mathbf{h}. \tag{5.17}$$

Since this choice of receiver maximizes the output SNR, it is also called the *matched filter*.

Output SNR and Diversity Order Given the weight vector in (5.17), the output SNR is given by

$$\begin{aligned}\gamma_{out} &= \frac{P_T|\mathbf{w}^H\mathbf{h}|^2}{\sigma^2||\mathbf{w}||^2} = \frac{P_T|\mathbf{h}^H\mathbf{h}|^2}{\sigma^2||\mathbf{h}||^2} = \frac{P_T}{\sigma^2}||\mathbf{h}||^2 \\ &= \sum_{m=0}^{M-1} \frac{P_T|h_m|^2}{\sigma^2} = \sum_{m=0}^{M-1} \gamma_m,\end{aligned} \tag{5.18}$$

i.e., the output SNR is the sum of the SNRs at each element. Furthermore, since $\mathbb{E}[\gamma_m] = \Gamma$, we have $\mathbb{E}[\gamma_{out}] = M\Gamma$, i.e., an average the output SNR scales by the number of antenna elements. However, as we have seen, the performance gains are not due to SNR, but due to diversity. To analyze the diversity, we first obtain the PDF of the output SNR. To do so, we recognize that the output SNR is a sum of independent (and identically distributed) exponentially random variables. The resulting PDF is the convolution of the individual PDFs; equivalently, the characteristic function of the output is the product of the individual characteristic functions. We have:

$$f_{\gamma_m} = \frac{1}{\Gamma} e^{-\gamma_m/\Gamma} \Rightarrow F_{\gamma_m}(s) = \mathbb{E}[e^{-s\gamma_m}] = \frac{1}{1+s\Gamma}$$

$$\Rightarrow F_{\gamma_{\text{out}}} = \prod_{m=0}^{M-1} F_{\gamma_m}(s) = \left[\frac{1}{1+s\Gamma}\right]^M. \quad (5.19)$$

Using the inverse Laplace transform, we have:

$$f(\gamma_{\text{out}}) = \frac{1}{(M-1)!} \frac{\gamma_{\text{out}}^{M-1}}{\Gamma^M} e^{-\gamma_{\text{out}}/\Gamma}, \quad (5.20)$$

and the resulting outage probability corresponding to a threshold of Γ_0 is given by

$$P_{\text{out}} = \int_0^{\Gamma_0} f(\gamma_{\text{out}}) d\gamma_{\text{out}} = 1 - e^{-\Gamma_0/\Gamma} \sum_{m=0}^{M-1} \left(\frac{\Gamma_0}{\Gamma}\right)^m \frac{1}{m!} \quad (5.21)$$

Diversity Order One would expect the SNR-optimal MRC approach to also achieve a diversity order of M. To see this, note that $e^x = \sum_{m=0}^{\infty} x^m/m!$ and, therefore,

$$P_{\text{out}} = 1 - e^{-\Gamma_0/\Gamma} \left[e^{\Gamma_0/\Gamma} - \sum_{m=M}^{\infty} \left(\frac{\Gamma_0}{\Gamma}\right)^m \frac{1}{m!}\right] = e^{-\Gamma_0/\Gamma} \sum_{m=M}^{\infty} \left(\frac{\Gamma_0}{\Gamma}\right)^m \frac{1}{m!}$$

For large Γ, the $m = M$ term dominates the sum. Further, $e^{-\Gamma_0/\Gamma} \simeq (1 - \Gamma_0/\Gamma)$. We have

$$P_{\text{out}} \simeq \left(1 - \frac{\Gamma_0}{\Gamma}\right) \times \left(\frac{\Gamma_0}{\Gamma}\right)^M \frac{1}{M!} \simeq \frac{C}{\Gamma^M},$$

where C is some constant, i.e., the diversity order achieved is M.

Using a similar analysis, it is also possible to bound the BER. As mentioned earlier, given SNR γ_{out}, most practical modulations suffer a BER of $aQ(\sqrt{b\gamma_{\text{out}}})$. Since $Q(x) \leq e^{-x^2/2}$, the BER is upper bounded by $ae^{-b\gamma_{\text{out}}/2}$. Given the PDF of the output SNR in (5.20), we have:

5.2 Receive Diversity

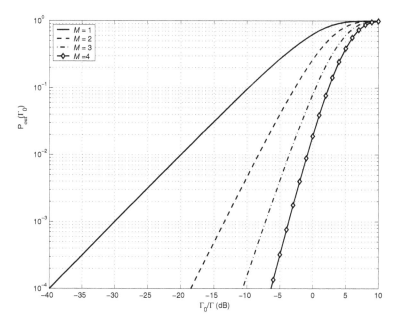

Fig. 5.5 Performance of a maximal ratio combining system

$$\text{BER} \leq \int_0^\infty ae^{-b\gamma_{\text{out}}} f(\gamma_{\text{out}}) d\gamma_{\text{out}} = a\left[\frac{1}{1+b\Gamma/2}\right]^M.$$

Again, as Γ grows asymptotically large, we see a diversity order of M.

Figure 5.5 plots the outage probability as a function of Γ_0/Γ, the ratio of the threshold SNR to the average SNR per element. As with selection combining, we see the linear relationship between outage and SNR (in log scale) at high SNR. Importantly, both selection combining and MRC achieve the same diversity order. However, as we have seen, using MRC results in a higher output SNR. This gain manifests as a *shift* in the outage curve—as seen in Fig. 5.6 with $M = 4$. We see that using MRC provides approximately a 3-dB gain in outage probability over using selection combining.

Our results show that there is a trade-off between complexity and performance in choosing between selection and maximum ratio combining. On the one hand, MRC provides the maximum possible SNR and, so, the lowest outage probability. However, implementing MRC requires accurate channel state information (CSI) at the receiver. Furthermore, the processor must be able to scale and phase shift (multiply by a complex weight) the signal received

(continued)

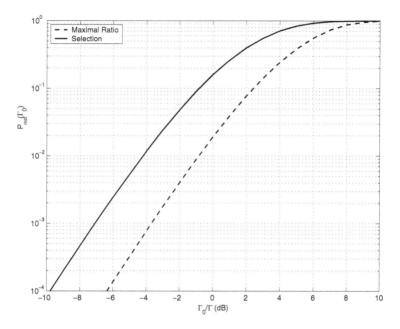

Fig. 5.6 Comparing selection and maximal ratio combining (M=4)

at each element. Since the scale factor is to match the channel, its value may have to fluctuate over many 10s of dB. Selection combining, on the other hand, requires only a simple measurement of SNR (a power measurement) and achieves the same diversity order—at the cost of a loss in SNR (and, so, an increased error rate).

5.2.3 Equal Gain Combining

There are, now, many alternative methods combining the M receive signals, each balancing implementation complexity and performance. Here, to illustrate these alternatives, we mention one alternative popular approach known as equal gain combining (EGC). To appreciate EGC, we revisit the MRC weight vector in (5.17). As seen in Fig. 5.3, x_m, the signal on antenna m is multiplied by the weight w_m^*. The key benefit from using the MRC weight $w_m = h_m$ is that the multiplication (since the weight uses a complex conjugate) eliminates the phase of the channel. The signal term $\mathbf{w}^H \mathbf{h}$ is, therefore, a sum of positive numbers; we use this fact in designing equal gain combining (EGC). EGC sidesteps the problem of large fluctuations in channel magnitude by choosing $w_m = e^{j\angle h_m}$, i.e., it only captures the phase of the channel. Specifically,

5.2 Receive Diversity

$$w_m^* h_m = e^{-j\angle h_m} h_m = |h_m| \quad \Rightarrow \quad \mathbf{w}^H \mathbf{h} = \sum_{m=0}^{M-1} |h_m| \qquad (5.22)$$

Using this choice of weight vector, we have the output noise power as $\sigma^2 \mathbf{w}^H \mathbf{w} = M\sigma^2$. The output SNR is, therefore,

$$\gamma_{\text{out}} = \frac{\left[\sum_{m=0}^{M-1} |h_m|\right]^2}{M\sigma^2}, \qquad (5.23)$$

i.e., the output SNR is due to the square of the sum of channel magnitudes, as opposed to the MRC case in (5.18).

Using the fact that $|h_m|$ has a Rayleigh PDF and the assumption of statistically independent fading, it is not difficult to show (see Problem 6) that the average output SNR is given by

$$\mathbb{E}[\gamma_{\text{out}}] = \Gamma\left[1 + (M-1)\frac{\pi}{4}\right].$$

Comparing this result to (5.18), we see only a small loss in average SNR. While closed-form solutions for outage probability and BER are difficult to obtain, EGC, as expected, also achieves a diversity order of M.

5.2.4 Impact of Spatial Correlation

A final note on the impact of spatial correlation on the performance of diversity systems. We have seen that the key benefits of such systems arise because of statistically independent fading from one receive element to the next. We also considered the opposite extreme of perfectly correlated fading. The real world is somewhere in between.

We have used Rayleigh fading to model wireless channels; this model assumes that the signal propagates over many (an asymptotically large number of) paths. In this case, the resulting small-scale fading changes rapidly with receiver position and any two receivers see statistically independent fading. In practice, the number of paths is finite and fading is correlated across space.

Figure 5.7 plots the outage probability for a two-element receive array wherein the small-scale fading is correlated with correlation coefficient ρ. Our previous results correspond to $\rho = 0$ (uncorrelated, diversity order 2) and $\rho = 1$ (perfectly correlated, diversity order 1). Importantly, even a correlation as high as $\rho = 0.5$ has minimal impact on performance and diversity order. In practice, we consider correlations lower than $\rho = 0.5$ as effectively uncorrelated.

The level of correlation is a function of the propagation environment. In general, the larger the number of propagation paths, the lower the correlation. This result

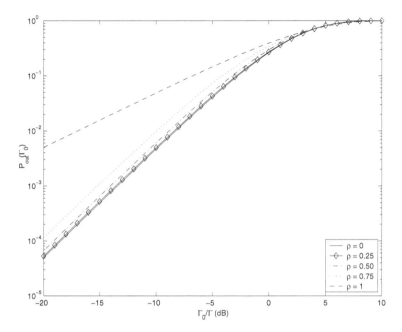

Fig. 5.7 Probability of outage for changing correlation

manifests itself differently at a mobile or at a basestation. Mobiles/users are generally surrounded by many objects which act as scatterers providing a rich multipath environment. The small-scale fading component, therefore, decorrelates over relatively short distances—as short as half a wavelength [3]. On the other hand, basestations are generally elevated and are, likely, not surrounded by scatterers. In this case, the multipath propagation is due to scattering by objects near the user. Depending on the severity of the local scattering, the decorrelation distance may become as long as 4–10 wavelengths [4].

5.3 Transmit Diversity

In the previous section, we analyzed receive diversity wherein multiple receive antennas were used to enhance the reliability of decoding a single symbol. This allowed us to write the receive signal as a vector, as in (5.8). We then processed this length-M vector using a weight vector $\mathbf{w}$. We now switch our focus to the multiple-input, single-output (MISO) case where a transmitter (usually the basestation) equipped with multiple (M) antennas communicates with a single-antenna receiver (usually a user or mobile), as illustrated in Fig. 5.8.

We now have two very distinct possible situations: when CSI is available at the transmitter (CSIT) and when it is not.

5.3 Transmit Diversity

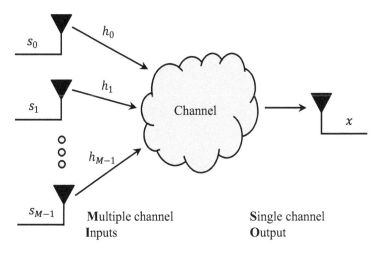

Fig. 5.8 Multiple-input, single-output system

5.3.1 With CSIT

As before, the transmitter wishes to send symbol s to the receiver. To do so using its M antennas, it multiplies the symbol with a weight vector $\mathbf{w}$. The transmitted signal is the length-M vector $\mathbf{w}s$. The received signal, a scalar, is the superposition of the M transmissions through the length-M channel vector $\mathbf{h}$, i.e.,

$$x = \mathbf{h}^T \mathbf{w} s + n,$$

where n is the noise. Since CSI is available at the transmitter, i.e., the transmitter knows the channel vector $\mathbf{h}$, it can tailor its choice of the weight vector to the channel. Using an analysis similar to the MRC case, it is straightforward to show that the weight vector that maximizes the receive SNR is given by $\mathbf{w} \propto \mathbf{h}^*$, i.e., the matched filter. This is also known as maximum ratio transmission (MRT).[2]

The proportionality constant is determined by allowed transmit power. If, as before, we assume $\mathbb{E}[|s|^2] = 1$, we need that $||\mathbf{w}||^2 \leq P_T$. Since increasing transmit power increases receive SNR, we will set $||\mathbf{w}||^2 = P_T$. Therefore, the weight vector

$$\mathbf{w} = \sqrt{P_T} \mathbf{h} / ||\mathbf{h}||.$$

---

[2] In the MRC case, we had $\mathbf{w} = \mathbf{h}$ because the weighting process included the complex conjugate. Note that in both cases, the symbol s, after processing, is multiplied with $||\mathbf{h}||^2$, i.e., it is scaled with the channel power.

> **Problem** Using an analysis similar to the MRC case, show that the diversity order achieved by MRT is also M. Effectively show that $P_{\text{out}} \propto \Gamma^{-M}$.

5.3.2 Without CSIT

If the transmitter does not know the channel vector, it has no way to choose a weight vector to match the channel. If one were to just transmit the same symbol s from all M antennas, this is equivalent to choosing the weight vector of all ones. The received signal would then be $x = \sqrt{P_T}(\sum_m h_m)s + n$. The effective channel is then $\sum_m h_m$; this term is zero mean and no diversity gains are available.

> **Problem** Show that a system with received signal given by
> $$x = \sqrt{P_T}\left(\sum_m h_m\right)s + n$$
> achieves a diversity order of only 1.

A crucial advance in achieving diversity gains was the introduction of the *time dimension*. The simplest example would be to create a receive vector by transmitting the symbol s over M transmit antennas and M symbol periods. The resulting received signal model is effectively the same as (5.8) (as is the further processing). However, this scheme pays a massive penalty by reducing the transmission rate by a factor of $(1/M)$. Even for $M = 2$, no reasonable system design would be willing to accept this penalty. Thankfully, alternative formulations are possible.

5.3.2.1 Alamouti's Scheme

In [5], Alamouti proposed a remarkably simple transmission scheme for the case of two transmit antennas. In this scheme, the transmitter transmits *two* symbols in *two* timeslots (Fig. 5.9).

Specifically, in the initial timeslot (indexed as 0), it transmits symbol s_0 from the first antenna ($m = 0$) and symbol s_1 from the second ($m = 1$). In this timeslot, the received signal is

$$x_0 = \sqrt{P_T/2}h_0 s_0 + \sqrt{P_T/2}h_1 s_1 + n_0.$$

5.3 Transmit Diversity

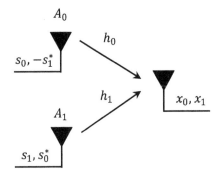

Fig. 5.9 Illustration of the Alamouti transmission scheme with two transmit antennas and one receive antenna

In the next timeslot, the transmitter transmits $-s_1^*$ from the first antenna and s_0^* from the second. The received signal is

$$x_1 = -\sqrt{P_T/2}h_0 s_1^* + \sqrt{P_T/2}h_1 s_0^* + n_1.$$

We note that by transmitting two symbols over two timeslots, there is no rate penalty. Now, writing the received signals over two timeslots as a matrix equation, but after conjugating x_1, we have:

$$\mathbf{x} = \begin{bmatrix} x_0 \\ x_1^* \end{bmatrix} = \sqrt{P_T/2} \begin{bmatrix} h_0 & h_1 \\ h_1^* & -h_0^* \end{bmatrix} \begin{bmatrix} s_0 \\ s_1 \end{bmatrix} + \begin{bmatrix} n_0 \\ n_1^* \end{bmatrix}$$

$$= \sqrt{P_T/2}\mathbf{H}\mathbf{s} + \mathbf{n}, \qquad (5.24)$$

We can now process this received signal by

$$\mathbf{r} = \mathbf{H}^H \mathbf{x} = (\sqrt{P_T/2})\mathbf{H}^H \mathbf{H} \mathbf{s} + \mathbf{H}^H \mathbf{n},$$

$$\Rightarrow \mathbf{r} = \begin{bmatrix} r_0 \\ r_1 \end{bmatrix} = \sqrt{P_T/2} \begin{bmatrix} |h_0|^2 + |h_1|^2 & 0 \\ 0 & |h_0|^2 + |h_1|^2 \end{bmatrix} \begin{bmatrix} s_0 \\ s_1 \end{bmatrix} + \mathbf{H}^H \mathbf{n}. \quad (5.25)$$

Importantly, the resulting terms are exactly as if we had two receive antennas and used MRC! This is the beauty of the Alamouti scheme—with a simple arrangement of symbols over space and time, it achieves full diversity order without any loss in rate. The only penalty is that we split the available power across two symbols, i.e., each symbol receives only half the power—when compared to MRC with two antennas, the error rate, whether outage or BER, for the Alamouti scheme shows a 3-dB penalty in average SNR.

Figure 5.10 illustrates the performance of the Alamouti scheme and compares this performance to using receive diversity. The four plots show the baseline case of a single transmit and receive antenna—the diversity order of one is clear (a reduction in BER of a factor of 10 when the SNR grows by a factor of 10—or 10 dB). The red plot illustrates receive diversity and the diversity order of 2 is again clear. Between

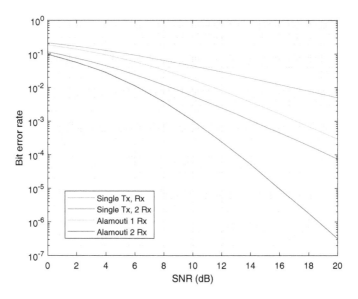

Fig. 5.10 Performance of the Alamouti scheme compared to receive diversity

Fig. 5.11 Illustration of the space-frequency block coding (SFBC) in 4G and 5G systems

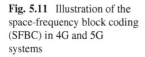 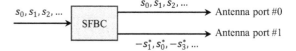

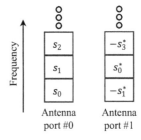

these two curves lies the performance of the Alamouti scheme—importantly, the BER curve is parallel to the case of receive diversity, i.e., it too achieves a diversity order of 2. However, also clear is the loss of 3dB due to the splitting of the available power between the two symbols transmitted simultaneously. Finally, the figure also shows that the transmit diversity of the Alamouti scheme can be combined with receive diversity (two receive antennas), achieving a diversity order of 4.

In 4G and 5G systems today, Almouti scheme is used in a form of space-frequency block coding (SFBC), where instead of using two different time instances, frequency resources are partitioned in equal blocks, as illustrated in Fig. 5.11. The details about how the frequency resources are partitioned will be covered in Chap. 7.

5.3 Transmit Diversity

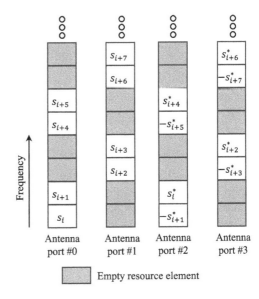

Fig. 5.12 Illustration of frequency-switched transmit diversity (FSTD) [6]

A variation of this scheme, where four transmit antennas are used, uses combination of SFBC and frequency-switched transmit diversity (FSTD), as shown in Fig. 5.12.

5.3.2.2 General Space-Time Coding

The Alamouti scheme is the simplest example of *space-time coding*—a careful arrangement of symbols across space and time (or frequency). However, it is limited to just two transmit antennas. In [7], the authors addressed the question of how to design good space-time codes and their limitations. A detailed discussion of the development is beyond the scope of this book; here, we mention the key results and limitations imposed.

In the general case, a transmitter with M antennas transmits symbols over $L \geq M$ timeslots. The channel, **h**, is assumed constant over the L timeslots, i.e., the channel coherence time is greater than L timeslots. The codebook is a set of valid transmissions, and each valid code is given by $M \times L$ matrix:

$$\mathbf{C} = \begin{bmatrix} c_{00} & c_{01} & \cdots & c_{0(L-1)} \\ c_{10} & c_{11} & \cdots & c_{1(L-1)} \\ c_{(M-1)0} & c_{(M-1)1} & \cdots & c_{(M-1)(L-1)} \end{bmatrix}$$

The receiver decodes the transmission to a *valid* codeword $\tilde{\mathbf{C}}$, denoted as $\mathbf{C} \rightarrow \tilde{\mathbf{C}}$. An error occurs if $\tilde{\mathbf{C}} \neq \mathbf{C}$; define the error matrix $\mathbf{E} = \mathbf{C} - \tilde{\mathbf{C}}$. The key is that the average error probability can be expressed as

$$P[\mathbf{C} \to \tilde{\mathbf{C}}] \leq \left(\prod_{m=1}^{R} \lambda_m\right) \Gamma^{RM} \qquad (5.26)$$

where R is the *rank* of the error matrix and $\lambda_m, m = 1, \ldots, R$ are the nonzero eigenvalues of the $M \times M$ matrix $\mathbf{EE}^H$. We now know what makes for a good code:

1. A good code has the highest possible rank in the error matrix between any two codewords. This is the rank criterion.
2. A good code has the highest product of non-zero eigenvalues of the error matrix.[3]

In their work on space-time codes, Tarokh et al. [7, 8] developed space-time trellis codes—a trellis code is specific to the modulation used. The output of the code, one per transmit antenna, is a function of a state and input. The trellis specifies the state transitions as well. A trellis does not terminate; a termination condition can be added to "finish" a codeword. Given their implementation complexity and relative lack of flexibility, trellis codes have been less popular than space-time block codes.

A space-time block code encodes a block of symbols at a time—the arrangement of symbols in space and time is independent of the modulation scheme used. Of importance are orthogonal space-time block codes (OSTBC) wherein each individual symbol can be independently decoded—the Alamouti scheme above is an example of an OSTBC since, after processing, each of the two symbols can be isolated from each other (the symbols are said to be orthogonal to each other). Orthogonality is a highly desirable condition because it eliminates the need for joint decoding across symbols—a complex process with exponential computational burden.

In general, one can encode N symbols in L timeslots; ideally we would want $N = L$, i.e., a rate-1 arrangement, as in the Alamouti code. We briefly summarize the key results here. For an OSTBC achieving full diversity order:

- For *real* symbols it is always possible to design a rate-1 OSTBC. For $M = 2, 4, 8$, it is possible for the code matrix ($\mathbf{C}$ above) to be square; else, the code matrix is rectangular, but can still be rate-1. However, real symbols are rarely used in wireless communications.
- For complex data symbols, such as most often used in wireless networks, a rate-1 OSTBC exists for $N = 2$ only. Essentially, the Alamouti code above is unique in being rate-1 and valid for orthogonal codes.
- Also for complex data symbols, a rate-1/2 code can be designed for any value of M (since a complex symbol can be thought of as two real symbols).
- For the special cases of $M = 3$ and $M = 4$, i.e., a transmitter with three or four antennas, rate-3/4 codes exist.

---

[3] This is not exactly the determinant since this is only the product of nonzero eigenvalues. However, since any decent space-time code design would have full rank, in practice this is often referred to as the determinant criterion.

As we have seen, the orthogonality in OSTBC codes allow for symbol-by-symbol decoding of the transmitted symbols. Clearly this is the simplest possible decoding arrangement. However, as we have seen, a rate-1 code is possible for only the case of $M = 2$, with lower rate if using more antennas. Any improvement in rate is possible only at the expense of decoding complexity.

> So far we have considered the case of multiple transmit antennas and a single receive antenna. It is not difficult to show that with multiple receive antennas, the diversity order is the product of the numbers of transmit and receive antennas.

5.4 MIMO Information Theory and Rate Considerations

Our discussion so far has focused on increasing the reliability of decoding data symbols transmitted at a pre-chosen, but fixed, data rate. We now investigate a complementary metric, the achievable rate—the maximum rate that can theoretically be achieved. A detailed development is beyond the scope of this book; here we provide the reader with a sufficient understanding of the role the rate analysis would play in system design. We will begin with a single point-to-point system (as in Sects. 5.2 and 5.3) wherein a single transmitter communicates with a single receiver.

In the point-to-point SISO case, no interference exists, and the signal-to-interference-plus-noise ratio (SINR) is replaced with γ. Based on the discussion in Chap. 3, the maximum achievable rate, in bits/s/Hz, is given by $\log_2(1 + \gamma)$.

We now extend this expression to the case of multiple antennas at both the transmitter (N_T) and receiver (N_R).[4]

5.4.1 Parallel Channels

Consider the case of $N_T = N_R$ and in parallel, i.e., transmitter antenna n communicates only with receive antenna n. Mathematically, if we denote as h_{nm} the channel between transmitter n and receiver m, we have $h_{nm} = 0 \ \forall n \neq m$, i.e., the parallel channels do not interact. This rather contrived scenario is surprisingly informative.

---

[4] It is worth noting the change in notation from previous sections. In Sects. 5.2 and 5.3, we used M to designate the number of antennas at the basestation assuming the user has only a single antenna. Here, for now, we do not distinguish between the basestation and user or whether we are in the uplink or downlink.

Denote as P_n the power allocated to the n-th parallel channel and by $P_N = \sigma^2$ the noise power, i.e., the normalized SNR on this n-th channel is given by $\gamma_n = |h_{nn}|^2 P_n/\sigma^2$ and the rate achievable on this channel is $\log_2(1+\gamma_n)$. Since the channels do not interact, the overall rate is given by

$$R = \sum_{n=0}^{N_T-1} \log_2(1+\gamma_n).$$

To obtain the maximum possible rate, assuming the transmitter knows the channels (or at least the channel powers), we can optimize the powers allocated to the individual channels. The overall maximum achievable rate is, therefore,

$$R_{\max} = \max_{\{P_n, n=0,1,\ldots,N_T-1\}} \left[\sum_{n=0}^{N_T-1} \log_2\left(1 + P_n \frac{|h_{nn}|^2}{\sigma^2}\right) \right], \quad (5.27)$$

such that $\quad P_n \geq 0, \quad (5.28)$

$$\sum_{n=0}^{N_T-1} P_n \leq P_T, \quad (5.29)$$

where the constraint in (5.28) ensures that power can only be nonnegative, while (5.29) ensures that the sum of all allocated powers meets the constraint of the total power available at the transmitter (set to P_T).

Solving this optimization problem uses the method of Lagrange multipliers [9] leading to the unconstrained problem (note the minimization and the negative on the rate expression!)

$$\mathcal{L}(\{P_n\}; \lambda) = \min_{\{P_n, n=0,1,\ldots,N_T-1\}} \left[-\sum_{n=0}^{N_T-1} \log_2\left(1 + P_n \frac{|h_{nn}|^2}{\sigma^2}\right) \right.$$

$$\left. + \lambda \left(\sum_{n=0}^{N_T-1} P_n - P_T \right) \right]$$

Differentiating, as usual, and setting to zero leads to

$$\frac{\partial \mathcal{L}}{\partial P_n} = \frac{|h_{nn}|^2}{\sigma^2} \frac{\log_2(e)}{\left(1 + P_n \frac{|h_{nn}|^2}{\sigma^2}\right)} + \lambda = 0$$

$$\Rightarrow \forall n = 0, 1, \ldots, N_T - 1 \quad \left(\frac{\sigma^2}{|h_{nn}|^2} + P_n \right) = \mu \text{(a constant)}.$$

5.4 MIMO Information Theory and Rate Considerations

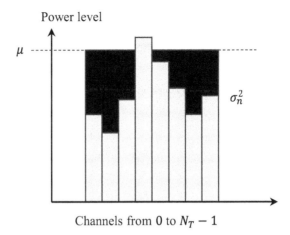

Fig. 5.13 Illustration of waterfilling

Equivalently, imposing the condition that powers must be nonnegative,

$$P_n = \left(\mu - \frac{\sigma^2}{|h_{nn}|^2}\right)^+, \qquad (5.30)$$

where $(x)^+ = \max(x, 0)$. This allocation of powers is known as *waterfilling*.

Figure 5.13 provides a geometric interpretation of the power allocation and illustrates why this process is known as waterfilling. For the n-th channel, $\sigma_n^2 = \sigma^2/|h_{nn}|^2$ sets the effective noise level and acts like a post over which "water" (power) is to be poured. Stronger channels have a lower effective noise level. *Water is poured* (power is allocated) until the water level reaches the constant μ. The value of μ is determined by the available power—after waterfilling, all the available power is used and we must have:

$$\sum_{n=0}^{N_T-1} P_n = P_T.$$

As the figure indicates, some channels may be so weak (large effective noise level σ_n^2) so as to not receive any power. However, interestingly, the optimal allocation scheme *does not* assign all available to the strongest channel. This is because there are diminishing marginal returns to allocating power to channels—at high SNR, $\log_2(1 + x) \simeq \log_2(x)$ and the rate grows only logarithmically with power; on the other hand, at low SNR, $\log_2(1 + x) \simeq x$ and the rate grows linearly with allocated power. Consequently, allocating power to weak (but, not too weak) channels can increase the overall rate.

Denoting as $P_n^*, n = 0, 1, \ldots, N_T - 1$ as the optimal power allocations, the overall achievable rate is

$$R_{\max} = \sum_{n=0}^{N_T-1} \log_2\left(1 + \frac{P_n^* |h_{nn}|^2}{\sigma^2}\right)$$

Finally, a note on the role of CSI: waterfilling requires knowledge of the channel state at the transmitter (CSIT). If the transmitter does not know the channels, it can only divide the available power equally leading to a total achievable rate of

$$R = \sum_{n=0}^{N_T-1} \log_2\left(1 + \frac{P_T}{N_T} \frac{|h_{nn}|^2}{\sigma^2}\right) \tag{5.31}$$

Comparing these two expressions, and using (5.30), we see that the gains from CSIT would be substantial when the channels have very different power levels. If all channels had approximately the same power, an equal power allocation would be close to optimal.

5.4.2 General MIMO Channels

The discussion in Sect. 5.4.1 considered the simplified scenario of parallel channels. In practice, the most likely scenario is when all transmit antennas communicate with all receive antennas. In this case, the transmitter transmits a length-N_T vector **s** received as a length-N_R vector **x** through the $N_R \times N_T$ full channel matrix **H** (and corrupted by noise), i.e.[5]

$$\mathbf{x} = \mathbf{Hs} + \mathbf{n}, \tag{5.32}$$

where $\mathbf{n} \sim \mathcal{CN}(\mathbf{0}, \sigma^2 \mathbf{I})$ denotes the additive white Gaussian noise (AWGN) (see Chap. 3).

Interestingly, this general system can be reduced to the case of parallel channels. To do so, consider the singular value decomposition[6] of **H** given by $\mathbf{H} = \mathbf{U}\boldsymbol{\Sigma}\mathbf{V}^H$. Rewriting (5.32), we have:

---

[5] In the case of parallel channels, $N_R = N_T$ and **H** is a diagonal matrix.

[6] Any $M \times N$ matrix **A** can be decomposed as $\mathbf{A} = \mathbf{U}\boldsymbol{\Sigma}\mathbf{V}^H$. The columns of **U** are the M eigenvectors of $\mathbf{AA}^H$ and the columns of **V** are the N eigenvectors of $\mathbf{A}^H\mathbf{A}$. The $M \times N$ matrix $\boldsymbol{\Sigma}$ is a diagonal matrix of singular values (with appropriate zero padding to match matrix dimensions). For example, if $M \leq N$, $\boldsymbol{\Sigma} = [\text{diag}(\sigma_1, \sigma_2, \ldots, \sigma_M) \mid \mathbf{0}_{M \times N-M}]$ where σ_m^2 are the M eigenvalues (all real) of $\mathbf{AA}^H$. Note that this is for an arbitrary rectangular matrix **A**, and these singular values should not be confused with the noise power. By convention, we choose the singular values as positive and in descending order. Since $\mathbf{AA}^H$ and $\mathbf{A}^H\mathbf{A}$ are positive semi-definite matrices, **U** and **V** are unitary matrices, i.e., $\mathbf{UU}^H = \mathbf{U}^H\mathbf{U} = \mathbf{I}_M$, $\mathbf{VV}^H = \mathbf{V}^H\mathbf{V} = \mathbf{I}_N$ and $\sigma_m \geq 0$. The matrix **U** (**V**) is known as the matrix of left (right) singular vectors.

5.4 MIMO Information Theory and Rate Considerations

$$\mathbf{x} = \mathbf{U}\mathbf{\Sigma}\mathbf{V}^H\mathbf{s} + \mathbf{n},$$
$$\Rightarrow \mathbf{U}^H\mathbf{x} = \mathbf{\Sigma}\mathbf{V}^H\mathbf{s} + \mathbf{U}^H\mathbf{n},$$
$$\Rightarrow \tilde{\mathbf{x}} = \mathbf{\Sigma}\tilde{\mathbf{s}} + \tilde{\mathbf{n}},$$

where we define $\tilde{\mathbf{x}} = \mathbf{U}^H\mathbf{x}$, $\tilde{\mathbf{s}} = \mathbf{V}^H\mathbf{s}$ and $\tilde{\mathbf{n}} = \mathbf{U}^H\mathbf{n}$.

> Since the matrix $\mathbf{U}$ is unitary, the new noise vector, $\tilde{\mathbf{n}}$ has the same statistics as the original vector $\mathbf{n}$. To see this, we note that $\mathbb{E}[\tilde{\mathbf{n}}] = \mathbf{U}^H \mathbb{E}[\mathbf{n}] = \mathbf{0}$. Further, the covariance of $\tilde{\mathbf{n}}$ is given by
>
> $$\mathbb{E}[\tilde{\mathbf{n}}\tilde{\mathbf{n}}^H] = \mathbb{E}[\mathbf{n}\mathbf{U}^H\mathbf{U}\mathbf{n}^H] = \mathbb{E}[\mathbf{n}\mathbf{n}^H] = \sigma^2\mathbf{I}.$$
>
> Importantly, since $\mathbf{n}$ and $\tilde{\mathbf{n}}$ follow the same statistics, we can analyze the system as before.

Consider the case of $N_R < N_T$, i.e., the receiver has fewer antennas than the transmitter. We have

$$\tilde{\mathbf{x}} = \begin{bmatrix} \tilde{x}_0 \\ \tilde{x}_1 \\ \vdots \\ \tilde{x}_{N_R-1} \end{bmatrix} = \begin{bmatrix} \sigma_0 & 0 & \cdots & 0 & 0 & \cdots & 0 \\ 0 & \sigma_1 & \cdots & 0 & 0 & \cdots & 0 \\ \vdots & \vdots & \ddots & \vdots & 0 & \cdots & 0 \\ 0 & 0 & \cdots & \sigma_{N_R-1} & 0 & \cdots & 0 \end{bmatrix} \begin{bmatrix} \tilde{s}_0 \\ \tilde{s}_1 \\ \vdots \\ \tilde{s}_{N_T-1} \end{bmatrix} + \tilde{\mathbf{n}}, \quad (5.33)$$

where, mathematically, the zeros to the right of the vertical line in the matrix denote $(N_T - N_R)$ columns of zeros to ensure the matrix dimensions are compatible. Remarkably, this is exactly like the case of N_R parallel channels (Fig. 5.14)!

From a communication point of view, the zeros show that we have only N_R ($<N_T$ in this case) parallel channels. The role of channel (h_{nn} in Sect. 5.4.1) is played by the singular values σ_n. The optimal power allocation to the n-th channel in the rotated (tilde) space is given by

$$P_n = \left(\mu - \frac{\sigma^2}{\sigma_n^2}\right)^+,$$

where, as before, μ is chosen to meet the power constraint, $\sum_n P_n = P_T$. Analogous to (5.31), the achievable rate is given by

$$R = \sum_{n=0}^{N_R-1} \log_2\left(1 + P_n \frac{\sigma_n^2}{\sigma^2}\right) \quad (5.34)$$

Fig. 5.14 Illustration of N_R parallel channels

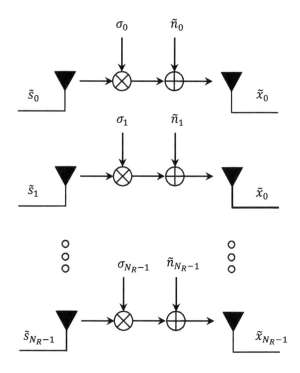

So far we have considered the case of $N_R < N_T$. We can show, analogously, that if $N_R > N_T$, we can transmit a maximum of N_T parallel data streams. In fact, the maximum number of parallel data streams is the *rank* of the channel **H**—the number of nonzero singular values of **H** (or the number of nonzero eigenvalues of $\mathbf{HH}^H$).

Figure 5.15 illustrates the processing chain for a channel matrix **H** with rank $R \leq \min(N_T, N_R)$.

It is instructive to consider the rate equation in some special cases where an analytical solution is available. We consider four examples:

Case 1: 1 transmitter and N_R receivers, $\mathbf{H} = \begin{bmatrix} h_0, h_1, \ldots, h_{N_R-1} \end{bmatrix}^T$, rank($\mathbf{H}$) = 1.

Since rank($\mathbf{H}$) = 1, only one singular value is nonzero; this singular value and the resulting achievable rate are given by

$$\sigma_1 = \sqrt{|h_0|^2 + |h_1|^2 + \ldots |h_{N_R-1}|^2}, \tag{5.35}$$

5.4 MIMO Information Theory and Rate Considerations

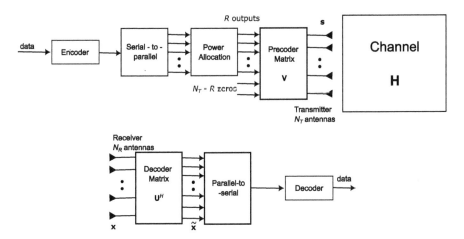

Fig. 5.15 A MIMO communication system with CSIT

$$R = \log_2\left(1 + \frac{P_T}{\sigma^2}\sum_{n=0}^{N_R-1}|h_n|^2\right). \tag{5.36}$$

Case 2: N_T transmitters and 1 receiver, $\mathbf{H} = \begin{bmatrix} h_0, h_1, \ldots, h_{N_T-1} \end{bmatrix}$, rank($\mathbf{H}$) = 1.

Since rank($\mathbf{H}$) = 1, again only one singular value is nonzero; this singular value and the resulting achievable rate are given by

$$\sigma_1 = \sqrt{|h_0|^2 + |h_1|^2 + \ldots |h_{N_T-1}|^2}, \tag{5.37}$$

$$R = \log_2\left(1 + \frac{P_T}{\sigma^2}\sum_{n=0}^{N_T-1}|h_n|^2\right), \tag{5.38}$$

As (5.36) and (5.38) show, in the case of rank-1 channels, the best we can do is to gather all the channel energy into one data stream.

Problem Show that in the case of one transmit and N_R receive antennas, the receiver is the same as the MRC receiver used to maximize diversity gains. Similarly, show that with N_T transmit and one receive antennas, the transmitter is the same as using MRT.

Case 3: N transmitters and M receivers with perfect line of sight (LOS), without multipath.

Let d_t be the distance between the transmit elements and d_r the distance between the receive elements. The transmitter transmits in direction ϕ_t with respect to its baseline while the receiver receives from angle ϕ_r with respect to its baseline. In this case, at an operating wavelength of λ

$$h_{mn} = C \exp(j(2\pi/\lambda)d_r n_r \cos\phi_r) \exp(j(2\pi/\lambda)d_t n_t \cos\phi_t),$$

$$n_r = 0, 1, \ldots, N_R - 1, n_t = 0, 1, \ldots, N_T - 1,$$

where C is some constant capturing the path loss. Note that even though the channel matrix $\mathbf{H}$ is of size $N_R \times N_T$, it is still rank-1 and $d_1 = \sqrt{N_T N_R}$. The achievable rate is given by

$$R = \log_2\left(1 + |C^2| N_T N_R \frac{P_T}{\sigma^2}\right), \tag{5.39}$$

i.e., in line-of-sight conditions, the arrays at the transmitter and receiver only provide a power gain.

Case 4: $N_T = N_R$ and the channel has full rank with equal singular values.

Since the square of the singular values of $\mathbf{H}$ are the eigenvalues of $\mathbf{HH}^H$, using the fact that the sum of eigenvalues is the trace of the matrix

$$\sum_{n=0}^{N_T-1} \sigma_n^2 = \text{trace}\left(\mathbf{HH}^H\right) = \sum_{n_t=0}^{N_T-1} \sum_{n_r=1}^{N_R-1} |h_{n_r n_t}|^2,$$

which is the total power in the channel matrix. Let us denote this as P_h. Since the channel has equal singular values, $\sigma_n^2 = P_h/N_T, \forall n = 0, \ldots, N_T - 1$. Since all singular values are equal, the energy allocation is also uniform ($P_n = P_T/N_T$) and

$$R = \sum_{n=0}^{N_T-1} \log_2\left(1 + \frac{P_T \sigma_n^2}{N_T \sigma^2}\right) = \sum_{n=0}^{N_T-1} \log_2\left(1 + \frac{P_T}{N_T} \frac{P_h}{N_T \sigma^2}\right)$$

$$= N_T \log_2\left(1 + \frac{P_T}{N_T} \frac{P_h}{N_T \sigma^2}\right). \tag{5.40}$$

Note the significant difference in the capacities described in Eqs. (5.39) and (5.40). Under perfect line-of-sight conditions, or if either the transmitter or receiver uses only one antenna, the channel is rank-1 and we only achieve *power gains*—the rate increases as the logarithm of the number of elements. However, when the channel is set up such that each eigenchannel is has equal power (the singular values are all equal), the rate gains *are linear*.

This final statement is crucial; from (5.34), the linear gains in rate arise when the rank of the channel is greater than one—the rate is maximized with a full-rank channel, i.e., min(N_R, N_T). In this case, the parallel channels allow us to transmit *independent* data streams (N_T in the final example above), thereby increasing rate.

5.5 Trade-Off Between Rate and Reliability

We have now considered two opposing metrics: in Sects. 5.2 and 5.3, our focus was on maximizing the reliability given a fixed transmission rate. However, as we saw, good space-time codes exist only for a small number of antennas (with the best rate-1 code for two transmit antennas only—the Alamouti scheme). In Sect. 5.4, on the other hand, we used information theoretic concepts to maximize rate assuming perfect reliability. Intuitively, a higher rate should result in lower reliability. In this section, we explore a trade-off between these two viewpoints, trading off rate for reliability.

5.5.1 Simultaneous Multiplexing and Space-Time Coding

In [8], Tarokh et al. present a scheme that represents a trade-off between data throughput and diversity order. The scheme is based transmitting multiple data streams while limiting the coding complexity but yet exploiting all available spatial degrees of freedom. Consider a transmitting array of N_T elements divided into Q groups of subarrays, each with N_q elements, i.e., $\sum_{q=1}^{Q} N_q = N_T$. At the input, B bits are divided into Q blocks of B_q bits each, i.e., we will have Q parallel data streams. Using a low-complexity space-time encoder (denoted as C_q, the B_q bits are encoded, yielding N_q symbols per timeslot for transmission using N_q elements. The overall code can be represented as $C_1 \times C_2 \times \ldots \times C_Q$. Note that each code may itself be a trellis or block code. Let $\mathbf{c}_q$ denote the data transmitted using the qth subarray.

The receiver decodes each block of data successively. Without loss of generality, we start with decoding the first block. The data received at the N_R receiving elements is given by

$$\mathbf{x} = \mathbf{Hc} + \mathbf{n}, \tag{5.41}$$

$$= \begin{bmatrix} h_{00} & h_{01} & \cdots & h_{0(N_1-1)} & h_{0N_1} & \cdots & h_{0(N_T-1)} \\ h_{10} & h_{11} & \cdots & h_{1(N_1-1)} & h_{1N_1} & \cdots & h_{1(N_T-1)} \\ \vdots & \vdots & \cdots & \vdots & \vdots & \ddots & \vdots \\ h_{(N_R-1)0} & h_{(N_R-1)1} & \cdots & h_{(N_R-1)(N_1-1)} & h_{(N_R-1)N_1} & \cdots & h_{(N_R-1)(N_T-1)} \end{bmatrix} \begin{bmatrix} \mathbf{c}_1 \\ \mathbf{c}_2 \\ \vdots \\ \mathbf{c}_Q \end{bmatrix} + \mathbf{n}, \tag{5.42}$$

where the channel matrix $\mathbf{H}$ is partitioned to isolate the channel from the first subarray to the receiver. We assume the channel $\mathbf{H}$ is full-rank, i.e., min(N_T, N_R).

We will start by recovering c_1, i.e., for now, only N_1 of the N_T transmitters are of interest. The data transmissions c_2 to c_Q (through the channels from their respective transmitters) act as interference and degrade the decoding of the data in the first block, c_1.

Denote as $\mathbf{\Omega}(C_1)$ the $N_R \times N_1$ channel matrix from the N_1 elements in the first subarray to the N_R receive elements (the matrix to the left of the partition). Similarly denote as $\mathbf{\Lambda}(C_1)$ the $N_R \times (N_T - N_1)$ matrix of the *interfering* channel,[7] i.e.

$$\mathbf{\Omega}(C_1) = \begin{bmatrix} h_{00} & h_{01} & \cdots & h_{0(N_1-1)} \\ h_{10} & h_{11} & \cdots & h_{1(N_1-1)} \\ \vdots & \vdots & \cdots & \vdots \\ h_{(N_R-1)0} & h_{(N_R-1)1} & \cdots & h_{(N_R-1)(N_1-1)} \end{bmatrix},$$

$$\mathbf{\Lambda}(C_1) = \begin{bmatrix} h_{0N_1} & \cdots & h_{0(N_T-1)} \\ h_{1N_1} & \cdots & h_{1(N_T-1)} \\ \vdots & \ddots & \vdots \\ h_{(N_R-1)N_1} & \cdots & h_{(N_R-1)(N_T-1)} \end{bmatrix}.$$

Since $\mathbf{\Lambda}(C_1)$ has $N_T - N_1$ columns, if $N_R > N_T - N_1$, $\mathrm{rank}(\mathbf{\Lambda}(C_1)) \leq N_T - N_1$ and there exist at least $N_R - (N_T - N_1) = (N_R - N_T + N_1)$ linearly independent vectors that are orthogonal to all $(N_R - N_1)$ vectors in $\mathbf{\Lambda}(C_1)$. Denote as $\mathbf{\Theta}(C_1)$ as the $N_R \times (N_R - N_T + N_1)$ matrix whose columns are orthogonal to $\mathbf{\Lambda}(C_1)$. In addition, we ensure the columns of $\mathbf{\Theta}(C_1)$ are mutually orthonormal. Note that due to the constraint $N_R > N_T - N_1$, it is always possible to create such a matrix $\mathbf{\Theta}(C_1)$ given the channel matrix $\mathbf{H}$.

Now consider a new set of *received* data:

$$\tilde{\mathbf{x}}_1 = \mathbf{\Theta}(C_1)^H \mathbf{y}, \qquad (5.43)$$

$$= \mathbf{\Theta}(C_1)^H \mathbf{\Omega}(C_1) \mathbf{c}_1 + \mathbf{\Theta}(C_1)^H \mathbf{\Lambda}(C_1) \begin{bmatrix} \mathbf{c}_2 \\ \vdots \\ \mathbf{c}_Q \end{bmatrix} + \mathbf{\Theta}(C_1)^H \mathbf{n}, \qquad (5.44)$$

$$= \mathbf{\Theta}(C_1)^H \mathbf{\Omega}(C_1) \mathbf{c}_1 + \mathbf{\Theta}(C_1)^H \mathbf{n} \qquad (5.45)$$

The noise term $\mathbf{\Theta}(C_1)^H \mathbf{n}$ is also zero mean and satisfies $\mathrm{E}\left\{\mathbf{\Theta}(C_1)^H \mathbf{n}\mathbf{n}^H \mathbf{\Theta}(C_1)\right\} = \sigma^2 \mathbf{\Theta}(C_1)^H \mathbf{\Theta}(C_1) = \sigma^2 \mathbf{I}_{N_R-N_T+N_1}$, i.e., the noise term is still white. Note that, importantly; all the interference has been eliminated. This is because of how we chose the matrix $\mathbf{\Theta}(C_1)$.

---

[7] The notation is in a large part from Tarokh et al. [8].

5.5 Trade-Off Between Rate and Reliability

This final equation is that of an equivalent space-time coded communication system with N_1 transmitters and $(N_R - N_T + N_1)$ receivers. On the data in $\mathbf{c}_1$, we therefore get a diversity order of $N_1(N_R - N_T + N_1)$, which is significantly lower than the potential diversity gain for a single data stream of $N_R N_T$. Note that the achieved diversity order makes sense since the data stream is transmitted using N_1 antennas and the N_R receivers must use $(N_T - N_1)$ degrees of freedom to suppress the $(N_T - N_1)$ interfering transmissions, leaving $(N_R - N_T + N_1)$ degrees of freedom to enhance reliability (diversity) on the N_1 transmissions of interest.

Obviously, one could repeat the process and achieve $N_q(N_R - N_T + N_q)$ diversity order on the q-th data in $\mathbf{c}_q$. However, there is a more intelligent way of processing the same data. Assume the data in $\mathbf{c}_1$ is accurately decoded. At the receiver, we now know the transmitted data from the first subarray and the channel $\mathbf{\Omega}(C_1)$, and so when decoding $\mathbf{c}_2$, one can *eliminate* this source of interference from the data. Let

$$\mathbf{y}_2 = \mathbf{y} - \mathbf{\Omega}(C_1)\mathbf{c}_1 = \mathbf{\Lambda}(C_1)\begin{bmatrix}\mathbf{c}_2\\\mathbf{c}_3\\\vdots\\\mathbf{c}_Q\end{bmatrix} + \mathbf{n}. \tag{5.46}$$

This is equivalent to a communication system with N_R receivers and $N_T - N_1$ transmitters (and $N_T - N_1 - N_2$ interfering sources). To decode $\mathbf{c}_2$, only N_2 of these transmitters are of interest. Using the same scheme as described above, one can achieve $N_2(N_R - (N_T - N_1 + N_2)) = N_2(N_R - N_T + N_1 + N_2)$-order diversity. This is *greater* than the $N_2(N_R - N_T + N_2)$ order diversity attained using the "obvious" approach. Repeating this process for each subarray, to decode $\mathbf{c}_q$ one can achieve diversity order $N_q\left(N_R - N_T + \sum_{p=1}^{q} N_p\right)$, i.e., each successive data block can be decoded with greater diversity order.

The trade-off is, therefore, as follows: if we were to transmit a single data stream, we could achieve a diversity order of $N_R N_T$. On the other hand, if we allowed to transmit (multiplex) as many streams as possible (highest rate), we would transmit $\min(N_R, N_T)$ streams. Here, we are choosing a balance between the two extremes—we are transmitting Q streams, but at a diversity order of $N_q\left(N_R - N_T + \sum_{p=1}^{q} N_p\right)$ for stream q.

We note that since each block gets a different order of diversity, one can transmit these blocks with different power levels to achieve somewhat equal error rates. In [8] the authors suggest a power allocation in inverse proportion to diversity order. Finally, as a special case, the famous Bell Labs Layered Space-Time (BLAST) scheme [10] uses $N_q = 1$, with the maximum data throughput but minimum diversity.

The scheme presented in [8] is flexible in that it allows for a trade-off between throughput (multiplexing) and required reliability. However, it suffers from one significant drawback—the need for an adequate number of antennas $N_R > N_T - N_1$ such that a null space can be formed to cancel the interference. As mentioned earlier, in cellular communications, one could expect multiple antennas at the basestation; however, a mobile device would not have many antennas. Thinking in terms of the downlink, expecting $N_T > N_T - N_1$ may not be realistic (note that BLAST requires $N_R \geq N_T$).

5.5.2 Diversity-Multiplexing Trade-Off

We end this diversity section by stating the *fundamental trade-off*, stated by Zheng and Tse, between diversity order and data throughput (also multiplexing) [11]. We have seen that given a diversity order of d, the probability of error in Rayleigh channels is given by $P_e \propto \text{SNR}^{-d}$ at high SNR. Similarly, with system using N_T transmit and N_R receive antennas, the capacity of the system raises (in the high SNR limit) as $R \propto \min(N_R, N_T)\log(\text{SNR})$.

Zheng and Tse define a scheme $\mathcal{R}(\text{SNR})$ to achieve diversity gain d and *multiplexing gain r* if

$$\lim_{\text{SNR}\to\infty} \frac{\log P_e(\text{SNR})}{\log \text{SNR}} = -d \tag{5.47}$$

$$\lim_{\text{SNR}\to\infty} \frac{R(\text{SNR})}{\log \text{SNR}} = r \tag{5.48}$$

where $R(\text{SNR})$ is the data throughput of the scheme. Note that the scheme assumes that as the SNR increases, the data rate rises as well (possibly through adaptive modulation). Any scheme with a fixed data rate, however high, is said to have a *multiplexing gain of zero*. In this regard, all schemes we have developed so far have a multiplexing gain of zero. The thinking behind such a definition is that as the SNR increases, one could either gain reliability (reduce P_e) or increase throughput or part of each. The definition determines how much we gain in each metric SNR increases.

Let $d^\star(r)$ be the supremum of all possible schemes with multiplexing gain r. Clearly $d_{\max} = d^\star(0)$ and $r_{\max} = \sup(r : d^\star(r) > 0)$. The authors state the most interesting result:

Diversity-Multiplexing Trade-Off Consider the case of N_T transmit and N_R receive antennas. For a block length $L > N_R + N_T - 1$, the optimal tradeoff curve, $d^\star(r)$ is given by the piecewise-linear function connecting the points $(r, d^\star(r))$, $r = 0, 1, \ldots, min(N_R, N_T)$, where

$$d^\star(r) = (N_R - r)(N_T - r). \tag{5.49}$$

Figure 5.16 illustrates the fundamental trade-off between reliability (diversity order) and data rate (multiplexing). The curve shows linearly joined points $(r, d^\star(r))$, with $d^\star(r)$ defined in Eq. (5.49). The theorem states that it is not possible

5.5 Trade-Off Between Rate and Reliability

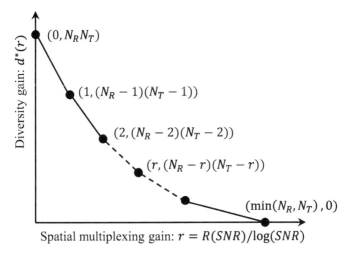

Fig. 5.16 Illustrating the diversity-multiplexing trade-off

to use the available SNR in any manner better than this curve. Note that as expected at zero multiplexing gain (fixed transmission rate), the diversity order that can be achieved is $N_R N_T$, whereas if one were increasing the data throughput as $\min(N_R, N_T) \log(\text{SNR})$, there is no diversity, i.e., the error rate *does not* fall with SNR.

Equation (5.49) suggests that at the integer points, when the multiplexing rate is r, the system communicates r parallel data streams. The transmitter and receive each use r degrees of freedom to eliminate the inter-stream interference leaving a diversity order of $(N_R - r)(N_T - r)$. Another interpretation uses the eigenvalues of the channel matrix [12]. A multiplexing rate of r says that the raw data rate of transmission is $R = r \log(SNR)$. An outage occurs when the achievable rate falls below this chosen transmission rate. The achievable rate is given by

$$C = \sum_{m=1}^{\min(N_R, N_T)} \log_2 \left(1 + \frac{P_m}{\sigma^2} \sigma_m^2 \right), \tag{5.50}$$

where σ_m, $m = 1, \ldots \min(N_R, N_T)$ are the nonzero singular values of $\mathbf{H}$ and P_m is the power allocated to stream m.

To interpret this equation, we replace P_m/σ^2 with the SNR. At high SNR, we can ignore the "1 +." For an outage to occur, we need:

$$\sum_{m=1}^{\min(N_R, N_T)} \log_2 \left(\text{SNR} \sigma_m^2 \right) < r \log_2(SNR). \tag{5.51}$$

The outage events are therefore controlled by the singular values of **H**. For an outage to happen, these singular have to be "bad" (small):

- If the chosen multiplexing gain, r, is close to zero, an outage occurs if *all eigenvalues are poor*. This happens rarely (and yields diversity order $N_R N_T$).
- If $r = \min(N_R, N_T)$, we need *all singular values to be large* to avoid an outage (note that the number of the "large" singular values effectively provides the linear term in front of the $\log_2(SNR)$.)
- Somewhere in between, to avoid an outage, we need r of the eigenvalues to be large enough, resulting in a diversity order of $(N_R - r)(N_T - r)$.

5.6 Multiuser MIMO Systems

The discussion so far has focused on the MIMO single-user case, wherein one BS communicates with a single-user terminal (which may or may not possess multiple antennas). However, a wireless network must satisfy the needs of multiple spatially distributed users. In a traditional approach to network design, each active user in a cell is allocated its own resource block, a slice of time, and/or frequency, for its communication needs. The use of multiple antennas at the BS, however, allows for multiple users to be served on the *same* time-frequency resource. This capability, also called space-division multiple access (SDMA), recovers users' signals using their individual spatial signatures—their channels.

5.6.1 Interference Mitigation

In communicating to multiple users simultaneously, the signals from (uplink) or to (downlink), the users can interfere with each other. A key question is how to cancel (or mitigate) this inter-user interference. The popular approaches are easiest to appreciate in the uplink—consider a multiuser system in the uplink wherein K single-antenna users transmit symbols to a single BS with M antenna elements. For one symbol period, designating the symbol transmitted by the k-th user as s_k, the received signal is the length-M vector:

$$\mathbf{x} = \sum_{k=1}^{K} \mathbf{h}_k s_k + \mathbf{n} = \mathbf{H}\mathbf{s} + \mathbf{n}, \tag{5.52}$$

where $\mathbf{s} = [s_1, s_2, \ldots, s_K]^T$ denotes the vector of transmitted symbols, $\mathbf{h}_k$ denotes the length-M channel of user k to the BS and $\mathbf{H} = [\mathbf{h}_1, \mathbf{h}_2, \ldots, \mathbf{h}_K]$ denotes the $M \times K$ matrix of channels of all K users to the BS; finally, $\mathbf{n} \sim \mathcal{CN}(0, \sigma^2 \mathbf{I})$ denotes noise. As we have done earlier, we assume the channel matrix **H** is known. Given

5.6 Multiuser MIMO Systems

the received signal $\mathbf{x}$, we wish to retrieve the symbol vector $\mathbf{s}$ for the K users. To do so optimally, however, requires us to jointly process all users signals—a process that is exponentially complex in the number of users. We now explore some popular practical alternatives; however, it is worth noting that this list is not comprehensive.

5.6.1.1 Successive Interference Cancellation (SIC)

The SIC approach creates a series of (seemingly) single-user problems from the multiuser problem above. We rewrite the received signal, focusing on user 1, as

$$\mathbf{x} = \mathbf{h}_1 s_1 + \sum_{k=2}^{K} \mathbf{h}_k s_k + \mathbf{n} = \mathbf{h}_1 s_1 + \mathbf{n}_1,$$

where $\mathbf{n}_1$ is the overall noise and multiuser interference seen by user 1. Notice the similarity between this equation and that in (5.8). The key difference is that, due to the randomness of the symbols s_k arising from the symbol set and the fact that the channels are known, $\mathbf{n}_1$ is *not* white and Gaussian. As a *simplification*, when using SIC, we ignore this fact and treat $\mathbf{n}_1$ as being so. Now, to process the signal in (5.8), we used MRC—we do the same here and process $\mathbf{x}$ to obtain

$$y_1 = \mathbf{h}_1^H \mathbf{x},$$

which is used for symbol decoding.

Now, clearly, one repeats this process for each user. However, the SIC approaches attempt to do one better—since we have decoded the symbol of user 1 (s_1, and know its channel $\mathbf{h}_1$), we can subtract its contribution to the signal $\mathbf{x}$. So, consider:

$$\mathbf{x}_2 = \mathbf{x} - \mathbf{h}_1 s_1 = \mathbf{h}_2 s_2 + \sum_{k=3}^{K} \mathbf{h}_k s_k + \mathbf{n} = \mathbf{h}_2 s_2 + \mathbf{n}_2.$$

Here, we are *assuming* that the symbol s_1 has been correctly recovered. If so, we note that the noise vector $\mathbf{n}_2$ has one fewer contribution, i.e., the noise power seen by user 2 is lower than that seen by user 1. We now apply MRC to user 2.

The process continues, successively subtracting the contribution of each user as its symbol is decoded. Each user, therefore, sees one less interference source than the previous user decoded.

A crucial aspect of this SIC procedure is that it depends on accurate recovery of the users' symbols. If improperly decoded, the incorrect symbol would add to, not subtract from, interference seen by the next user. This *error propagation* limits the performance of the SIC approach. To minimize its impact, it is important to order users such that early users have the lowest probability of errors, i.e., we order users such that

$$||\mathbf{h}_1||^2 \geq ||\mathbf{h}_2||^2 \geq \cdots \geq ||\mathbf{h}_K||^2.$$

5.6.1.2 Zero Forcing

Unlike the SIC approach, zero forcing (ZF) processes all K symbols at once. Considering (5.52), the ZF criterion is to find a vector $\mathbf{s}_{ZF}$ which comes closest to the received signal in Euclidean sense, i.e.,

$$\mathbf{s}_{ZF} = \arg\min_{\mathbf{s}} \left[||\mathbf{x} - \mathbf{Hs}||^2 \right].$$

Using the theory of differentiating with respect to a vector [13], the solution to this optimization is

$$\mathbf{s}_{ZF} = \left(\mathbf{H}^H \mathbf{H} \right)^{-1} \mathbf{H}^H \mathbf{x}.$$

The length-K vector $\mathbf{s}_{ZF}$ is then processed on a per-user basis. Clearly, for an effective solution, we need more observations (M at the M antennas) than unknowns (K for the K user symbols), i.e., we need $M \geq K$.

The reason this approach is called zero forcing is clear when considering what the resulting vector is. We have:

$$\mathbf{s}_{ZF} = \left(\mathbf{H}^H \mathbf{H} \right)^{-1} \mathbf{H}^H \mathbf{x} = \left(\mathbf{H}^H \mathbf{H} \right)^{-1} \mathbf{H}^H (\mathbf{Hs} + \mathbf{n}),$$
$$= \mathbf{s} + \mathbf{n}_{ZF},$$

i.e., in the symbol vector, the multiuser interference has been eliminated (forced to zero). If the noise vector $\mathbf{n}_{ZF} = \left(\mathbf{H}^H \mathbf{H} \right)^{-1} \mathbf{H}^H \mathbf{n}$ were white and Gaussian, per-user decoding would be optimal; unfortunately, it is not. In Problem 1, you are asked to derive the statistics of this resulting noise vector.

It is worth noting that the output vector $\mathbf{s}_{ZF}$ is not a vector of symbols, but is a vector of complex numbers—symbols corrupted by noise—we would use a minimum distance decoding process for each user separately.

5.6.1.3 Minimum Mean Squared Error (MMSE) Receiver

The MMSE receiver also searches for a vector of symbol estimates, but takes a different starting point. It looks for a $M \times K$ combining matrix $\mathbf{W}$ which would bring the received signal, $\mathbf{x}$, as close to the true symbols as possible. However, since we do not know the true symbols, it can only do this on average. Specifically,

5.6 Multiuser MIMO Systems

$$\mathbf{W}_{\text{MMSE}} = \arg \min_{\mathbf{W}} \mathbb{E}\left[\underbrace{||\mathbf{s} - \mathbf{W}^H \mathbf{x}||^2}_{\text{error}}\right].$$

$$\underbrace{\phantom{\mathbf{W}_{\text{MMSE}} = \arg \min_{\mathbf{W}} \mathbb{E}[||\mathbf{s} - \mathbf{W}^H \mathbf{x}||^2]}}_{\text{mean squared error}}$$

Assuming that all users transmit at the same power and transmit independent symbols such that $\mathbb{E}[\mathbf{ss}^H] = \sigma_s^2 \mathbf{I}$ and using $\mathbb{E}[\mathbf{nn}^H] = \sigma^2 \mathbf{I}$, we can show that

$$\mathbf{W}_{\text{MMSE}} = \sigma_s^2 \left(\sigma_s^2 \mathbf{H}\mathbf{H}^H + \sigma^2 \mathbf{I}\right)^{-1} \mathbf{H}$$

$$\Rightarrow \mathbf{s}_{\text{MMSE}} = \mathbf{W}_{\text{MMSE}}^H \mathbf{x} = \sigma_s^2 \mathbf{H}^H \left(\sigma_s^2 \mathbf{H}\mathbf{H}^H + \sigma^2 \mathbf{I}\right)^{-1} \mathbf{H}\mathbf{s} + \mathbf{W}_{\text{MMSE}}^H \mathbf{n}.$$

In the final equation, we expanded only the signal term to illustrate a very interesting point. Using the matrix inversion lemma,[8] we have for the signal component:

$$\mathbf{s}_{\text{MMSE}} = \left(\mathbf{H}^H \mathbf{H} + \frac{\sigma^2}{\sigma_s^2}\mathbf{I}\right)^{-1} \mathbf{H}^H \mathbf{x}.$$

This expression is surprisingly similar to the signal component from zero forcing. At high SNR (large σ_s^2/σ^2), the two are essentially the same. The explanation is as follows: in ZF, if the channel matrix has small singular values ($\mathbf{H}^H\mathbf{H}$ has small eigenvalues), the inversion process increases the noise power. So, in completely eliminating the interference, ZF results in *noise enhancement*. The MMSE approach, on the other hand, balances the noise component with any residual interference. At high SNR, the noise is less important and so the two approaches converge.

5.6.1.4 Downlink Interference Cancellation

Our discussion so far has focused on the uplink; similar schemes have also been formulated for the downlink. In this case, as shown in Fig. 5.17, the transmitting BS attempts to form beams toward multiple users simultaneously while minimizing the power leakage between users (though, it is worth noting that, with multipath propagation, beampatterns would not "look as nice") The BS precodes the vector of symbols, using a $M \times K$ precoding matrix $\mathbf{W}$. The transmitted signal and signal received at the K users are given by

Transmitted Signal: $\quad \mathbf{x} = \mathbf{W}\mathbf{s}$

Received Signal: $\quad \mathbf{y} = \mathbf{H}^T \mathbf{x} + \mathbf{n}$

---

[8] $\mathbf{A}\mathbf{B}^H \left(\mathbf{B}\mathbf{A}\mathbf{B}^H + \mathbf{C}\right)^{-1} = \left(\mathbf{B}^H \mathbf{C}^{-1} \mathbf{B} + \mathbf{A}^{-1}\right)^{-1} \mathbf{B}^H \mathbf{C}$. We use $\mathbf{A} = \mathbf{I}$, $\mathbf{B} = \mathbf{H}$, $\mathbf{C} = (\sigma^2/\sigma_s^2)\mathbf{I}$.

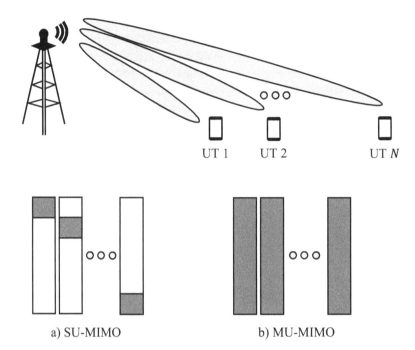

Fig. 5.17 Differences between single-user MIMO (SU-MIMO) and multiuser MIMO (MU-MIMO) in terms of resource utilization. MU-MIMO reuses all the resources, thus increasing the data rate per user.

In extending ZF [14] and MMSE [15] to the downlink, a crucial consideration is that the transmitting BS must meet a power constraint. In the case of a sum power constraint, the precoders for the ZF and MMSE cases are given by

$$\mathbf{W}_{ZF} = \alpha \mathbf{H}^*(\mathbf{H}^T\mathbf{H}^*)^{-1}$$
$$\mathbf{W}_{\text{MMSE}} = \mathbf{H}^*(\mathbf{H}^T\mathbf{H}^* + \alpha \mathbf{I})^{-1}$$

where, in both cases, the parameter α is chosen to meet the power constraint $\mathbb{E}[||\mathbf{W}\mathbf{s}||] = P_T$, where P_T is the available power. Finally, it is worth noting the MMSE approach has been extended to account for inter-cell interference as well [16].

5.6.2 Performance Metrics

This ability to communicate with multiple user raises many questions for system design; important ones are which users to serve with what priority levels? A BS with M antennas can serve at most M users, but usually will try to serve fewer in a single time/frequency resource. Multiple publications, e.g., [16, 17], have shown that choosing the correct combination of users (from the larger set of active users) has a significant impact on performance.

Another question is: "what metric is to drive system design?" Consider a system where K users are scheduled indexed as $k = 1, \ldots, K$. Two obvious choices of design criteria are to maximize one of

$$\text{Max Sum-Rate: } \sum_k R_K \qquad \text{Max-min rate: } \min_k R_k,$$

where R_k denotes the achievable rate to user k. Optimizing the sum-rate metric maximizes use of the available resources, but favors users with strong channels at the expense of users with weak channels (low achievable rate). This is consistent with the discussion on waterfilling wherein stronger eigenchannels are allocated more power than weak eigenchannels. For the max-min rate metric, on the other hand, resources are allocated to ensure all users achieve the same rate. This rate achieved is, therefore, limited by the weakest user and results in poor overall rates. As a balance between these two extremes, a popular metric is the *proportionally fair* sum-rate defined as [17, 18]

$$\text{Proportionally Fair: } \sum_k \log(R_k) \simeq \sum_k \frac{1}{\bar{R}_k} R_k,$$

where $\bar{R}_k$ is the *long-term* average rate achieved for user k. The benefit of this scheme is that strong users do get priority, but if a user gets very low rate, its weight $(1/\bar{R}_k)$ increases. The final expression also explains why the proportionally fair metric is one form of a *weighted* sum-rate.

For multiuser MIMO networks, optimizing user scheduling and power allocation for any of these metrics is an NP-hard problem and requires careful algorithm design with solutions only guaranteed to a local optimum [16].

5.6.3 Massive MIMO Systems

We conclude this chapter with a brief discussion of an important trend in industry—the use of a large number of antennas at the BS, so-called massive MIMO systems. The concept of massive MIMO systems came out of an asymptotic analysis of MIMO systems letting the number of BS antennas, M, grow toward infinity [19]. The author showed that as long as $M/K \to \infty$, the users' channels are effectively orthogonal. In turn, this implies that multiuser schemes, such as ZF and MMSE, reduce to the single-user MRC/MRT approach in Sects. 5.2 and 5.3. Indeed, in this case, the rate achievable to each user grows without bound!

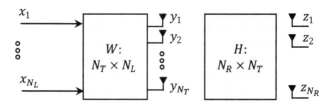

Fig. 5.18 Illustrating massive MIMO system and channel

Importantly, however, this result is true only if the users' channels are known perfectly. Throughout this chapter, we have assumed this to be true—beginning with MRC (channels state information at the receiver or CSIR) or with MRT (CSI at the transmitter or CSIT), we have channel values for the processing. Channel estimation, therefore, plays a fundamentally important role in wireless networks, especially MIMO and multiuser MIMO networks.

5.6.3.1 Channel View from the Device Side

In 5G deployments, the term massive MIMO is used for systems that deploy $N_T = 32$ or $N_T = 64$ antenna ports at the basestation side, while the number of the antennas at the device side is still limited to $N_R = 4$. We have already stated that the number of transmit layers, N_L, is equal to the channel rank can be up to $\min(N_T, N_R)$, but it is instructive to see what kind of channel does the device *see*. To illustrate this, we will slightly change the notation and denote the received signal with z_n, where $n = 1, 2, \ldots, N_R$, as shown in Fig. 5.18.

Let us assume that in Fig. 5.18, we have the following parameters: $N_T = 64$, $N_L = N_R = 4$ layers of data, precoding matrix W has dimensions $N_T \times N_L = 64 \times 4$ and channel H's dimensions are $N_R \times N_T = 4 \times 64$. Therefore, we can write:

$$y_1 = w_{1,1}x_1 + w_{1,2}x_2 + w_{1,3}x_3 + w_{1,4}x_4 \tag{5.53}$$
$$y_2 = w_{2,1}x_1 + w_{2,2}x_2 + w_{2,3}x_3 + w_{2,4}x_4 \tag{5.54}$$
$$\vdots \tag{5.55}$$
$$y_{64} = w_{64,1}x_1 + w_{64,2}x_2 + w_{64,3}x_3 + w_{64,4}x_4 \tag{5.56}$$

which can be written in a compact form as

$$\mathbf{y} = \mathbf{W}\mathbf{x} \tag{5.57}$$

The received signal vector $\mathbf{z}$ consist of only $N_R = 4$ components:

$$\mathbf{z} = \mathbf{H}\mathbf{W}\mathbf{x} = \mathbf{H}'\mathbf{y} \tag{5.58}$$

5.6 Multiuser MIMO Systems

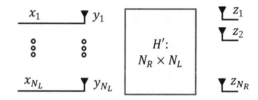

Fig. 5.19 Equivalent precoding and channel matrix, as seen by a device on the downlink receive side

where the channel seen by the receiver, **H'**, has dimensions $N_R \times N_L = 4 \times 4$. Therefore, the device does not *see* the **W** matrix; it only sees the effective *channel* **H'** (Fig. 5.19).

5.6.4 Impact of Massive MIMO on the Link Budget

Massive MIMO impacts the link budget in two ways:

1. Higher antenna/beamforming gain due to larger number of antennas
2. Lower interference margin as the inter-cell interference is reduced due to narrower beams point only to intended users

We can illustrate this with the following example. Let the regular 4×4 MIMO system have antenna gain of 18 dBi and interference margin of 13 dB. Also, let a massive MIMO system, with 64 transmit/receive units have an antenna gain of 24 dBi (18 dBi from the array gain and 6 dBi of individual antenna element gain) and interference margin of 7 dB.

The useful signal power reading difference at the UE comes from the antenna gain only, and is equal to 6 dB (24 dBi–18 dBi) (Fig. 5.20).

However, the actual link budget difference also comes from the reduced interference margin (additional 6 dB delta), so the total MAPL difference for the same data rate is $6 + 6 = 12$ dB, as shown in Fig. 5.21.

As a result, the coverage distance can be more than doubled with massive MIMO for the same target data rate (e.g., solid line associated with massive MIMO achieves 200 Mbps at 270 meters, whereas 4×4 based MIMO system achieves that data rate at 132 meters) (Fig. 5.22).

Problems

1. Find the noise vector mean and covariance of the noise vector in zero forcing, $\mathbf{n}_{ZF}$.
2. Calculate the bit error rate (BER) values for BPSK modulation scheme for $\gamma_{B|dB} = [6, 12]$ dB in AWGN scenario and compare to the average BER

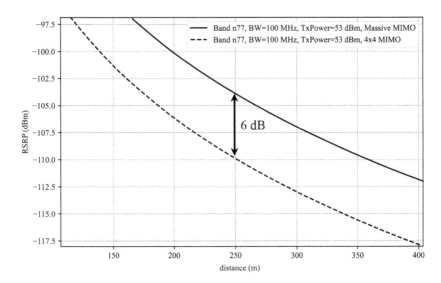

Fig. 5.20 Reference signal received power (RSRP), to be discussed in Chaps. 10 and 11, differs by 6 dB as a result of different maximal antenna gain

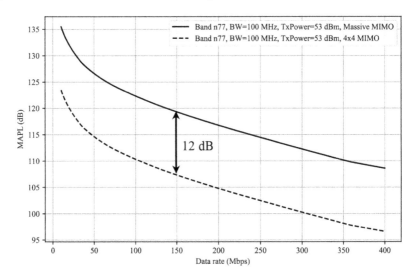

Fig. 5.21 Maximal allowable pathloss is 12 dB higher in case of massive MIMO due to lower interference and higher antenna gain

under Rayleigh fading for $\Gamma_{|dB} = [6, 12]$ dB. The BER for BPSK is BER $= Q(\sqrt{2\gamma_B}) = 0.5\,\text{erfc}\left(\sqrt{\gamma_B}\right)$, where γ_B is in linear scale.

3. In a selection combining scheme, if the threshold for reliable communication, Γ_0, is 8 dB below the average SNR Γ, determine the following:

5.6 Multiuser MIMO Systems

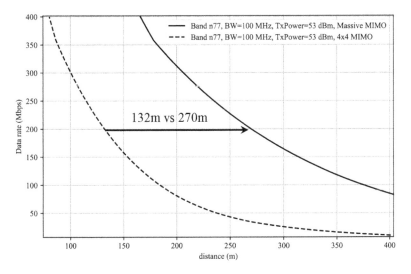

Fig. 5.22 Impact of massive MIMO on data rate over a distance

   a. Probability that the signal level at a single antenna falls below the threshold Γ_0.
   b. Outage probability for $M = 2, 3, 4$.
4. In selection combining, if the outage probability is set to 1%, determine:
   a. How does the ratio Γ_0/Γ change for $M = 2, 3,$ and 4.
   b. How do the changes calculated in a) affect the fading margin?
   c. How does the average output SNR change for different values of M specified in a).
5. Repeat the calculation in Problem 4 for the MRC combining scheme.
6. Derive the expression for the equal gain combining scheme under Rayleigh fading.
7. Show that $N \log_2 \left(1 + \frac{SNR}{N}\right) > \log_2 (1 + SNR)$ for $N \geq 2$ and $SNR > 0$. What is the significance of this inequality?
8. A 2×2 MIMO system is subject to the so-called keyhole effect: that is, the signal passes through a narrow opening, as illustrated in Fig. 5.23. Determine the overall 2×2 channel matrix H and determine its rank.
9. A selection combining receiver consists of $M = 4$ diversity combining branches, with the average per-branch SNR of $\Gamma = 10$ dB. Assuming Rayleigh fading, determine the following:

   a. Threshold SNR value Γ_0, if the probability that the received signal's SNR will fall below the Γ_0 threshold is 1%.

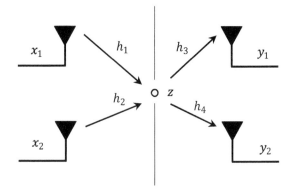

Fig. 5.23 Keyhole effect illustration

b. Channel capacity of the SISO system when the signal operates at the threshold E_b/N_0 of Γ_0 if the baseband channel bandwidth is $W = 10$ MHz and the bit rate is $R_b = 20$ Mbps.

10. Consider a receive diversity system with $N = 2$ antennas. We have seen that selection combining always selects the branch with the greatest SNR. Another solution is to use switch-or-stay combining (SSC). In this scheme, the receiver initially picks a branch at random and then sticks with its choice until the branch SNR falls below a threshold γ_T. Once the branch SNR falls below this threshold, the receiver switches to the other branch.

    Derive the outage probability, $P(\gamma < \gamma_0)$, and the PDF of the output SNR γ. You can assume that the fading to each element is Rayleigh and i.i.d. (*Hint:* Think in terms of two ranges, $\gamma < \gamma_T$ and $\gamma \geq \gamma_T$).

11. Consider a MIMO system with N_T transmitters and N_R receivers. Both the transmitter and receiver are assumed to know the $N_R \times N_T$ channel $\mathbf{H}$. The transmitter transmits only a single data stream, s. The transmitter uses weights $\mathbf{w}_t$ and the receiver weights $\mathbf{w}_r$ to maximize the output SNR. Show that these weight vectors are given by $\mathbf{w}_t = \sqrt{P_T}\mathbf{v}_1$ and $\mathbf{w}_r = \mathbf{u}_1$ where $\mathbf{u}_1$ and $\mathbf{v}_1$ are the left and right singular vectors of $\mathbf{H}$ corresponding to the largest singular value (σ_1).

    If the receiver sees an average noise power of σ^2 on each element, what is the SNR achieved on this data stream?

    Note that the transmitter must meet an energy constraint $||\mathbf{w}_t||^2 \leq P_T$.

12. The channel to an M-element receiver array is measured as $\mathbf{h}$. However, due to some power amplifier problems, the odd numbered elements get attenuated by 3dB (no phase change). What weights should the receiver use to maximize output SNR?

    You can assume M is even and the numbering is from zero to $M - 1$.

13. Consider a two-element receive diversity system with independent, but not identical, Rayleigh fading. The mean signal-to-noise ratio levels of the two channels are Γ_0 and Γ_1, respectively. The SNR of each element is distributed exponentially $\gamma_i \sim (1/\Gamma_i)e^{-\gamma_i/\Gamma_i}$. The system uses selection diversity. Derive

5.6 Multiuser MIMO Systems

the outage probability of such a two-element system, i.e., the probability that the output SNR is lower than a threshold γ_s. Do you expect such a system to achieve full diversity order? If yes, is there a penalty to pay for having one channel weaker than the other?

Develop an approach to repeating this question for the case with maximal ratio combining at the receiver. Do not worry about obtaining the final answer, but understand how you would approach this analysis.

14. In Q1, you obtained the covariance of $\mathbf{n}_{ZF}$, the noise vector after zero forcing (ZF). Let us use this covariance matrix to investigate the noise enhancement with ZF.

    The original received signal, before ZF, is given by

    $$\mathbf{x} = \mathbf{Hs} + \mathbf{n}.$$

    Here, $\mathbf{H}$, the $M \times K$ channel matrix, is known and the noise is white, i.e., $\mathbf{n} \sim \mathcal{CN}(0, \sigma^2 \mathbf{I})$. Since the symbols in the vector $\mathbf{s}$ are for K different users, we model the symbols as independent random variables. Specifically, $\mathbb{E}[\mathbf{ss}^H] = P_T \mathbf{I}$ where P_T is the average power of each symbol.

    For convenience, let us assume that $K = M$, i.e., $\mathbf{H}$ is a square matrix. We define the SNR as

    $$\gamma_{\text{orig}} = \frac{\mathbb{E}[\|\mathbf{Hs}\|^2]}{\mathbb{E}[\|\mathbf{n}\|^2]}.$$

    Show that

    $$\gamma_{\text{orig}} = \frac{P_T}{\sigma^2} \frac{\sum_{m=1}^{M} \lambda_m}{M},$$

    where λ_m, $m = 1, \ldots, M$ denotes the M eigenvalues of the matrix $\mathbf{H}^H \mathbf{H}$. This implies the SNR is proportional to the *arithmetic average* of these eigenvalues.
    Hint: $\|\mathbf{x}\|^2 = \text{tr}[\mathbf{xx}^H]$ where $\text{tr}[\cdot]$ denotes the trace of a matrix.

    Now consider the signal after ZF. We have:

    $$\mathbf{s}_{ZF} = \mathbf{s} + \mathbf{n}_{ZF}.$$

    We define the resulting SNR as

    $$\gamma_{ZF} = \frac{\mathbb{E}[\|\mathbf{s}\|^2]}{\mathbb{E}[\|\mathbf{n}_{ZF}\|^2]}.$$

    Using your covariance matrix from Q1, show that

    $$\gamma_{ZF} = \frac{P_T}{\sigma^2} \frac{1}{\frac{1}{M}\sum_{m=1}^{M} \frac{1}{\lambda_m}},$$

i.e., the post-ZF SNR is proportional to the *harmonic mean* of the eigenvalues. Since the harmonic mean of a set of positive numbers is always lower than its arithmetic mean, the post-ZF SNR is lower than the pre-ZF SNR.[9]

References

1. A.F. Molisch, *Wireless Combinations* (IEEE Press by Wiley, 2011)
2. S. Verdu, *Multiuser Detection* (Cambridge University Press, Cambridge, 1998)
3. J.C. Liberti, T.S. Rappaport, *Smart Antennas for Wireless Communications: IS-95 and Third Generation CDMA Applications* (Prentice-Hall, Englewood Cliffs, 1997)
4. S. Chen, S. Sun, G. Xu, X. Su, Y. Cai, Beam-space multiplexing: practice, theory, and trends, from 4G TD-LTE, 5G, to 6G and beyond. IEEE Wirel. Commun. **27**(2), 162–172 (2020)
5. S.M. Alamouti, A simple transmit diversity technique for wireless communications. IEEE J. Select Areas Commun. **16**(8), 1451–1458 (1998)
6. E. Dahlman, S. Parkvall, J. Sköld, *4G, LTE-Advanced Pro and The Road to 5G* (Academic Press, New York, 2016)
7. V. Tarokh, N. Seshadri, A.R. Calderbank, *Space-Time Codes for High Data Rate Wireless Communication: Performance Criterion and Code Construction*, vol. 44, No. 2 (1998), pp. 744–765
8. V. Tarokh, A. Naguib, N. Seshadri, A.R. Calderbank, Combined array processing and space-time coding. IEEE Trans. Inform. Theory **45**(4), 1121–1128 (1999)
9. S. Boyd, L. Vandenberghe, *Convex Optimization* (Cambridge University Press, Cambridge, 2004)
10. G.D. Golden, G.J. Foschini, R.A. Valenzuela, P.W. Wolniansky, Simplified processing for high spectral efficiency wireless communication employing multi-element arrays. J. Sel. Areas Commun. **17**, 1841–1852 (1999)
11. L. Zheng, D. Tse, Diversity and multiplexing: a fundamental tradeoff in multiple-antenna channels. IEEE Trans. Inform. Theory **49**(5), 1073–1096 (2003)
12. H. Bolcskei, D. Gesbert, C.B. Papadias, A. van der Veen, *Space-Time Wireless Systems* (Cambridge University Press, Cambridge, 2006)
13. S. Haykin, *Adaptive Filter Theory*, 4th edn. (Prentice Hall, Englewood Cliffs, 2002)
14. Q.H. Spencer, A.L. Swindlehurst, M. Haardt, Zero-forcing methods for downlink spatial multiplexing in multiuser MIMO channels. IEEE Trans. Signal Process. **52**(2), 461–471 (2004)
15. S. Shi, M. Schubert, H. Boche, Downlink MMSE transceiver optimization for multiuser MIMO systems: duality and sum-MSE minimization. IEEE Trans. Signal Process. **55**(11), 5436–5446 (2007)
16. A.A. Khan, R.S. Adve, W. Yu, Optimizing downlink resource allocation in multiuser MIMO networks via fractional programming and the Hungarian algorithm. IEEE Trans. Wirel. Commun. **19**(8), 5162–5175 (2020)
17. W. Yu, T. Kwon, C. Shin, Multicell coordination via joint scheduling, beamforming, and power spectrum adaptation. IEEE Trans. Wirel. Commun. **12**, 1–14 (2013)
18. S. Zimmermann, U. Killat, Resource marking and fair rate allocation, in *2002 IEEE International Conference on Communications. Conference Proceedings. ICC 2002 (Cat. No.02CH37333)*, vol. 2 (2002)
19. T.L. Marzetta, Massive MIMO: an introduction. Bell Labs Tech. J. **20**, 11–22 (2015)

---

[9] The penalty in SNR provides the ability to decode each user separately, a significant advantage.

Chapter 6
Introduction to Multicarrier Transmission

6.1 Orthogonal Frequency Division Multiplexing (Multicarrier Transmission)

We will now examine an effective way of transmitting high data rates over wireless channels using digital modulation schemes. As we have seen, the higher the symbol rate, the larger the bandwidth needed for transmission. If W is the bandpass channel bandwidth for the signal we intend to transmit, for high data rates, it is likely that $W \gg B_C$, where B_C is the channel coherence bandwidth which is in the range:

$$\frac{1}{50\tau_{RMS}} \leq B_C \leq \frac{1}{5\tau_{RMS}}, \tag{6.1}$$

where τ_{RMS} denotes the RMS delay spread (see Chap. 2). Since $W \gg B_c$, the symbol period, $T_s < \tau_{RMS}$, i.e., a sequential transmission of symbols, would suffer significant inter-symbol interference (ISI) which is the time domain interpretation—equivalently, frequency selective fading, which is the frequency domain interpretation. Decoding when suffering from ISI requires combinatorial processing since a single measurement is due to multiple transmitted symbols, i.e., ISI makes recovering the transmitted symbols exponentially complex.

The multicarrier transmission approach is to divide the channel into N parallel subchannels, each $\Delta f = W/N < B_c$ wide, and transmit N symbols simultaneously over the N subchannels. The value of N is selected to guarantee that each subchannel experiences flat fading, i.e., no ISI. Within each subchannel, we can

---

Supplementary Information The online version contains supplementary material available at (https://doi.org/10.1007/978-3-031-76455-4_6).

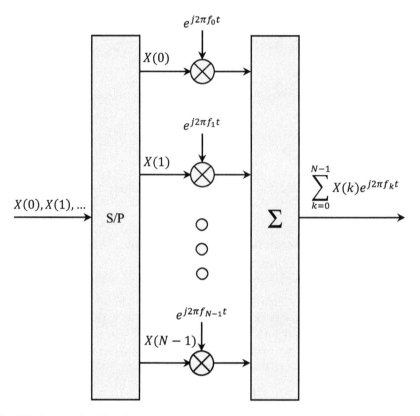

Fig. 6.1 A nominal multicarrier transmitter

transmit digitally modulated symbols with a duration of $T_U \sim 1/\Delta f > 1/B_C$, which also means that the symbol duration is much larger than the delay spread, or $T_U \gg \tau_{RMS}$. By doing so, we have significantly reduced the impact of ISI between the adjacent symbols (only the edges of each symbol are affected).

In Fig. 6.1, we see a nominal multicarrier transmission implementation, where a stream of data symbols

$$X = \begin{bmatrix} X(0), X(1), \ldots, X(N-1) \end{bmatrix} \quad (6.2)$$

is transmitted via N parallel subcarriers, nominally at frequencies $f_k = f_0 + k\Delta f$, where $k = 0, 1, \ldots, N-1$.
resulting in a transmit signal over one symbol duration given by

$$s(t) = \sum_{k=0}^{N-1} X(k) e^{j2\pi f_k t} = e^{j2\pi f_0 t} \sum_{k=0}^{N-1} X(k) e^{j2\pi k \Delta f t}, \ 0 \le t \le T_U \quad (6.3)$$

6.1 Orthogonal Frequency Division Multiplexing (Multicarrier Transmission)

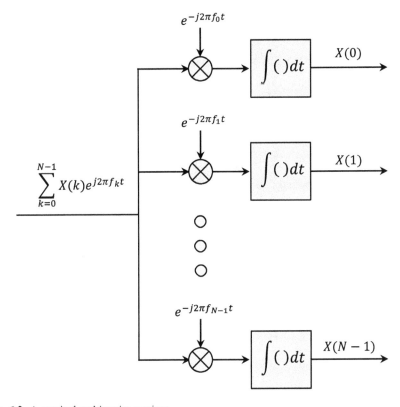

Fig. 6.2 A nominal multicarrier receiver

The serial-to-parallel (S/P) block takes the serial stream of symbols and parallelizes them before they are modulated by N local oscillators. If we did not place any constraints on the parameters, specifically Δf, the N data symbols would interfere with each other, requiring a complex decoding process.

The key to multicarrier transmission is to choose Δf such that the data symbols do not interact with each other, i.e., they are *orthogonal*. But, what is the subcarrier spacing Δf that ensures orthogonality between different subcarriers at the receiver? To determine this, let us look at the receiver structure, which, as shown below, estimates each symbol *individually*.

As seen in Fig. 6.2, the received signal (for now, we ignore fading) is demodulated using each of the N subcarrier frequencies, followed by integration. The integration is over a symbol period, i.e., T_U in the case where rectangular pulses are used. Orthogonality is ensured if the following condition is met:

$$\int_0^{T_U} e^{j2\pi f_k t} e^{-j2\pi f_\ell t} dt = 0, \quad k \neq \ell \tag{6.4}$$

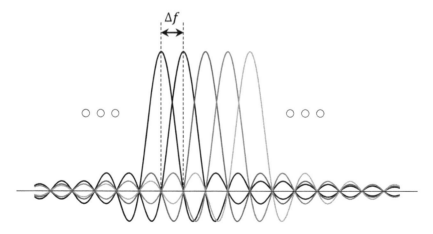

Fig. 6.3 Illustrating multicarrier transmission in the frequency domain

and we obtain:

$$\int_0^{T_U} e^{j2\pi(f_k-f_\ell)t} dt = 0 \quad (6.5)$$

$$\frac{1}{j2\pi(f_k-f_\ell)} e^{j2\pi(f_k-f_\ell)t}\Big|_0^{T_U} = e^{j2\pi(f_k-f_\ell)T_U} - 1 = 0$$

$$\Rightarrow 2\pi(f_k - f_\ell)T_U = 2\pi m$$

$$\Rightarrow f_k - f_\ell = \frac{m}{T_U}$$

i.e., for orthogonality, the frequency spacing must be an integer multiple of $1/T_U$ with the smallest spacing given by $\Delta f = 1/T_U$. The multicarrier signal, using this smallest spacing, is therefore

$$s(t) = \sum_{k=0}^{N-1} X(k) e^{j2\pi \frac{k}{T_U} t}, \quad 0 \le t \le T_U, \quad (6.6)$$

where, for now, we have eliminated the common term of $e^{j2\pi f_0 t}$. Since the time-domain signal uses rectangular pulses of duration T_U, in the frequency domain, as shown in Fig. 6.3, $S(f)$ is a sum of sinc pulses spaced Δf apart (for purposes of illustration, the plot uses $X(k) = 1, \forall k$).

As developed so far, the transmitter (receiver) requires N modulators (demodulators), a remarkably inefficient and expensive architecture. The key to making multicarrier transmissions practical is to consider what happens if we sample this signal with the sampling interval $T_{samp} = T_U/N$. The N samples over one symbol interval T_U can be expressed as

6.1 Orthogonal Frequency Division Multiplexing (Multicarrier Transmission)

Fig. 6.4 IFFT block used in Example 6.1

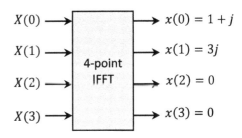

$$s(nT_{samp}) \equiv s(n) = \sum_{k=0}^{N-1} X(k)e^{j2\pi \frac{kn}{N}}, \ 0 \le n \le N-1. \quad (6.7)$$

This expression in (6.7) for the sampled signal $s(n)$ bears striking resemblance to the Inverse Discrete Fourier Transform (IDFT) for a length-N sequence $X(k)$ [1]:

$$\text{IDFT}[X(k)] = x(n) = \frac{1}{\sqrt{N}} \sum_{k=0}^{N-1} X(k)e^{j2\pi \frac{kn}{N}}, \quad (6.8)$$

except for the scaling factor $1/\sqrt{N}$ which can easily be compensated for. Importantly, a DFT/IDFT can be efficiently implemented using a Fast Fourier Transform (FFT).

The implication is that a multicarrier signal with minimum subcarrier spacing that still ensures orthogonality can be implemented via an FFT process/chipset, hence ensuring a cheap and efficient design of the transmitter. Given the similar structure, the receiver also requires an FFT process/chipset (i.e., complex symbols $X(k)$ can be obtained from $x(n)$ via a DFT, that is, by inverting the IDFT operation):

$$\text{DFT}[x(n)] = X(k) = \frac{1}{\sqrt{N}} \sum_{n=0}^{N-1} x(n)e^{-j2\pi \frac{kn}{N}}, \ 0 \le k \le N-1 \quad (6.9)$$

Example 6.1 A simple multicarrier system uses a 4-point IDFT (IFFT) module to encode 4 complex input symbols at a time. If the outputs of the IDFT are $x(0) = 1+j, x(1) = 3j, x(2) = x(3) = 0$, determine the originally encoded sequence.

Solution Input symbol values $X(k)$ can be determined using DFT (Fig. 6.4):

$$X(k) = \frac{1}{\sqrt{4}} \sum_{n=0}^{3} x(n) e^{-j2\pi kn/4} \tag{6.10}$$

$$X(0) = \frac{1}{2}[x(0) + x(1) + x(2) + x(3)] = \frac{1}{2}[1 + j + 3j + 0 + 0] = 1/2 + 2j \tag{6.11}$$

$$X(1) = \frac{1}{2}\left[x(0) + x(1)e^{-j2\pi \times 1/4}\right] = \frac{1}{2}[1 + j + 3j(-j)] = 2 + j/2 \tag{6.12}$$

$$X(2) = \frac{1}{2}\left[x(0) + x(1)e^{-j2\pi \times 2/4}\right] = \frac{1}{2}[1 + j + 3j(-1)] = 1/2 - j \tag{6.13}$$

$$X(3) = \frac{1}{2}\left[x(0) + x(1)e^{-j2\pi \times 3/4}\right] = \frac{1}{2}[1 + j + 3j(j)] = -1 + j/2 \tag{6.14}$$

■

Each of the substreams carries $\log_2(M)$ bits per digital modulated symbol, where M is the digital modulation order[1] (e.g., $M = 64$ for 64QAM), and the data rate of each substream is $R_k = \log_2(M)/T_U$. Given that there are N data substreams, substituting $T_U = 1/\Delta f = N/W$, the total data rate of the multicarrier signal is

$$R_{OFDM} = N \times R_k = N\frac{\log_2(M)}{T_U} = \log_2(M)W \tag{6.15}$$

which is the same as if we used a single-carrier transmission with the much shorter symbol duration $T_{SC} = 1/W = T_U/N$. Therefore, in theory, we achieve the same data rate using multicarrier transmission as that in the case of a single carrier. However, crucially, in using multicarrier transmission, the benefit is that we eliminate ISI.

> In our discussion so far, the entire bandwidth W is dedicated to a single user. In this case, we are *multiplexing* the user's signals on the N subcarriers, and the scheme is known as orthogonal frequency division multiplexing or OFDM. However, the data symbols can belong to *different* users providing a *multiple access* scheme; in this case, the transmission scheme is known as orthogonal frequency division multiple access or OFDMA. OFDMA is the basis of 4G and 5G wireless networks.

As stated above, the OFDM signal in the frequency domain comprises a sum of sinc functions that are spaced at the minimum spacing required for orthogonality. As

---

[1] It is not necessary that the same modulation scheme be used in all subcarriers. However, this is the most obvious approach, and we will focus on this case.

6.1 Orthogonal Frequency Division Multiplexing (Multicarrier Transmission)

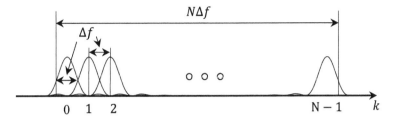

Fig. 6.5 Multicarrier signal, with sinc²() representation of overlapping subcarriers that are spaced Δf apart from each other, k is the subcarrier index

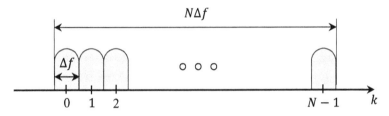

Fig. 6.6 Multicarrier signal representation equivalent to the one in Fig. 6.5 with nonoverlapping subcarriers, each Δf wide

a result, the bandwidth of an OFDM signal is approximately the product of spacings between the sinc pulses and the number of subcarriers:

$$BW_{OFDM} \approx N\frac{1}{T_U} = N\Delta f. \tag{6.16}$$

While the spectral content of individual subcarriers overlaps, as shown in Fig. 6.5: the interpretation for bandwidth calculation (and for the guardband calculation, as will be seen in Chap. 8) is that individual subcarriers are Δf wide and are not overlapping, as illustrated in Fig. 6.6.

Example 6.2 Consider a multicarrier system with a total bandwidth of $W = 20$ MHz. Suppose the system operates in a suburban area with channel rms delay spread $\tau_{RMS} = 5\,\mu s$. How many subcarriers can be supported if we need to ensure approximately flat fading for each subcarrier?

Solution The coherence bandwidth is in the range:

$$\frac{1}{50\tau_{RMS}} \leq B_C \leq \frac{1}{5\tau_{RMS}} \tag{6.17}$$

After substituting $\tau_{RMS} = 5\,\mu s$, we get:

$$4 \text{ kHz} \leq B_C \leq 40 \text{ kHz}. \tag{6.18}$$

If we pick $B_C = 40$ kHz, the subcarrier spacing Δf needs to satisfy $\Delta f \leq B_C$, so that the number of subcarriers is

$$N_{\text{SUB}} = \frac{W}{\Delta f} \geq \frac{W}{B_C} = \frac{1 \times 10^6}{40 \times 10^3} = 500. \tag{6.19}$$

For practical reasons, the FFT implementation is easier if the number of points is a power of 2, so we can chose the closest power of 2 values that satisfies the flat fading requirement, i.e., $N_{\text{SUB}} = N_{\text{FFT}} = 512 > 500$. In that case, the total bandwidth W is divided into 512 subbands with subcarrier spacing $\Delta f = W/N_{\text{FFT}} = 39.0625$ kHz.

The corresponding per subcarrier symbol duration is $T_U = 1/\Delta f = 25.6\,\mu s$. If we compare it with the $\tau_{\text{RMS}} = 5\,\mu s$, we can see that ISI affects about $5/25.6 = 19.53\%$ of each symbol. Alternatively, we could have selected $N_{\text{FFT}} = 1024$, resulting in $\Delta f = 19.53125$ kHz, which means the delay spread affects less than 10% of the symbol. ∎

It is possible to refine the solution by selecting a more convenient subcarrier spacing Δf. For example, if we set $\Delta f = 15$ kHz, it will still satisfy the flat fading condition $\Delta f \leq B_C$, but we need to decouple the number of FFT points (power of 2) from the number of data subcarriers if we want to ensure that the signal spectrum is contained within the given channel bandwidth. For example, if we set $N_{\text{SUB}} = 1200$, where N_{SUB} is the number of parallel subcarriers, while the number of FFT points is the next power of 2, which in this case is $N_{\text{FFT}} = 2048$, the spectrum used for transmission is $N_{\text{SUB}} \times \Delta f = 18$ MHz, which is less than the channel bandwidth $W = 20$ MHz. 4G LTE uses 2048 subcarriers and a 15 kHz subcarrier bandwidth.

> One important consideration in selecting the number of FFT points and symbol duration is the impact of motion. Any movement in the channel, but especially when applied to the user causes, a time-varying channel (time domain interpretation), equivalently, a Doppler shift (frequency domain interpretation). Coupled with multipath, the Doppler shift causes frequency spreading and, consequently, subcarrier orthogonality is lost. This results in inter-carrier interference (ICI; see Sect. 6.3.3). Ideally, the Doppler shift should be less than 1% of the subcarrier spacing to guarantee negligible ICI [2] for users with high SNR and high modulation order. In practice, around 2% is considered acceptable (e.g., $f_D = 300$ Hz for $\Delta f = 15$ kHz in LTE).
>
> Note that $f_D = v/\lambda$, where v is the radial velocity and λ is the operating wavelength. Hence, for an operating frequency of 3 GHz, we have $\lambda = 0.1$m, i.e., $f_D = 300$ Hz $\equiv v = 30$ m/s or approximately 110 kmph (highway

(continued)

speeds). Essentially, for most applications, meeting the target of 2% Doppler shift is quite feasible.

6.2 Cyclic Prefix and Channel Impact

As illustrated in Example 6.2, an OFDM symbol duration T_U is chosen such that $T_U \gg \tau_{RMS}$, which means that the ISI is limited to a fraction of the symbol. We can do more, and almost completely eliminate the impact of ISI, by introducing a guard time at the beginning of each symbol such that the duration of this guard time is $T_g \geq \tau_{RMS}$. By doing so, we introduce a small overhead as the added guard time carries no information. This is considered a good trade-off as the added guard time eliminates ISI. Depending on the content of the guard time, we can have different implementations of OFDM systems. The widely used approach is to copy the last μ time domain samples from the end of an OFDM symbol at the beginning; called a *cyclic prefix*, this approach is shown in Fig 6.7.

A more detailed explanation of why the cyclic prefix helps with the ISI closely follows the methodology outlined in [3]. An OFDM symbol consists of N_{FFT} samples that span the duration of T_U, and the new, extended symbol, consists of $N_{FFT} + \mu$ samples, where $\mu < N_{FFT}$ is the number of cyclic prefix samples. The total duration of the extended OFDM symbol is

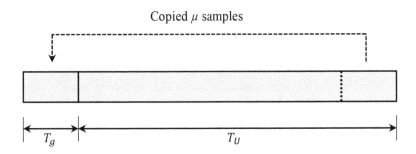

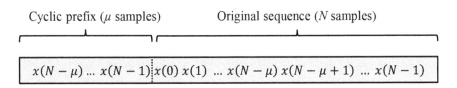

Fig. 6.7 Illustrating the cyclic prefix

$$T_S = T_U + T_g = \frac{N_{FFT} + \mu}{N_{FFT}} T_U. \tag{6.20}$$

Analytically, if

$$\boldsymbol{x}(n) = [x(0), x(1), \ldots, x(N-1)], \tag{6.21}$$

the newly created symbol sequence with the prepended cyclic prefix is

$$\tilde{x}(n) = \left[\underbrace{x(N-\mu)}_{=x(-\mu)}, \underbrace{x(N-\mu+1)}_{=x(-(\mu-1))}, \ldots, \underbrace{x(N-1)}_{=x(-1)}, x(0), x(1), x(2), \right. $$
$$\left. \ldots, x(N-1) \right]. \tag{6.22}$$

> The addition of a prefix comes at the cost of losing a fraction of the power on added samples. In addition to the power loss, cyclic-prefix insertion also introduces a spectral efficiency penalty as the OFDM symbol rate is reduced without a corresponding reduction in the overall signal bandwidth. Why are we willing to pay this penalty?

Our discussion so far has ignored the impact of the fading channel. However, we formulated the multicarrier approach specifically to deal with the channel delay spread. Let us now consider a discrete channel with the impulse response $h(n)$ with an rms delay spread of τ_{RMS} equivalent to $\nu(<\mu)$ sampling intervals (note that $T_{samp} = T_U/N$ is the sampling interval). The signal at the output of the channel, $y(n)$, is the *linear* convolution of the input signal and the channel impulse response. The convolution output can be written as

$$\tilde{x}(n) \longrightarrow \boxed{h(n)} \longrightarrow y(n) = \tilde{x}(n) * h(n)$$

$$y(n) = h(n) * \tilde{x}(n) = \sum_{k=0}^{\nu} h(k)\tilde{x}(n-k) \tag{6.23}$$

$$y(0) = h(0)x(0) + h(1)x(-1) + \ldots + h(\nu)x(-\nu) \tag{6.24}$$

$$= h(0)x(0) + h(1)x(N-1) + \ldots + h(\nu)x(N-\nu) \tag{6.25}$$

6.2 Cyclic Prefix and Channel Impact

$$y(1) = h(0)x(1) + h(1)x(0) + \ldots + h(\nu)x(N - \nu + 1) \quad (6.26)$$

$$\vdots \quad (6.27)$$

$$y(N-1) = h(0)x(N-1) + h(1)x(N-2) + \ldots + h(\nu)x(N-\nu-1) \quad (6.28)$$

Over the range from $0 \leq n \leq N - 1$, this is equivalent to

$$y(n) = \sum_{k=0}^{\nu} h(k)x((n-k) \mathrm{mod}_N) = h(n) \otimes x(n) \quad (6.29)$$

where $()\mathrm{mod}_N$ is the modulo N operator (i.e., $x(n)$ has been periodically expanded, with a period N) which is the expression for *circular* convolution (denoted by $\otimes$). Thus, by adding the cyclic prefix, we have converted the linear convolution of the channel into the circular convolution. The total duration of the linear channel output $y(n)$ is $N_{\text{FFT}} + \mu + \nu + 1 - 1 = N_{\text{FFT}} + \mu + \nu$ samples.

At the receiver, the first μ symbols are discarded, thus eliminating the ISI from the previous symbol, and the last ν samples are absorbed into the ISI of the next symbol, leaving N_{FFT} samples for DFT operation. The most important benefit of (effectively) achieving a circular convolution is as follows: in the discrete domain, if $\mathrm{DFT}[x(n)] = X(k)$ and $\mathrm{DFT}[h(n)] = H(k)$, then,

$$\mathrm{DFT}[y(n)] = Y(k) = \mathrm{DFT}[x(n) \otimes h(n)] = X(k)H(k). \quad (6.30)$$

This Is a Crucial Result It shows that, upon using the cyclic prefix, the received signal after the frequency selective channel acts just like frequency-flat fading on each *individual* subcarrier! Put another way, the use of the prefix has created N parallel frequency-flat channels. If the channel $H(k)$ is known for $k = 0, \ldots, N-1$, we can recover the symbols using $X(k) = Y(k)/H(k)$.

To illustrate this result, consider the following example: let $x(n)$, expressed as

$$x(n) = \frac{1}{\sqrt{N}} \sum_{k=0}^{N-1} X(k) e^{j\frac{2\pi k n}{N}}$$

be the transmitted OFDM symbol, and the channel provides two multipath components with delays τ_1 and τ_2, as illustrated in Fig. 6.8.

Ignoring noise to focus on the OFDM concepts, the received signal $y(n)$ is

$$y(n) = y_1(n) + y_2(n), \quad (6.31)$$

where we assume that the processing window is relative to τ_1 delay and the number of samples corresponding to $\tau_2 - \tau_1$ is m

$$y_1(n) = x(n) \quad (6.32)$$

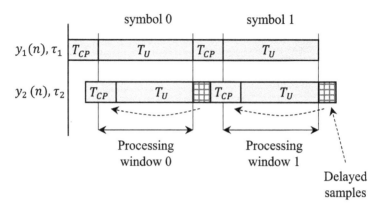

Fig. 6.8 Illustration of two delayed OFDM signals at the receiver. The delayed samples from $y_2(n)$ appear in the cyclic prefix

$$y_2(n) = x\left((n-m) \bmod N\right) \tag{6.33}$$

At the receiver, we operate over a window of N-samples duration, and we perform the DFT to demodulate the original sequence $X(k)$:

$$\begin{aligned}
Y(k) &= \frac{1}{\sqrt{N}} \sum_{k=0}^{N-1} y(n) e^{-j\frac{2\pi kn}{N}} \\
&= \frac{1}{\sqrt{N}} \sum_{k=0}^{N-1} y_1(n) e^{-j\frac{2\pi kn}{N}} + \frac{1}{\sqrt{N}} \sum_{k=0}^{N-1} y_2(n) e^{-j\frac{2\pi kn}{N}} \\
&= \frac{1}{\sqrt{N}} \sum_{k=0}^{N-1} x(n) e^{-j\frac{2\pi kn}{N}} + \frac{1}{\sqrt{N}} \sum_{k=0}^{N-1} x\left((n-m) \bmod N\right) e^{-j\frac{2\pi kn}{N}} \\
&= \frac{1}{\sqrt{N}} \sum_{k=0}^{N-1} x(n) e^{-j\frac{2\pi kn}{N}} + \frac{1}{\sqrt{N}} e^{-j\frac{2\pi km}{N}} \sum_{k=0}^{N-1} x\left((n-m) \bmod N\right) e^{-j\frac{2\pi k(n-m)}{N}} \\
&= X(k) + e^{-j\frac{2\pi km}{N}} X(k) \\
&= X(k)\left(1 + e^{-j\frac{2\pi km}{N}}\right) = X(k) H(k) \tag{6.34}
\end{aligned}$$

To recover the signal, we simply divide $Y(k)$ with the channel at subcarrier k:

$$X(k) = \frac{Y(k)}{H(k)}. \tag{6.35}$$

The discrete OFDM transmitter model, including the cyclic prefix addition and RF modulation, is illustrated in Fig. 6.9, and the receiver model is as shown

6.2 Cyclic Prefix and Channel Impact

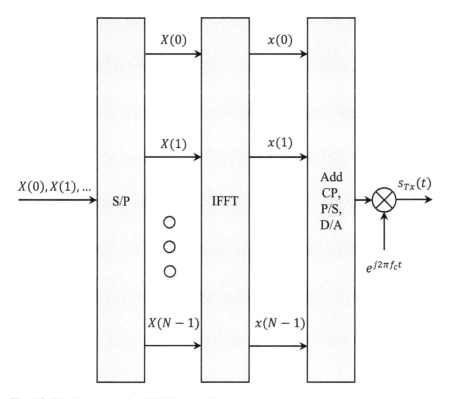

Fig. 6.9 Block diagram of an OFDM transmitter

in Fig. 6.10. Besides the already introduced functional blocks like IFFT and CP addition, the transmitter block diagram incorporates a serial to parallel (S/P) block, whereby the input sequence is parallelized to generate inputs into the IFFT block, and the inverse serial to parallel (S/P) block that takes a parallel block out the IFFT outputs and puts them back in a series of samples. The digital-to-analog (D/A) conversion is performed before the signal upconversion. Some details are omitted, like the QAM modulation that generates symbols $X(k)$ at the transmitter, and receive filtering, digital-to-analog (D/A) conversion, and QAM demodulation at the receiver.

The carrier frequency f_c corresponds to the center of one of the blocks within cellular spectrum introduced in Chap. 4. It can be related to f_0 in (6.3) as

$$f_c = f_0 + N\Delta f/2 \qquad (6.36)$$

for practical systems, where f_0 is the center RF frequency of the first subcarrier.

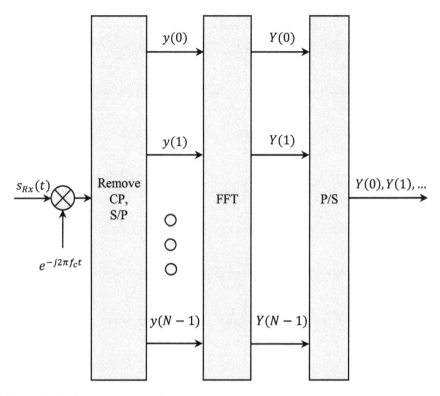

Fig. 6.10 Block diagram of an OFDM receiver

Example 6.3 (Impact of CP on Data Rate) Consider an OFDM system with total passband bandwidth $W = 20$ MHz. The channel has a maximum delay spread of $\tau_{RMS} = 5\,\mu s$, and assume an OFDM system with 16-QAM modulation applied to each subcarrier. To keep the overhead small, the OFDM system uses $N_{FFT} = 2048$, $N_{SUB} = 1200$ subcarriers, and the subcarrier spacing of $\Delta f = 15$ kHz. The length of the cyclic prefix is set to $\mu = 160$ samples to ensure no ISI between OFDM symbols. For these parameters, find the total transmission time associated with each OFDM symbol, the overhead of the cyclic prefix, and the data rate of the system.

Solution Since $\Delta f = 15$ kHz, the useful symbol duration is $T_U = 1/\Delta f = 66.67\,\mu s$. Evidently, $T_U \gg \tau_{RMS}$, which means that ISI is small. The OFDM samples are spaced $T_{samp} = T_U/N_{FFT} = 32.55$ ns apart, and additional $\mu = 160$ samples are added to form a guard time $T_g = T_{CP} = \mu T_{samp} = 5.21\,\mu s$ long. Therefore, the total OFDM symbol duration is $T_{OFDM} = T_g + T_U = 5.21 + 66.67 = 71.87\,\mu s$. The cyclic prefix is sufficiently long to completely eliminate the ISI, and the overhead due to CP is defined as

$$\mathrm{CP}_{OH} = \frac{T_g}{T_{OFDM}} = \frac{160}{160 + 2048} = 7.24\%. \tag{6.37}$$

Each subcarrier is 16-QAM modulated, and therefore carries $\log_2(16) = 4$ bits/symbol. The total OFDM signal data rate is:

$$R_{OFDM} = \underbrace{1200}_{N_{SUB}} \times \underbrace{4}_{\text{bits/symbol}} \times \underbrace{\frac{1}{71.87\,\mu s}}_{\text{symbol rate}} = 66.78\,\text{Mbps}. \tag{6.38}$$

Without a cyclic prefix, the total OFDM signal data rate would have been

$$R_{OFDM} = \underbrace{1200}_{N_{SUB}} \times \underbrace{4}_{\text{bits/symbol}} \times \underbrace{\frac{1}{66.67\,\mu s}}_{\text{symbol rate}} = 72\,\text{Mbps}. \tag{6.39}$$

Note that if we used single-carrier digital transmission, the data rate would be $18 \times 10^6 \times 4 = 72$ Mbps for the same bandwidth of 18 MHz ($N_{SUB} \times \Delta f = 1200 \times 15$ kHz), assuming that the transmission rate is the inverse of the transmission bandwidth.

The 7.24% loss due to CP overhead can be observed from the ratio $66.78/72 = 92.754\% = 100\% - 7.246\%$. ∎

6.3 OFDM Challenges

So far, we have seen many advantages of OFDM communication systems, including resistance to frequency selective fading and ease of implementation via FFT chipsets. An additional advantage, bandwidth scalability, will be discussed in Chap. 8. However, OFDM has its challenges as well, and we now some key associated issues.

6.3.1 Peak-to-Average Power Ratio

As the name indicates, the peak-to-average power ratio (PAPR) is defined as the ratio between the peak power and the average signal power. This is an important metric for transmitters because we would like the power amplifiers within the transmit chain to act linearly (we had briefly discussed amplifier linearity in Chap. 3). If the PAPR is too high, the amplifier can saturate and cause a distortion in the transmitted signal. Figure 6.11 illustrates the nonlinear behavior of practical power amplifiers.

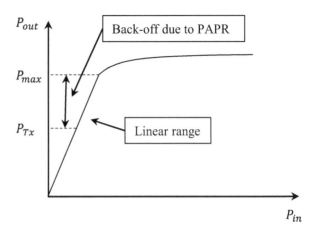

Fig. 6.11 Power amplifier amplification curve, with linear and saturation ranges

Typical values in conventional digital modulation schemes are in the 2–5 dB range, depending on the pulse shape, and the smallest values correspond to the rectangular pulse shape. In OFDM, we normally use a rectangular pulse shape, but as will be shown next, the PAPR values are significant.

Consider the OFDM signal given by

$$x_{OFDM}(n) = \frac{1}{\sqrt{N}} \sum_{k=0}^{N-1} X(k) e^{j\frac{2\pi kn}{N}} \qquad (6.40)$$

Let us assume that symbols $X(k)$ are normalized such that their average power is set to 1. We then have

$$\begin{aligned} P_{AV} &= \mathbb{E}\left[|x_{OFDM}(n)|^2\right] \\ &= \mathbb{E}\left[\left(\frac{1}{\sqrt{N}} \sum_{k=0}^{N-1} X(k) e^{j\frac{2\pi kn}{N}}\right)\left(\frac{1}{\sqrt{N}} \sum_{k=0}^{N-1} X^*(k) e^{-j\frac{2\pi kn}{N}}\right)\right] \\ &= \frac{1}{N} \sum_{k=0}^{N-1} \mathbb{E}[|X(k)|^2] = \mathbb{E}[|X(k)|^2] = 1, \end{aligned} \qquad (6.41)$$

where we make the common assumption that individual symbols are independent and, hence, the cross-terms go to zero.

The peak power for the OFDM signal occurs when all the symbols $X(k)$ have both the highest amplitude and their phases align.[2] If the constellation points for $X(k)$ symbols has an intrinsic PAPR of α, the peak power is given by

---

[2] For example, for a 16-QAM constellation covering the set $\{\pm\sqrt{3}A \pm j\sqrt{3}A, \pm A \pm jA, \pm A \pm \sqrt{3}A, \pm\sqrt{3}A \pm A\}$ with A is chosen to ensure unit average power, the largest value of $x_{OFDM}(n)$

6.3 OFDM Challenges

$$\max\left(|x_{OFDM}(n)|^2\right) = \frac{1}{N}\left[\sum_{k=0}^{N-1} \max(|X(k)|)\right]^2 = N\alpha P_{AV}. \quad (6.42)$$

This expression indicates that OFDM suffers from high PAPR values, linearly proportional to N since

$$PAPR = \frac{P_{peak}}{P_{AV}} = N\alpha. \quad (6.43)$$

For example, if the number of subcarriers is $N = 1024$, the peak-to-average power ratio is $PAPR_{|dB} = 30.1 + 10\log_{10}(\alpha)$ dB. From Fig. 6.11 we see that, to avoid distortion, the peak signal power should be within the linear range of the PA, i.e., we need $P_{peak} < P_{max}$. Consequently, the average transmit power P_{Tx} is backed off to ensure this condition is met.

In practice, it is extremely unlikely that perfect amplitude alignment of OFDM signal components would happen, and 30 dB PAPR values are unrealistic. Peak values of different OFDM symbols vary and are difficult to determine analytically, so we treat the PAPR statistically. Instead of an absolute peak amplitude/power value, we look at the distribution of peak values. If we set P_t as an instantaneous target power that corresponds to statistical peak-to-average power ratio (we assumed the average power to be 1), we can calculate the probability that at least one instantaneous sample power per OFDM symbol is larger than P_t as [1, 3]

$$\text{Prob}\,(PAPR > P_t) = 1 - \left(1 - e^{-P_t}\right)^N \quad (6.44)$$

This type of probability is known as a complementary cumulative distribution function (CCDF). The PAPR value is considered to correspond to the CCDF point where the instantaneous power P_t is exceeded with a probability of 0.0001 (or even 0.001). That means that only one in a ten-thousand samples will exceed the chosen P_t value and cause distortion; such a low probability is considered tolerable.

Figure 6.12 plots both the theoretical and simulated PAPR CCDF for the cause of QPSK symbols and $N = 1024$. The two curves are in agreement, as one would expect. The plot indicates that the PAPR exceeds 10 dB approximately one-tenth of the time, and also exceeds 12 dB in more than in one in ten-thousand OFDM symbols.

Practical systems employ various PAPR reduction techniques (e.g., surveyed in [4]) to bring PAPR values below 10 dB even the instantaneous power P_t is exceeded with a probability of 0.0001.

---

occurs when all N symbols are all equal to each other and one of $\pm\sqrt{3}A \pm j\sqrt{3}A$. This is clearly a rare occurrence, but defines the worst-case (peak power) scenario.

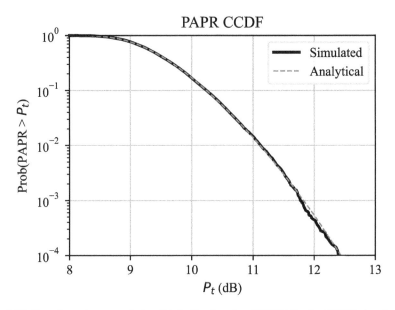

Fig. 6.12 Complementary cumulative distribution function (CCDF) for the OFDM signal peak-to-average distribution

6.3.2 Phase Noise

Wireless communication receivers are coherent in that they track the phase of the received signal; however, there is always a slight error in the phase tracking. This (random) phase error is also called phase noise. Let $y(n) = x_{OFDM}(n)e^{j\phi(n)}$ be a received OFDM symbol, where $\phi(n)$ is a random process representing the phase noise [5] and $x(n)$ is an OFDM symbol given by

$$x_{OFDM}(n) = \frac{1}{\sqrt{N}} \sum_{k=0}^{N-1} X(k) e^{j\frac{2\pi kn}{N}}. \tag{6.45}$$

Typically, the amplitude of the phase noise is small compared to subcarrier spacing, so we have $|\phi(n)| \ll 1$, and the following linear approximation applies:

$$e^{j\phi(n)} \approx 1 + j\phi(n). \tag{6.46}$$

Therefore, the received signal $y(n)$ can be written as

$$y(n) \approx x_{OFDM}(n)[1 + j\phi(n)], \tag{6.47}$$

and the output of the FFT module at the receiver is

6.3 OFDM Challenges

$$Y(k) = \frac{1}{\sqrt{N}} \sum_{n=0}^{N-1} y(n) e^{-j\frac{2\pi kn}{N}} \quad (6.48)$$

$$= \frac{1}{\sqrt{N}} \sum_{n=0}^{N-1} x_{OFDM}(n) e^{-j\frac{2\pi kn}{N}} + \frac{1}{\sqrt{N}} \sum_{n=0}^{N-1} j\phi(n)$$

$$\times \left[\frac{1}{\sqrt{N}} \sum_{l=0}^{N-1} X(l) e^{j\frac{2\pi ln}{N}} \right] e^{-j\frac{2\pi kn}{N}}$$

$$= X(k) + j\frac{X(k)}{N} \underbrace{\sum_{n=0}^{N-1} \phi(n)}_{l=k} + \frac{j}{N} \underbrace{\sum_{n=0}^{N-1} \sum_{\substack{l=0 \\ l \neq k}}^{N-1} \phi(n) X(l) e^{j\frac{2\pi (l-k)n}{N}}}_{l \neq k}$$

$$= X(k)(1 + \Phi) + N^{ICI}(k)$$

where $1 + \Phi = 1 + j/N \sum_{n=0}^{N-1} \phi(n)$ is the common phase error (CPE) and $N^{ICI}(k)$ is the inter-carrier interference (ICI) that is dependent on subcarrier k. While the CPE can be compensated for by using pilot signals, the ICI acts as an additive noise that cannot be compensated for. The impact of phase noise decreases as the subcarrier spacing Δf increases because $\phi(n)$ becomes relatively smaller. The phase noise is much more pronounced at higher frequencies [5]; this is in part[3] the reason why, in 5G networks, the subcarrier spacing at higher frequencies is larger than at the lower frequencies (e.g., 120 kHz and 240 kHz in FR2 range vs 15 kHz and 30 kHz in FR1 range), as we will discuss in Chap. 8.

6.3.3 Carrier Frequency Offset

As with phase noise, there is always an error between the carrier frequency used at the receiver (for downconversion) and that used at the transmitter (for upconversion). The mismatch may be due to a Doppler frequency shift or just because the two systems use different frequency sources. Frequency tracking helps reduce the error, but it cannot be eliminated.

Let the continuous time received signal subcarrier component carrier be offset by some frequency error f_D away from the transmitted frequency f_k (e.g., due to Doppler frequency shift). Focusing on a single baseband subcarrier frequency f_k in the OFDM symbol, we have:

---

[3] The other reason is that with the limited number of subcarriers, wider channel bandwidths in FR2 range can only be supported if Δf is sufficiently large.

$$s_k(t) = e^{j2\pi(f_k+f_D)t}$$
$$= e^{j2\pi \frac{k}{T_U}t} e^{j2\pi f_D t}$$
$$= e^{j2\pi \frac{k}{T_U}t} e^{j2\pi \frac{\delta}{T_U}t} \tag{6.49}$$

where $\delta = f_D T_U$ is the normalized carrier frequency offset (CFO). The received OFDM signal is subjected to the same CFO, so the sampled version of the received OFDM signal, after downconversion to baseband, is

$$y(n) = x_{OFDM}(n) e^{j2\pi \frac{n\delta}{N}} \tag{6.50}$$

At the output of the FFT module, the demodulated signal is

$$Y(k) = \frac{1}{\sqrt{N}} \sum_{n=0}^{N-1} y(n) e^{-j2\pi \frac{kn}{N}}$$

$$= \frac{1}{\sqrt{N}} \sum_{n=0}^{N-1} x_{OFDM}(n) e^{j2\pi \frac{n\delta}{N}} e^{-j2\pi \frac{kn}{N}}$$

$$= \frac{1}{N} \sum_{n=0}^{N-1} \left[\sum_{l=0}^{N-1} X(l) e^{j2\pi \frac{ln}{N}} \right] e^{j2\pi \frac{n\delta}{N}} e^{-j2\pi \frac{kn}{N}}$$

$$= \underbrace{\frac{1}{N} \sum_{n=0}^{N-1} X(k) e^{j2\pi \frac{n\delta}{N}}}_{k=l} + \underbrace{\frac{1}{N} \sum_{n=0}^{N-1} \sum_{\substack{l=0 \\ l \neq k}}^{N-1} X(l) e^{j2\pi \frac{n(\delta+l-k)}{N}}}_{k \neq l} \tag{6.51}$$

As in the case of phase noise, we have the complex coefficient that scales and rotates the constellation point $X(k)$, and we have the inter-carrier interference term:

$$Y(k) = X(k) \frac{1}{N} \sum_{n=0}^{N-1} e^{j\frac{2\pi \delta n}{N}} + ICI(k) \tag{6.52}$$

$$= X(k) \frac{1}{N} \frac{1 - e^{j\frac{2\pi \delta N}{N}}}{1 - e^{j\frac{2\pi \delta}{N}}} + ICI(k)$$

using the expression for the sum of a geometric progression. The expression in (6.52) can be simplified to

$$Y(k) = X(k) \frac{\text{sinc}(\delta)}{\text{sinc}\left(\frac{\delta}{N}\right)} e^{j\pi \frac{N-1}{N} \delta} + ICI(k) \tag{6.53}$$

6.3 OFDM Challenges

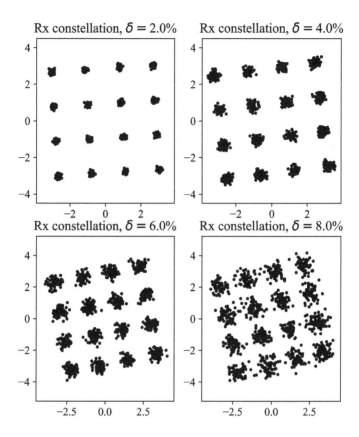

Fig. 6.13 Impact of carrier frequency offset (CFO) on received constellation points for different δ values. Common phase error rotates all the constellation points equally, whereas the inter-carrier interference causes noise-like behavior

where $e^{j\pi \frac{N-1}{N}\delta}$ is the common phase error and $\alpha = \frac{\text{sinc}(\delta)}{\text{sinc}\left(\frac{\delta}{N}\right)}$ is the attenuation factor. The signal to interference (SIR) value per subcarrier k is defined as [5]

$$SIR_k = \frac{\alpha^2 \mathbb{E}\left[|X(k)|^2\right]}{\mathbb{E}\left[|ICI(k)|^2\right]}. \tag{6.54}$$

Some examples of the impact of CFO on the received signal constellation are shown in Fig. 6.13.

6.4 Single-Carrier FDMA

Prior to the development of 4G cellular networks, the DL and UL modulation schemes were the same for any given technology. That changed in 4G, where the DL modulation scheme is cyclic prefix-based OFDMA (CP-OFDMA), whereas the UL is based on singe-carrier frequency division multiple access (SC-FDMA). In 5G, it is possible to have either SC-FDMA or CP-OFDMA for the UL, while the DL is CP-OFDMA based (Fig. 6.14). The reason for this difference is the high PAPR value of the OFDMA modulation scheme, which would limit the maximal transmit power in the UL due to a large power back-off.

SC-FDMA is similar to CP-OFDMA in a sense that there are parallel subcarrier transmissions (via the OFDMA Tx block) and resistance to multipath is retained, with the main difference being an additional DFT spreading module in front of the OFDMA transmitter, and an additional IDFT module after the OFDMA receiver, as shown in Fig. 6.15.

The series of L-point DFT and N-point IDFT at the transmitter side makes the UL signal single-carrier like as will be shown in the next subsection, which significantly reduces the PAPR. This reduction in PAPR is the key benefit of SC-FDMA. At the receiver, an additional IDFT operation is required to recover the transmitted data. During the LTE study phase within the 3GPP, two subcarrier mapping schemes have been considered (localized and distributed [6]), and the localized scheme has been adopted for the LTE UL, and is presented next.

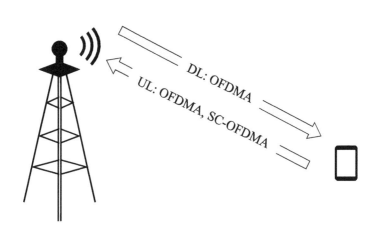

Fig. 6.14 Downlink and uplink modulation schemes in 4G and 5G cellular systems

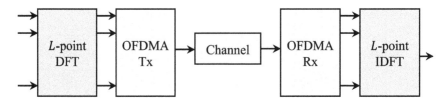

Fig. 6.15 High-level differences between the OFDMA and SC-FDMA transmitter and receiver

6.4 Single-Carrier FDMA

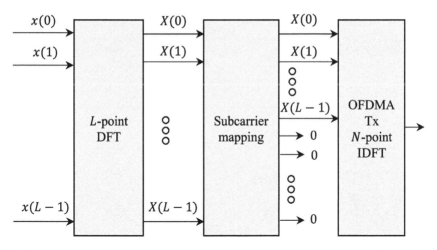

Fig. 6.16 Example of localized subcarrier mapping. In practical systems, the mapping does not always start from 0; this is just for illustration purposes

6.4.1 Localized Subcarrier Mapping

A complex sequence of L uplink symbols, $X(l)$ where $0 \leq l \leq L-1$, are outputs of an L-point DFT block and serve as an input to a localized subcarrier mapping module (the mapping module is omitted in Fig. 6.15). The output of the subcarrier mapping module consists of $N = L \times Q$ samples, where L consecutive samples are the same as the input, and the remaining $N - L$ output samples are set to 0, as illustrated in Fig. 6.16.

In a fully loaded uplink transmission, different users use different contiguous subcarrier mappings until all N inputs of the OFDMA transmitter are filled with nonzero values. For example, the first L subcarriers are assigned to the first user (the next L_i subcarriers could be assigned to the next user where L_i can be different from L, and so on). The output of the subcarrier mapping block, for the first user, is

$$\tilde{X}(k) = \begin{cases} X(k), & 0 \leq k \leq L-1 \\ 0, & L \leq k \leq N-1 \end{cases} \quad (6.55)$$

To determine the output of the OFDMA transmit block (after the N point IDFT), let $n = Q \times l + q$, where $0 \leq l \leq L-1$ and $0 \leq q \leq Q-1$. The time domain $\tilde{x}(n) = \text{IDFT}[\tilde{X}(k)]$ is

$$\tilde{x}(n) = \frac{1}{\sqrt{N}} \sum_{k=0}^{N-1} \tilde{X}(k) e^{j\frac{2\pi nk}{N}} = \frac{1}{\sqrt{QL}} \sum_{k=0}^{L-1} X(k) e^{j\frac{2\pi(Ql+q)k}{QL}}. \quad (6.56)$$

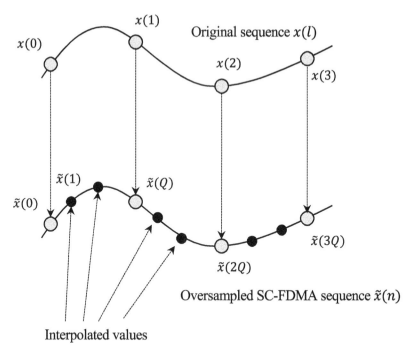

Fig. 6.17 Illustration of SC-OFDMA signal and the relationship between the input time sequence and the single-carrier like output sequence at the transmitter (before D/A conversion)

When $q = 0$, the time domain sequence, at time instances $n = 0, Q, 2Q, \ldots, (L-1)Q$, is

$$\tilde{x}(n) = \tilde{x}(lQ) = \frac{1}{\sqrt{Q}} \frac{1}{\sqrt{L}} \sum_{l=0}^{L-1} X(k) e^{j\frac{2\pi lk}{L}} = \frac{1}{\sqrt{Q}} x(l), \quad 0 \le l \le L-1. \quad (6.57)$$

Evidently, $\tilde{x}(n)$ is an oversampled version of $x(l)$, with Q as an oversampling factor. The $\tilde{x}(n)$ samples between these equally distanced samples are obtained when $q \ne 0$:

$$\tilde{x}(n) = \frac{1}{\sqrt{QL}} \sum_{k=0}^{N-1} \left[\frac{1}{\sqrt{L}} \sum_{p=0}^{L-1} x(p) e^{-j\frac{2\pi pk}{L}} \right] e^{j\frac{2\pi(Ql+q)k}{QL}}. \quad (6.58)$$

Therefore, the samples between the $\tilde{x}(lQ)$ points are also obtained from the $x(l)$ samples (but not through simple copying), and represent interpolated values, as shown above (Fig. 6.17).

The oversampled/interpolated signal $\tilde{x}(n)$ is a single carrier-like signal and as such has better peak-to-average power ratio.

Problems

1. One portion of the channel bandwidth uses subcarrier spacing of $\Delta f_1 = 1/T_1$, and another portion of the band uses subcarrier spacing of $\Delta f_2 = 1/T_2$, which is referred to as mixed numerology. Two different numerologies are not mutually orthogonal and a guard band of several subcarriers is needed between them. Show that numerology 1 subcarriers, located at $e^{j2\pi k \Delta f_1 t}$ (solid lines in the illustration below) and numerology 2 subcarriers, located at $e^{j2\pi m \Delta f_2 t}$ (dashed lines) are not mutually orthogonal. Assume that k and m are integers and that the first subcarrier with Δf_2 spacing starts after the last subcarrier with Δf_1 spacing.

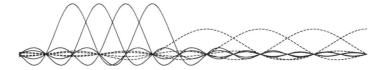

2. Consider an OFDM signal $x(n)$ and a discrete channel impulse response $h(n) = [1, 0, 0.5]$. Determine the following:

   a. Signal at the output of the DFT module at an OFDMA receiver and show that each subcarrier value at the output is a scaled value of the input subcarrier signal.
   b. Maximal and minimal channel gain using the result from (a).

3. For the given 16-QAM mapping table and an input integer sequence is $[2, 4, 6, 8]$ (*hint:* convert integers to binary), calculate the following:

   a. 16-QAM mapped output sequence $X(k)$ and time domain sequence $x(n)$ at the output of the 4-point IFFT module.
   b. Sequence $x(n)$ from (a), with an added cyclic prefix consisting of $\mu = 2$ samples, goes through a channel with an impulse response $h(n) = [1, 2]$. Determine the channel output $y(n)$.
   c. After dropping the cyclic prefix from $y(n)$ and truncating the remaining vector to length 4, determine the received sequence $r(n)$.
   d. $R(k)$ at the output of a four-point FTT module at the receiver.
   e. Frequency domain channel impulse response $H(k)$ based on the knowledge of the input sequence $X(k)$ and output $R(k)$.
   f. Time domain response $h(n)$ from $H(k)$ determined in e). Verify that it matches the impulse response specified in (b).

4. In OFDMA communication systems, peak-to-average power ratio (PAPR) is defined in a statistical sense as a probability that a certain threshold value P_0 is exceeded:

$$\text{Prob}(PAPR \geq P_0) = 1 - (1 - e^{-P_0})^{N_{FFT}}$$

Table 6.1 16-QAM constellation mapping table

Binary sequence	16-QAM symbol
0000	$1+j$
0001	$1+3j$
0010	$1-j$
0011	$1-3j$
0100	$3+j$
0101	$3+3j$
0110	$3-j$
0111	$3-3j$
1000	$-1+j$
1001	$-1+3j$
1010	$-1-j$
1011	$-1-3j$
1100	$-3+j$
1101	$-3+3j$
1110	$-3-j$
1111	$-3-3j$

If $\text{Prob}(PAPR \geq P_0) = 0.0001$, calculate the P_0 threshold for the cases when $N_{FFT} = 2048$ and $N_{FFT} = 4096$. Express the results in dB scale and comment on the differences.

5. Generate symbols using 16-QAM constellation points provided in Table 6.1 as an input sequence to an OFDM modulator and by setting the $N_{FFT} = 4096$. Of each symbol, select the peak signal level and determine the CCDF curve for the PAPR distribution assuming the constellation points are designed such that the average signal power is 1.
6. Generate OFDM symbols using 16-QAM constellation points provided in Table 6.1. If the channel is $h(n) = [1.0, 0.0, 0.7, -0.6]$ and the cyclic prefix consists of $\mu = 8$ samples, simulate the impact of channel and determine the received constellation points before and after equalization. Assume a complex AWGN is added, with zero mean and variance $\sigma^2 = 0.0016$
7. Consider an OFDM system with various system parameters that are a function of frequency range. Populate the missing values. If the channel rms delay is 0.1 µs, is there any subcarrier spacing that cannot absorb the ISI caused by the channel?

OFDM parameters	Up to 3 GHz	Up to 7 GHz	Up to 24 GHz	Above 24 GHz
Subcarrier spacing	15 kHz	30 kHz	60 kHz	120 kHz
Samples per OFDM symbol	4096	4096	4096	4096
OFDM symbol duration in µs				
CP samples	320	320	320	320
CP duration in µs				

6.4 Single-Carrier FDMA

8. The IEEE 802.22 standard, intended for TV white spaces (unlicensed vacated TV channels in the range from 54 to 698 MHz), uses OFDM both for uplink and downlink transmission, and its parameters are listed in the table below. Calculate and populate the missing values in the second (value) column.

Parameter	Value				Symbol
Channel bandwidth	6 MHz				
FFT size	2048				N_{FFT}
Subcarrier spacing					Δf
Useful symbol duration	298.76 μs				T_U
Cyclic prefix	$\frac{T_U}{32}$	$\frac{T_U}{16}$	$\frac{T_U}{8}$	$\frac{T_U}{4}$	T_{CP}
Data subcarriers	1440				N_{SUB}
Pilot subcarriers	240				N_P
Occupied bandwidth					
Modulation	QPSK, 16-QAM, 64-QAM				
Coding rate	1/2, 2/3, 3/4, 5/6				
Max and min rate for $T_{CP} = T_U/16$					

9. A simple multicarrier system uses SC-FDMA modulation for the uplink communication, where the DFT block consists of two points ($M = 2$), and an IDFT block consist of four points ($N = 4$). If the input sequence into the SC-FDMA DFT block consists of $x(0) = 1 + j$ and $x(1) = 3 - j$, determine the output SC-FDMA sequence (time domain representation of the four output symbols) for the localized subcarrier mapping. How many subcarrier mappings are possible?

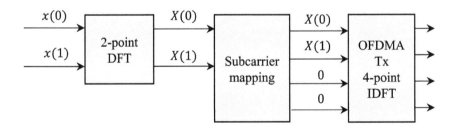

10. A simple multicarrier system uses SC-FDMA modulation for the uplink communication, where the DFT block consists of four points ($M = 4$), and an IDFT block consist of eight points ($N = 8$). If the input sequence into the

SC-FDMA DFT block consists of $x(0) = j, x(1) = 1 - j, x(2) = x(3) = 1$, determine the output SC-FDMA sequence (time domain representation of the eight output symbols) for the distributed subcarrier mapping.

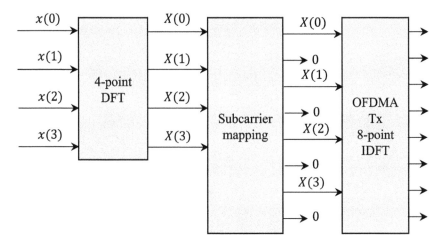

How does the output pattern relate to input pattern?

References

1. A. Goldsmith, *Wireless Communications* (Cambridge University Press, Cambridge, 2005)
2. E. Dahlman, S. Parkvall, J. Sköld, *4G LTE/LTE-Advanced for Mobile Broadband* (Academic Press, New York, 2011)
3. J.G. Andrews, A. Ghosh, R. Muhamed, *Fundamentals of WiMAX: Understanding Broadband Wireless Networking* (Prentice Hall, Englewood Cliffs, 2007)
4. Y. Rahmatallah, S. Mohan, Peak-to-average power ratio reduction in OFDM systems: a survey and taxonomy. IEEE Commun Surv Tutorials **15**(4), 1567–1592 (2013)
5. A. Zaidi, F. Athley, J. Medbo, U. Gustavsson, G. Durisi, X. Chen, *5G Physical Layer: Principles, Models and Technology Components* (Academic Press, New York, 2018)
6. F. Khan, *LTE for 4G Mobile Broadband: Air Interface Technologies and Performance* (Cambridge University Press, Cambridge, 2009)

Chapter 7
Radio Access Network Architecture

7.1 Radio Access and Core Network

So far, we have mainly talked about user terminals/equipments (UTs/UEs) and basestations (BSs) as essential elements to provide data connectivity to users. UTs are colloquially referred to as mobiles or mobile phones. These two elements, UEs and BSs, are the starting and/or terminating points for the wireless link in data connectivity. In 4G technology, as specified by the 3GPP starting from Release 8 (Rel-8), the basestation is denoted as eNodeB, or eNB for short, and together with the UT, they constitute the radio access network (RAN). Given that the 4G technology represents an evolution of the Universal Terrestrial RAN network developed in 3G, the 4G RAN network is called Enhanced Universal Terrestrial Radio Access Network (E-UTRAN).

To operate a cellular network, operators also need to deploy specialized network elements and connection interfaces between these elements to support the RAN. For example, users need to be authenticated into the operator's network, and their connectivity is required to follow policies based on the plan they subscribe to, their traffic needs to be routed, billed, etc. These functions are fulfilled with another important part of the wireless network, called the core network.

The core network stores important information about operator's customers, like the International Mobile Subscriber Identity (or IMSI, a unique number that identifies each customer), authentication keys, location, and other customer profile data. Based on these data, the mobile core network sets policies on the types of services users have access to, and tracks their usage for billing purposes.

Supplementary Information The online version contains supplementary material available at (https://doi.org/10.1007/978-3-031-76455-4_7).

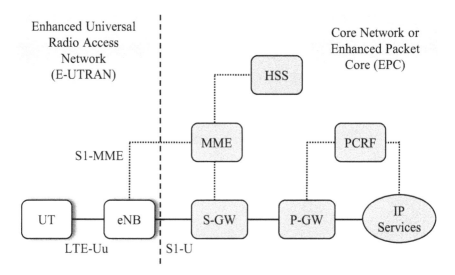

Fig. 7.1 4G RAN and core networks

Additionally, the core network routes the traffic to and from a mobile device, allocates an IP address, tracks which basestation users are connected to, etc. In 4G terminology, the core network is known as the Enhanced Packet Core (EPC). The E-UTRAN and EPC parts of a network are illustrated in Fig. 7.1.

In the RAN, functionality within the eNB includes admission control, mobility control, scheduling and resource allocation, security via data encryption, quality of service (QoS) mapping, etc. Basestations are typically interconnected via the so-called X2 interface, which helps with handover management as well as with interference coordination, along with other required inter-basestation information exchanges.

Connection interfaces between different elements in the network are well-defined by 3GPP. The UT is connected to an eNB via an LTE-Uu interface (the air interface) which includes not only the data transmission but also signaling between the eNB and UE as well as the signaling between the UT and the Mobility Management Entity (MME). The control plane signaling between the UT and eNB is called the access stratum (AS) signaling, whereas the signaling between the UT and the core network is called the non-access stratum (NAS) signaling. The interface that connects the eNB to the core network is called the S1 interface, and it is divided into the user plane S1-U interface and the control plane S1-MME interface.

In the core network, the Serving Gateway (S-GW) acts as a local mobility anchor for data bearers (it is a fixed point in the network, while the corresponding eNB could be anywhere in the network). It also collects billing information for individual users. The Packet Data Network Gateway (P-GW) allocates IP addresses to individual users and enforces QoS rules based on the settings from the Policy and Charging Rules Function (PCRF). The Home Subscriber Server (HSS) holds

7.1 Radio Access and Core Network

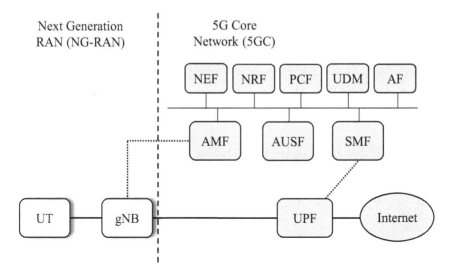

Fig. 7.2 5G RAN and core networks

subscriber profile data, service access restrictions, and roaming capabilities. It also plays a role in authentication. The MME is the key node that processes NAS signaling between the UE and the core network. The MME manages establishment and release of data bearers and plays a vital role in ensuring security between the UE and an outside network (Internet). It also is in charge of delivering core network paging (covered in Chap. 10) to eNBs in the same tracking area (explained in Chap. 13).

The 5G architecture is quite similar to that of 4G, with a different naming convention, as shown in Fig. 7.2. The User Plane Function (UPF) node takes a role of S-GW and P-GW and serves as a data gateway between the RAN and the Internet. The control plane functionality is divided among more nodes than in 4G. The Access and Mobility Management (AMF) node has functionality to similar the MME in 4G; it handles control signaling between the 5G core network and a UT via NAS signaling, mobility, authentication, etc. As with the MME, the AME also provides core network paging to gNBs. The Session Management Function (SMF) node allocates IP addresses to individual UEs, controls policy enforcement, and provides additional session management functions. The Policy Control Function (PCF) is responsible for policy rules; the Unified Data Management (UDM) handles authentication credentials and access authorization, etc.

The NG-RAN, standardized from 3GPP Rel-15 and higher, provides both NR and LTE/E-UTRA radio access. If a basestation provides 5G radio access, the node is denoted gNB. If, on the other hand, a basestation provides enhanced LTE access (Rel-15 and higher), it is denoted as ng-eNB. The interface between gNBs and ng-eNBs is called the Xn interface (Fig. 7.3).

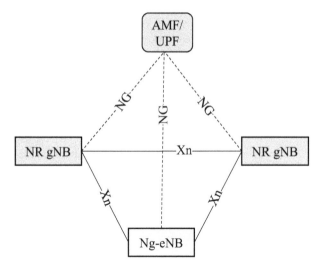

Fig. 7.3 5G RAN and core interconnection

While, normally, a 4G RAN is connected to the 4G core in an operator's network, and 5G RAN to 5G core, it is possible to mix and match different generations of RAN and core networks. In fact, most of the early 5G networks do not include a 5G core. Instead, in the transitional stage, there is a mix of 4G and 5G access by dual connectivity to both a gNB and an eNB. This is known as a Non-Standalone (NSA) deployment, and it is intended to leverage the existing 4G network infrastructure while providing 5G air interface. 3GPP defines several architectural options for 4G and 5G networks [1], and only three of them are illustrated in Fig. 7.4 (other options, 4, 5, and 7, are not shown).

In option 3, or the NSA mode, the 4G eNB plays a role of a Master Node that provides control plane connectivity to MME, whereas the 5G gNB is a secondary node (referred to as en-gNB in this configuration) that is intended to augment the data plane (i.e., data rate) using a 5G radio. 3GPP specs are known for its many acronyms, and there is no exception here. The NSA mode, in which **E-UTRAN** and **NR** RAN are connected, is referred to as a **D**ual **C**onnectivity mode, or EN-DC configuration. Details about the DC connectivity are provided in Chap. 12.

7.2 Basestation Architecture

Basestations typically comprise two main components: the baseband and the RF parts. They may be integrated together, as is usually the case with small cells, or consist of two or more hardware units, and interconnected via well-defined interfaces.

7.2 Basestation Architecture

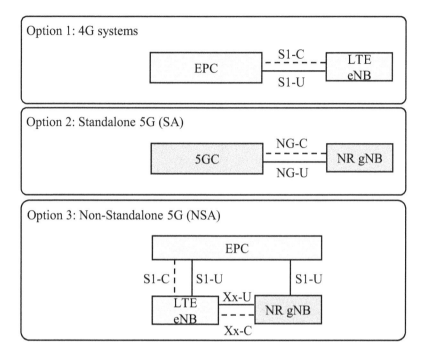

Fig. 7.4 4G and 5G deployment options

In a conventional basestation implementation that uses dedicated hardware, the baseband component of a macro basestation is usually packaged in a shelf (e.g., with a 19" rack) and are stored indoors (e.g., inside a building), or in an outdoor cabinet. The baseband processing capability is typically dimensioned to support three or six MIMO-based[1] sectors. Within these shelves, there are also controllers for X2, S1, and fiber-optic interfaces, as well as a power supply module.

Radio functionality is implemented in a unit called a Radio Head (RH) which is connected to receive/transmit antenna ports. Up until the emergence of 5G, the RH unit was purely dedicated for RF processing, and was in a close proximity to the baseband unit (tens of meters), with which it was connected via optical fiber. At longer spatial separations, which can be as long as 100 meters, the RH was also known as a remote radio head (RRH) or remote radio unit (RRU).

Given their importance in processing RF signals, careful site design is required to ensure stability of each site. In low- and mid-frequency bands, these radios are equipped with 2 or 4 antenna ports, whereas in high TDD bands, there could be 64 or more antenna ports (as discussed in Chap. 5). The connection between the baseband and RRH units is based on the Common Public Radio Interface (CPRI), as illustrated in Fig. 7.5.

---

[1] Multiple input, multiple output. See Chap. 5.

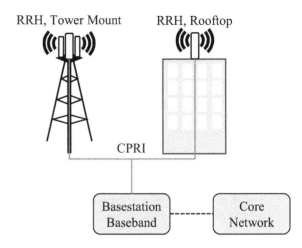

Fig. 7.5 Remote radio head deployment options example with CPRI connection between the baseband and RRH unit

7.2.1 Basestation Protocol Stack

The basestation protocol stack is divided into three layers, as illustrated in Fig. 7.6, with some small differences between 4G and 5G (4G lacks the SDAP; see below). While details on each of these layers will be covered in Chap. 8 (layer 1), Chap. 9 (layer 2), and Chap. 10 (layer 3), for completeness, here we cover these topics briefly:

- *Non-Access Stratum* (NAS) sublayer of layer 3 is the protocol between the UE and the core network whose purpose is to support the mobility of the UT and the session management procedures,
- *Radio Resource Control* (RRC) sublayer of layer 3 includes broadcast of system information related to the NAS, broadcast of system information related to the AS, paging, establishment, maintenance, and release of an RRC connection between the UE and E-UTRAN,
- *Service Data Application Protocol* (SDAP) maps QoS bearers to radio bearers according to their quality-of-service requirements and is defined for 5G only,
- *Packet Data Convergence Protocol* (PDCP) performs IP header compression, ciphering, reordering, and integrity protection,
- *Radio Link Control* (RLC) segments packets and handles retransmission,
- *Medium Access Control* (MAC) is responsible for the multiplexing of logical channels, hybrid-ARQ retransmissions, and scheduling,
- *Physical Layer* (PHY) processing includes coding/decoding, modulation/demodulation, multi-antenna mapping, and other physical-layer functions

7.2 Basestation Architecture

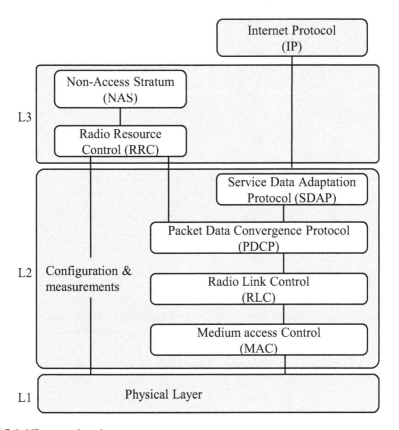

Fig. 7.6 NR protocol stack

7.2.2 CPRI Line Rates

The CPRI is a standard-based interface between the basestations, also known as radio equipment controller (REC), and RRH units (also referred to as radio equipment or RE). It provides a wide range of data rate options (e.g., option 3 can be used for 2×2 MIMO radios, option 5 can be sued for 4×4 MIMO supporting radios, etc). The first CPRI specification, version 1.0, was published in 2003, to support 3G cellular networks (UMTS networks). Subsequent releases were designed to support higher data rates, with the last version, v7.0, published in 2015 with support for 24 Gbps.

The mapping between the bit rate the CPRI interface supports, called line bit rate, and how many HSPA or LTE carriers are supported over the CPRI interface, called transport capacities, is listed in Table 7.1. The entries are calculated as follows: the clock rate for the HSPA (3G) radio is 3.84 MHz, and the data stream from the basestation to the radio consists of I (in-phase) and Q (quadrature) data represented by digital samples. Each sample consists of 16 bits (15 bits for D/A converter and 1 parity bit). Therefore, the uncoded data rate is

Table 7.1 CPRI line rates [2]

Option #	Line bit rate	Line coding	Transport capacity (# HSPA)	Transport capacity (# 20 MHz LTE 1x1)
1	614.4 Mbit/s	8B/10B	4	–
2	1228.8 Mbit/s	8B/10B	8	1
3	2457.6 Mbit/s	8B/10B	16	2
4	3072.0 Mbit/s	8B/10B	20	2
5	4915.2 Mbit/s	8B/10B	32	4
6	6144.0 Mbit/s	8B/10B	40	5
7	9830.4 Mbit/s	8B/10B	64	8
7A	8110.08 Mbit/s	64B/66B	64	8
8	10137.6 Mbit/s	64B/66B	80	10
9	12165.12 Mbit/s	64B/66B	96	12
10	24330.24 Mbit/s	64B/66B	192	24

Uncoded line rate $= 2$ (I and Q) $\times\, 16$ (bits/sample) $\times\, 3.84 \times 10^6 = 122.88$ Mbps.

The line coding column specifies the input/output number CPRI line coded bits. The purpose of line coding is to map input bit sequence into an output bit sequence in a way that achieves DC balance, that is, this mapping provides enough state changes between "1"s and "0"s to allow reasonable clock recovery. The notation 8B/10B means 8 bits are mapped to 10 bits. As a result, the coded line rate is

$$\text{Coded line rate} = 122.88 \times \frac{10}{8} = 153.6 \text{ Mbps}.$$

Given that, for instance, option 1 supports 614.4 Mbps, the total number of HSPA carriers that can be carried via a CPRI interface is $614.4/153.6 = 4$, as listed in the first row of Table 7.1. Therefore, CPRI Option 1 is sufficient to support up to 4 HSPA radios.

The LTE calculation is similar. The sampling rate of the LTE signal is $f_{samp} = 2048 \times 15 \times 10^3 = 30.72$ Msamples/second, where 2048 represents the number of FFT points used for OFDM modulation. The coded line rate for a single carrier is

$$\text{Coded line rate}_{LTE} = 2 \times 16 \times 30.72 \times 10^6 \times \frac{10}{8} = 1.2288 \text{ Gbps}.$$

If an LTE radio supports 4×4 MIMO, it needs to support up to 4 layers of data, which means that the required data rate for such radio is $4 \times 1.2288 = 4.9152$ Gbps. This data rate matches the Option 5 CPRI interface. Option 7A and higher use line coding rate of 66/64, thus imposing less overhead.

To put the numbers selected above in the right context, let us look at the dynamic range that the number of bits used for I and Q digital paths provide. Analog-to-digital converters (ADCs) as well as digital-to-analog converters typically use 15 bits, and 1 parity bit is added for the 16 bit representation of samples. The 15 bit ADCs have approximately $6 \times 15 = 90$ dB of signal to noise ratio.[2] That means that if the maximum signal level at the input of an ADC is -20 dBm, the noise floor due to the ADC device is $-20-90 = -110$ dBm. Received signal levels are usually much lower than -20 dBm.

7.3 Basestation Splits and the Midhaul

With a strong push toward the replacement of conventional dedicated hardware with general-purpose processors running customized software, known as virtualization, it became evident that parts of the baseband protocol stack, which operate in a relatively *slow* timeframe of a millisecond or so can easily be virtualized and stored in a more centralized location. Lower levels of the protocol stack, especially the high physical layer, and MAC layer may need to be closer to an RRH to reduce latency. Virtualization of the lower-level protocol stack is more gradual; the physical layer may still require some dedicated components. To facilitate this, 3GPP has defined a 5G gNB split into a central unit (gNB-CU) and one or more distributed units (gNB-DUs). This architecture, as defined in TS 38.401 [3], is illustrated in Fig. 7.7.

The CU physical entity provides support to higher layers of the basestation protocol stack (e.g., SDAP, PDCP, and RRC) and is used to control one or more DU entities. The DU, which can be located up to 20 km (see problem 4) away from the CU, provides support for lower layers of the basestation protocol stack (e.g., RLC, MAC, and parts of PHY layer). Each DU is dedicated to a single or multiple cells (e.g., 3 sectors), whereas a CU can control tens of DUs, thus controlling many

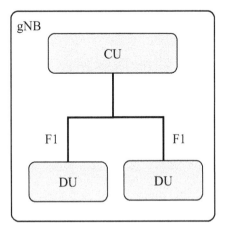

Fig. 7.7 Central unit (CU) and distributed unit (DU) connected via the midhaul interface F1

---

[2] Each bit adds approximately 6dB of improvement in SNR, where the noise is due to quantization. In addition, there a ADC-specific constant, but this analysis suffices for our purposes.

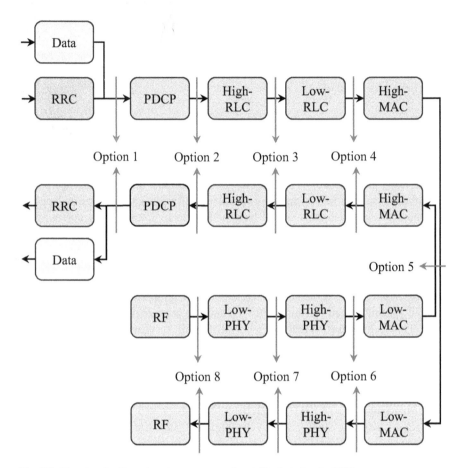

Fig. 7.8 Functional split options between central and distributed units ([1])

sites. 3GPP specifies eight possible splits between central and distributed unit, as shown in Fig. 7.8.

Typically, Option 2 is used in practice because the PDCP layer allows for traffic split and aggregation between different cells via dual connectivity. The throughput of the PDCP layer is similar to that of the MAC layer, so the midhaul capacity is mostly determined by the number of supported cells.

7.4 eCPRI and a Physical Layer Split

A new interface specification was developed, called enhanced CPRI (eCPRI), to address the CPRI data rate *explosion* associated with 5G bandwidths and large number of antennas. The latest version, v2.0, was released in May 2019 [4]. The

7.4 eCPRI and a Physical Layer Split

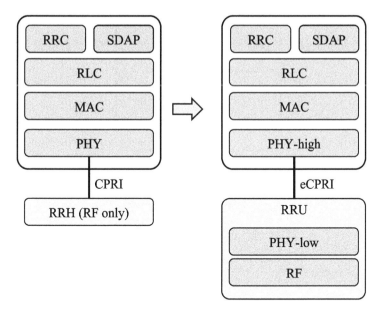

Fig. 7.9 Example of split 7 option where PHY layer is split into PHY-high and PHY-low to enable lower fronthaul data rate via eCPRI

main goal of eCPRI is to decrease the data rate between the basestation (DU) and the RRU.

To illustrate the differences between the CPRI data rates and 5G eCPRI data rates, let us consider a 100 MHz channel and a subcarrier spacing of $\Delta f = 30$ kHz (details on 5G subcarrier spacing will be discussed in Chap. 8). The sampling rate for this specific example is

$$f_{samp} = N_{FFT} \times \Delta f = 4096 \times 30 \times 10^3 = 122.88 \text{ Mbps}$$

Assuming a massive MIMO configuration with 64 antennas ($N_T = 64$), and 16 bits each for I and Q samples, the total CPRI data rate required would be

$$R_{CPRI} = 64 \times 2 \times 16 \times 122.88 \times 10^6 \approx 252 \text{ Gbps}.$$

Sustaining such a data rate is not feasible given that most optical interfaces only support rates up to 100 Gbps. To reduce the required data rate, the distribution of functionality between the basestation and a radio head has been changed in order to allow for lower data rates. One of the options from Fig. 7.8, split 7, is illustrated in Fig. 7.9. The eCPRI-based interface between the baseband and the new RRU is called the *fronthaul*. Split 7 is well suited for the data rate reduction over the fronthaul.

For added flexibility, 3GPP has defined a number of PHY layer splits that allow for a lower data rate transfer between the basestation and RRU unit with

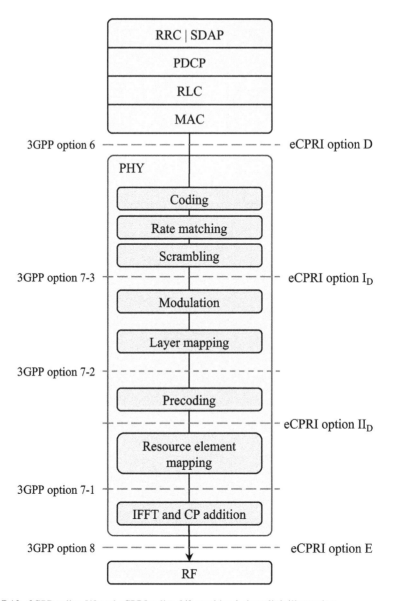

Fig. 7.10 3GPP splits [1] and eCPRI splits [4] combined, downlink illustration

various performance trade-offs. A more detailed view of different 3GPP and eCPRI specified splits is shown in Fig. 7.10. The eCPRI fronthaul data rate depends on which is the splitting point between the baseband and the RRU unit.

To illustrate possible eCPRI data rates associated with these splits, we will assume a spectral efficiency of 20 bps/Hz for 5G and look first at the eCPRI option D (3GPP option 6 split, which is intended for small cells). Ignoring the small overhead

7.4 eCPRI and a Physical Layer Split

of the MAC and IP layers, the data rate at the *Split D* point, just at the output of MAC layer is therefore

$$R_D = 100(\text{MHz}) \times 20(\text{bps}) = 2 \text{ Gbps}$$

This eCPRI option D data rate is significantly lower than the calculated CPRI data rate of 252 Gbps.

The next split, I_D (3GPP option 7-3), has a somewhat higher data rate as error control coding is added. For instance, with the code rate of $r_c = 5/6$ or higher, the data rate becomes:

$$R_{I_D} = 2 \times \frac{1}{r_c} = 2 \times \frac{6}{5} = 2.4 \text{ Gbps}$$

The II_D split happens after the modulation and MIMO precoding, so the data rate increases proportionally to the number of layers, but also because the constellation points are scaled due to precoding, and hence the need to be represented with more bits.

The calculation above can be refined if we account for the number data subcarriers where each subcarrier carries $\log_2(M)$ bits per OFDM sybmol, where M is the modulation order (e.g., $M = 16$ for 16-QAM). Calculation of approximate fronthaul data rates for different splits is described in [5]:

$$R_{7\text{-}3} = \frac{N_{SUB} N_{Layers} \log_2(M)}{T_{OFDM}} \frac{1}{r_c}$$

$$R_{7\text{-}2} = \frac{(I_{\text{bits}} + Q_{\text{bits}}) N_{SUB} N_{Layers}}{T_{OFDM}}$$

$$R_{7\text{-}1} = \frac{(I_{\text{bits}} + Q_{\text{bits}}) N_T N_{SUB} N_{Layers}}{T_{OFDM}} \quad (7.1)$$

where T_{OFDM} is the OFDM symbol duration, N_{Layers} is the number of MIMO layers, $r_c \approx 1$ is the code rate which can be set to 1 for the peak transmission rate, $I_{\text{bits}} = 16$ and $Q_{\text{bits}} = 16$ are in-phase and quadrature-phase components representing complex constellation points with some additional processing (e.g., MIMO precoding, gain) and N_{SUB} is the number of data subcarriers. The fronthaul throughput requirement is normally reduced by compressing the I and Q bits using the block floating point (BFP) compression [6].

Instantaneous data rates differ from the peak values that can be calculated using the formulas above, so the fronthaul may have variable data rates, typically lower than the peak rates, which allows for statistical multiplexing and improved link utilization over the fronthaul interconnection shared by several RRUs.

7.5 Backhaul Dimensioning

Backhaul dimensioning used to be an important part of network planning in cellular networks, where it was important to have the right backhaul throughput to support the demand on cellular sites. With the increase of supported data rate by the devices, as well as with the increase of the overall cell throughput via multiuser MIMO, network dimensioning has become somewhat simplified because most sites have fiber-optic connectivity with 10 Gbps links to the site. In a virtualized architecture, where many CUs are physically *stored* in the same central office location, the transport to the central office must have higher data rates.

On the other hand, there are still conventional sites, especially in remote areas, and this section explains the principles of dimensioning the backhaul transport and how to select the right values (Fig. 7.11).

The backhaul can be dimensioned based on a site peak rate, assuming that all sectors (three sectors in the example above) can achieve simultaneous peaks. Let us assume that each sector has 60 MHz spectrum deployed, with a following breakdown of speeds:

- 20 MHz ×10 bps/Hz, which is 200 Mbps
- 40 MHz ×20 bps/Hz, which is 800 Mbps
- 60 MHz combined has a total peak of $200 + 800 = 1000$ Mbps

Given that the backhaul may be run over an unsecured network, operators typically add an additional 20% overhead for the Internet Protocol Security (IPsec)

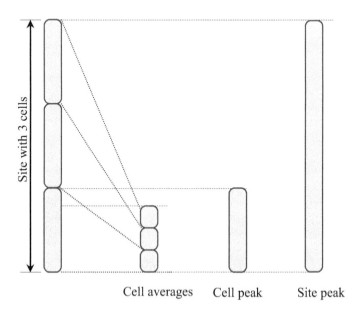

Fig. 7.11 An example of possible options for backhaul dimensioning

encryption that secures the network. Therefore, combining 3 sectors and applying the overhead, the backhaul link requires 3.6 Gbps to the site. However, it is statistically improbable that the site will support 3 peak rates simultaneously, and 3.6 Gbps link may be significantly overdimensioned.

The other approach may be to support site average data rates which are calculated as three sector averages. If, for instance, for the average spectral efficiency in 4G we assumed $SE_{avg,4G} = 2.5$ bps/Hz, the average sector throughput is $60 \times 2.5 = 150$ Mbps, and the total site average is $450 \times 3 = 450$ Mbps. After applying a 20% overhead, the backhaul data rate would be $450 \times 1.2 = 540$ Mbps.

This is a relatively low requirement, but it produces a different problem: the backhaul may not support the peak data rate that a UT supports. For this reason, a slightly more refined approach is to dimension the backhaul in a such way that it ensures one UT peak data rate. Therefore, the backhaul requirement is

$$R_{backhaul} = \max\left(3 \times R_{sector\_average}, R_{UT\_peak}\right) \times IPsec\_overhead \quad (7.2)$$

where R_{UT_peak} is a peak UT data rate. In the example above, the required backhaul data rate would be $R_{backhaul} = 1.2$ Gbps after accounting for 20% of IPsec overhead if the UT peak data rate is 1 Gbps.

7.6 Open RAN

Traditional basestation implementation was based on dedicated hardware and monolithic solutions that resulted in the so-called vendor lock-in situation where operators had to rely on the same vendor for the entire RAN solution, or, at best, deploy different vendor solutions in different geographic areas. A split of basestation functionality into CU, DU, and RRU, as shown in Fig. 7.12, has opened the door for multi-vendor solutions. However, it is worth emphasizing that the split itself does not guarantee interoperability as the solutions are still based on proprietary hardware and interfaces.

For a long time now, operators have been interested in the introduction of multi-vendor RAN solutions, where an operator can mix and match various basestation components from different vendors. Finally, in 2018 an Open RAN (a.k.a., O-RAN) alliance was founded by AT&T, China Mobile, Deutsche Telekom, NTT DOCOMO, and Orange. Many other vendors and operators have since joined in. The goal of the O-RAN alliance has been to define open interfaces between the split RAN

Fig. 7.12 Traditional RAN 4G and 5G networks

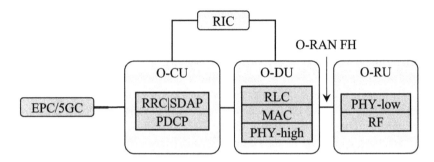

Fig. 7.13 Open RAN 4G and 5G functional split with O-RAN fronthaul (O-RAN FH) interface between the O-DU and O-RU

architecture units. The O-RAN alliance has adopted the 3GPP 7-2 split from [1]. The selection of the 7-2 split represents a trade-off between simplicity of RRU implementation, and the data rate and latency requirements for the fronthaul [7]. The O-RAN functional split allows for some variations for the PHY layer 7-2 split, where the precoding function may be part of O-CU or O-RU, and is called the 7-2x split by the O-RAN alliance [8]. A high-level view of the O-RAN functional split is shown in Fig. 7.13.

In addition to the CU, DU, and RU, a new RAN Intelligent Controller (RIC) unit has been added. The RIC unit is based on programmable components that perform optimization, using closed-loop control, to orchestrate the RAN. It is envisioned that the RIC will leverage artificial intelligence/machine learning (AI/ML) algorithms for better optimization and control. This RIC is divided into two subunits, one that operates in nearly real timescale of around 1s or more and another subunit that operates in a timescale from 10 ms to 1 s.

The O-RAN alliance aligns with the goal of operators to have access to the best-of-breed products from various vendors, thus avoiding a vendor-lock-in model. The question is: "why is O-RAN needed if 3GPP specifies various basestation splits and interfaces?" The answer is that while 3GPP does define most of the RAN interfaces, it does not specify the fronthaul interface. This is where the O-RAN alliance steps in and specifies an open fronthaul, which enables connectivity between the O-DU and O-RU units possibly designed by different vendors. Moreover, the X2 interface between 4G eNBs, while specified by 3GPP, does not interoperate between different vendors, so testing standards for interoperability are specified by O-RAN.

Another important aspect of O-RAN is that it is based on a Cloud-RAN (C-RAN) architecture, where virtualized cloud data centers are used instead of dedicated hardware [7]. One of the key advantages of having Cloud-RAN stored in a centralized location is that resources can be pooled together dynamically, thus reducing the cost by multiplexing the resources. To illustrate this benefit, consider that the amount of traffic processed by a RAN network nearly follows a Pareto distribution, where about 70% of the traffic is serviced by fewer than 30% of sites.

This means that over 70% of the sites have a very low load, and fewer sites are needed when using a centralized architecture with resource sharing.

Finally, the O-RAN is based on open and programmable networks following the principles of Software-Defined Networks (SDN).

7.6.1 The O-RAN Working Groups

Similar to 3GPP, the O-RAN specification development is divided into technical Work Groups (WG), which cover different parts of O-RAN architecture. There are 11 working groups as of the time of writing of this textbook [9]:

- **WG1: Use Cases and Overall Architecture Work Group** identifies key O-RAN use cases, deployment scenarios, and development of the overall O-RAN architecture. It is further subdivided into three task groups.
- **WG2: The Non-Real-Time RAN Intelligent Controller and A1 Interface Work Group** supports Non-Real-Time intelligent radio resource management, higher layer procedure optimization, and policy optimization in RAN and provides AI/ML models to Near-RT RIC.
- **WG3: The Near-Real-Time RIC and E2 Interface Work Group** defines the use cases and requirements related to near real-time RIC and E2 interface.
- **WG4: The Open Fronthaul Interfaces Work Group** specifies open fronthaul interfaces to enable multi-vendor O-DU and O-RU interoperability.
- **WG5: The Open F1/W1/E1/X2/Xn Interface Work Group** specifies enhancements to existing 3GPP profiles, defines test cases to insure multi-vendor interoperability and defines Operation and Management (O&M) models.
- **WG6: The Cloudification and Orchestration Work Group** drives the decoupling of RAN software from the underlying hardware, thus allowing commodity hardware platforms to be leveraged for all parts of RAN deployment.
- **WG7: The White-box Hardware Work Group** consists of nine task groups who work on specification of a complete hardware reference design for different deployment types (e.g., indoor, outdoor) taking into account energy efficiency.
- **WG8: Stack Reference Design Work Group** develops software architecture design and release plan for O-CU and O-DU following the O-RAN and 3GPP protocol stack.
- **WG9: Open X-haul Transport Work Group** focuses on transport equipment, physical media, and control/management protocols with the transport network.
- **WG10: Operation and Maintenance (OAM) Work Group** develops O1 interface and OAM architecture.
- **WG11: Security Work Group** focuses on defining the requirements and specifying the architecture and protocols for security of O-RAN systems.

Each working group is led by three or four co-chairs representing different companies.

Problems

1. Coaxial cable connecting the baseband part of the basestation and the radio unit has an attenuation of 2.1 dB per 30.5 meters. How much signal power is lost if distance between the baseband and radio is 72m? Explain the reason why the CPRI connection helps.
2. Assuming a single MIMO layer stream in 4G network with a data rate of 100 Mbps, how does the IFFT block affect the overall throughput over the CPRI interface?
3. The common purpose public interface (CPRI) supports ten different line bit rate options, supporting a wide range of HSPA and LTE channels. Calculate the missing values in the table below. Note that the LTE capacity relates to 4x4 MIMO case.

Option #	Line bit rate	Line coding	Transport capacity (# of 20 MHz LTE 4x4)
1	614.4 Mbit/s	8B/10B	
2	1228.8 Mbit/s	8B/10B	
3	2457.6 Mbit/s	8B/10B	
4	3072.0 Mbit/s	8B/10B	
5	4915.2 Mbit/s	8B/10B	
6	6144.0 Mbit/s	8B/10B	
7	9830.4 Mbit/s	8B/10B	
7A	8110.08 Mbit/s	64B/66B	
8	10137.6 Mbit/s	64B/66B	
9	12165.12 Mbit/s	64B/66B	
10	24330.24 Mbit/s	64B/66B	

4. CPRI and eCPRI specifications allow for a maximal delay of $100\mu s$ delay. Calculated the maximal distance between the baseband and remote unit if the optical fiber with refraction index of 1.5 is used, where the refractive index is the ratio of the speed of light in a vacuum and the speed of light in optical fiber.
5. CPRI allows for connection of RRHs to the centralized basestation location at distances up to 20 km. If the average RRH density is 3 radios per km^2, determine the following:
    a. Number of radios that can be covered from one centralized location.
    b. If the traffic distribution is such that 20% of the sites carry 50% of the traffic in a traditional distributed radio access network, how many baseband resources are needed if they are centralized?
6. The O-RAN alliance has adopted the 7-2x split. Assuming $N_{SUB} = 1200$, $r_c = 1$, $N_{Layers} = 4$, $T_{OFDM} = 1/14$ ms and block floating point compression where I and Q samples are compressed to 9 bits of I + 9 bits of Q, with 4 additional bits added for every 12 samples, calculate the required fronthaul throughput in the downlink.

7. Calculate the uplink fronthaul throughput using the same assumptions as in Problem 6, but with a maximal number of MIMO layers set to 2.
8. Repeat the calculation in Problem 6 for the downlink assuming $N_{SUB} = 3276$ and $T_{OFDM} = 1/28$ ms, the other parameters are the same. Compare the result with the value calculated in the chapter, equation (5).
9. If we modify the Problem 8 assumption by allowing for massive MIMO transmission, where the $N_{Layers} = 16$, calculate the required peak fronthaul requirement.
10. Repeat the calculation in Problem 9 for the UL, assuming $N_{Layers} = 8$ for the UL. What is the impact of TDD duty cycle of 0.2 in the UL?
11. User terminal supports a peak data rate of 2.5 Gbps and a cellular site, consisting of 4 sectors, deploys radios that have a combined bandwidth of 140 MHz, with average spectral efficiency of 3 bps/Hz and a peak spectral efficiency of 20 bps/Hz. Calculate the required backhaul for the cellular site by making sure that at least one user terminal peak data rate is supported.

References

1. 3rd Generation Partnership Project; Technical Specification Group Radio Access Network; Study on new radio access technology: Radio access architecture and interfaces (Release 14). https://www.3gpp.org/ftp//Specs/archive/38_series/38.801/38801-e00.zip. Cited May 14 2024
2. CPRI Specification V7.0, Common Public Radio Interface (CPRI); Interface Specification. http://www.cpri.info/downloads/CPRI_v_7_0_2015-10-09.pdf. Cited 13 May 2024
3. 3GPP TS 38.401, NG_RAN; Architecture description, Release 18. https://www.3gpp.org/ftp/Specs/archive/38_series/38.401/38401-i10.zip. Cited May 14 2024
4. Common Public Radio Interface: eCPRI Interface Specification, eCPRI Specification V2.0. http://www.cpri.info/downloads/eCPRI_v_2.0_2019_05_10c.pdf. Cited May 15 2024
5. V.Q. Rodriguez, F. Guillemin, A. Ferrieux, L. Thomas, *2020 International Wireless Communications and Mobile Computing (IWCMC): Cloud-RAN Functional Split for an Efficient Fronthaul Network* (2020)
6. S. Lagén, L. Giupponi, A. Hansson, X. Gelabert, *Modulation Compression in Next Generation RAN: Air Interface and Fronthaul Trade-offs* (2021)
7. P. Michele B. Leonardo D. Salvatore, B. Stefano M. Tommaso, Understanding O-RAN: architecture, interfaces, algorithms, security, and research challenges. IEEE Commun. Surv. Tutorials **25**(2), 1376–1411 (2023)
8. O-RAN alliance, O-RAN Working Group 4 (Open Fronthaul Interfaces WG) Control, User and Synchronization Plane Specification, v14.00. https://orandownloadsweb.azurewebsites.net/specifications. Cited May 14 2024
9. Open RAN alliance, About O-RAN ALLIANCE: O-RAN Technical Work Groups. https://www.o-ran.org/about. Cited May 15 2024

Chapter 8
3GPP-Based Air Interface, Layer 1

8.1 Time Domain Numerology

Now that we are familiar with the OFDM type of transmission, we are ready to discuss specific time and frequency domain physical layer parameters of 4G and 5G. First, we start with the time domain numerology.

8.1.1 4G Time Domain Numerology

In 4G, the subcarrier spacing is set to $\Delta f = 15^1$ kHz for both downlink and uplink. This value was chosen as the best value that meets several criteria such as being sufficiently wide to minimize the sensitivity to Doppler shift, and sufficiently narrow to experience flat fading for the majority of deployment scenarios, and for 4G operating bands. The symbol duration, as already discussed, is related to subcarrier spacing as follows: $T_U = 1/\Delta f = 66.67\,\mu s$. As we will see later, the 4G symbol duration is sufficiently long such that even a relatively small overhead of only 6.67% is sufficiently long to eliminate most of the delay spread encountered in the majority of deployment scenarios (e.g., urban, suburban deployments).

---

[1] Evolved Multimedia Broadcast Multicast (eMBMS) can have $\Delta f = 7.5$-kHz spacing and NarrowBand Internet of Things (NB-IoT), not covered in this textbook, can have $\Delta f = 3.75$-kHz subcarrier spacing, both part of 4G technology.

Supplementary Information The online version contains supplementary material available at (https://doi.org/10.1007/978-3-031-76455-4_8).

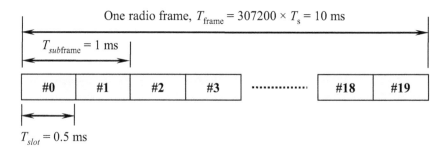

Fig. 8.1 Illustration of an LTE radio frame consisting for 10 subframes and 20 slots

The number of FFT/IFFT points is set to $N_{FFT} = 2048$ for a $W = 20$-MHz-wide channel, so that the sampling interval $T_{samp} = T_U/N_{FFT} = 1/(\Delta f N_{FFT}) = 32.55$ ns. Therefore, the sampling rate for a 4G system is $f_{samp} = 1/T_{samp} = \Delta f N_{FFT} = 30.72$ Msps. This sampling rate is 8 times larger than 3.84 Msps, which happens to be the clock rate for 3G communication systems. This integer ratio is by design, as it enables a common clock for implementation of both 3G and 4G systems using the same hardware.

Time domain resources of the LTE signal are divided into 10-ms frames, which are further subdivided into 1-ms subframes. Subframes are also referred to as Transmit Time Intervals (TTI) because they represent the processing unit over which channel estimation and data detection is conducted at the receiver. Given the sampling frequency of 30.72 Msps, there are exactly 307200 samples within a 10-ms frame, as illustrated in Fig. 8.1. Subframes are further divided into 2 slots of 0.5 ms, so that there are 20 slots within a frame.

Within 1 TTI, there is room for 15 OFDM symbols if no cyclic prefix is added:

$$15 \times T_U = \frac{15}{\Delta f} = \frac{15}{15 \times 10^3} = 1 \times 10^{-3} \text{ s} = 1 \text{ ms}. \tag{8.1}$$

However, due to channel delay spread and the associated inter-symbol interference, LTE specifications allocate some of the samples to the cyclic prefix. Specifically, 1/15th of the available samples within a 1-ms subframe, or 2048 samples in total, are divided and allocated between the remaining 14 symbols. However, since the ratio $2048/14 = 146.2857$ is not an integer number, it was decided that 2 symbols have a CP with 160 samples, and the remaining 12 symbols have a CP of 144 samples, thus using up to 2048 samples for the CP: $2 \times 160 + 12 \times 144 = 2048$. Given that a subframe is divided into two slots, each slot comprises seven symbols, as illustrated in Fig. 8.2, and then the whole structure repeats over the subsequent slot.

8.1 Time Domain Numerology

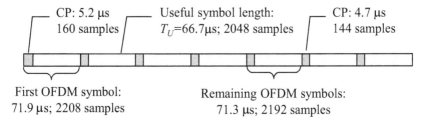

Fig. 8.2 Illustration of a single LTE slot that consists of seven symbols. The first symbol is longer than the remaining six symbols

For smaller LTE channel bandwidth (e.g., 10 MHz, 5 MHz, etc.), the number of IFFT/FFT points can be reduced (e.g., 1024 IFFT/FFT points for a 10-MHz channel, 512 IFFT/FFT points for a 5-MHz channel, etc.), although that is not necessary. In general, the number of CP samples per symbol for different number of IFFT points is calculated as follows:

$$\mu_{LTE} = \begin{cases} \frac{160 \times N_{FFT}}{2048}, & l = 0 \\ \frac{144 \times N_{FFT}}{2048}, & l = 1, 2, 3, 4, 5, 6 \end{cases} \tag{8.2}$$

where l is the symbol index within a slot.

Because of an uneven distribution of CP durations, the OFDM CP overhead is not calculated individually per symbol, but instead we use the overall overhead within one slot:

$$CP_{OH} = \frac{160 \times T_{samp} + 6 \times 144 \times T_{samp}}{0.5 \times 30720 \times T_{samp}} = 6.67\%. \tag{8.3}$$

8.1.1.1 Extended Cyclic Prefix

In addition to normal CP, the LTE standard also specifies an extended cyclic prefix. The extended CP is designed to eliminate multipath from channels with longer delay spread. In this case, only 12 T_U durations are used for data, and the remaining $3T_U$'s are used for CP. Unlike in the case of normal CP, in the extended case, it is possible to evenly divide the samples for the CP, so that we have:

$$\mu_{LTE, extended} = 3 \times 2048/12 = 512$$

samples per CP. With a sampling rate of 32.55 ns, this leads to an extended CP length of 16.6 μs.

Example 8.1 Calculate the CP overhead for the:

(a) Extended cyclic prefix of 16.6 μs with a regular subcarrier spacing of $\Delta f = 15\,\text{kHz}$
(b) For the extended cyclic prefix of 33.3 μs with the eMBMS subcarrier spacing of $\Delta f = 7.5\,\text{kHz}$

Solution Given that for the extended CP all the symbols are of equal length, the overhead can be calculated over a single symbol:

(a)

$$CP_{OH} = \frac{512 \times T_{samp}}{(512 + 2048) \times T_{samp}} = 0.2\ (20\%) \tag{8.4}$$

(b) $\Delta f = 7.5\,\text{kHz} \implies N_{FFT} = 4096$. The sampling interval remains the same: $T_{samp} = 1/(\Delta f N_{FFT}) = 32.55$ ns. The longer CP duration $T_{CP} = 33.3$ μs corresponds to twice as many samples as in a), or $\mu_{LTE} = 1024$, and the overhead remains the same:

$$CP_{OH} = \frac{1024 \times T_{samp}}{(1024 + 4096) \times T_{samp}} = 0.2\ (20\%) \tag{8.5}$$

where 4096 corresponds to the number of samples per useful symbol interval $T_U = \frac{1}{7.5\,\text{kHz}} = 133.33$ μs.

8.1.2 5G Time Domain Numerology

Time domain samples in 5G are expressed relatively to a basic unit called T_c, which is defined as (3GPP TS 38.211 [1, Section 4.1]):

$$T_c = \frac{1}{\Delta f \times N_{FFT}} \tag{8.6}$$

where $\Delta f = 480\,\text{kHz}$ and $N_{FFT} = 4096$. The ratio between the LTE sampling interval T_S and T_c in 5G systems is

$$\kappa = \frac{T_S}{T_c} = 64. \tag{8.7}$$

The main departure from LTE, where a single $\Delta f = 15\text{-kHz}$ solution for all bands is adopted, is that 5G has adopted a range of subcarrier spacings defined as

8.1 Time Domain Numerology

Table 8.1 5G supported numerologies

μ	$\Delta f = 2^\mu \times 15\,\text{kHz}$	Cyclic prefix
0	15	Normal
1	30	Normal
2	60	Normal, Extended
3	120	Normal
4	240	Normal
5	480	Normal
6	960	Normal

$$\Delta f^{NR} = 2^\mu \Delta f^{LTE}, \quad \mu = 0, 1, 2, 3, 4, 5, 6. \tag{8.8}$$

These NR subcarrier spacing values are listed in TS 38.211 [1, Table 4.2-1], and are reproduced in the Table 8.1:

Note that even though the largest $\Delta f = 960\,\text{kHz}$ (as of Rel-17), the 480-kHz value is still used to calculate the T_c value, which is approximately $T_c \approx 0.5\,\text{ns}$.

The number of samples per each NR symbol (excluding the CP part) is specified in TS 38.211:

$$N_U = 2048\kappa 2^{-\mu} \tag{8.9}$$

If we set $\kappa = 64$, the number of samples for different subcarrier spacing values is

$$N_U^\mu = \begin{cases} 131,072, & \mu = 0 \\ 65,536, & \mu = 1 \\ 32,768, & \mu = 2 \\ 16,384, & \mu = 3 \end{cases} \tag{8.10}$$

For most practical systems, this would result in significant oversampling. One possible implementation is to set the number of samples per OFDM symbol to $N_U^\mu = 4096$ regardless of the subcarrier spacing value. From (8.9), we determine that that means that $\kappa = 2^{\mu+1}$. Note that $\kappa = 64$ is still used for the definition of T_c. However, for implementation, the κ values that determine the number of samples per OFDM symbol can be modified. The NR sampling rate then becomes:

$$f_{samp}^{NR} = 4096 \times \Delta f^{NR} = 2^{\mu+1} f_{samp}^{LTE}. \tag{8.11}$$

For instance, if $\Delta f^{NR} = 15\,\text{kHz}$, the sampling rate for the NR will be 2× as high as for the LTE because the $N_{FFT} = 4096$ is twice as high as in the case of LTE. Lower sampling rates are possible for a lower number of FFT points, especially if the channels are narrow. For example, if the channel bandwidth is 10 MHz, the sufficient number of FFT points is 1024 because there are only $N_{RB} = 52$ resource blocks in a 10-MHz channel for $\Delta f^{NR} = 15\,\text{kHz}$, or $52 \times 12 = 624 < 1024$

subcarriers. In this case, a sufficient sampling rate is $f_{samp}^{NR} = 1024 \times 15 \times 10^3 = 15.36$ Msps. The implied value of κ is then $1024/2048 = 1/2$.

The cyclic prefix duration is also specified in TS 38.211 [1, Section 5.3.1], as

$$N_{CP,l}^{\mu} = \begin{cases} 512\kappa 2^{-\mu}, & \text{extended cyclic prefix} \\ 144\kappa 2^{-\mu} + 16\kappa, & \text{normal cyclic prefix, } l = 0 \text{ or } l = 7 \times 2^{\mu} \\ 144\kappa 2^{-\mu}, & \text{normal cyclic prefix, } l \neq 0 \text{ or } l \neq 7 \times 2^{\mu} \end{cases} \quad (8.12)$$

In practical systems, when $\kappa = 2^{\mu+1}$, the CP duration is

$$N_{CP,l}^{\mu} = \begin{cases} 1024, & \text{extended cyclic prefix} \\ 288 + 16 \times 2^{\mu+1}, & \text{normal cyclic prefix, } l = 0 \text{ or } l = 7 \times 2^{\mu} \\ 288, & \text{normal cyclic prefix, } l \neq 0 \text{ or } l \neq 7 \times 2^{\mu} \end{cases} \quad (8.13)$$

For example, for $\mu = 0$, the number of cyclic prefix samples is

$$N_{CP,l}^{0} = \begin{cases} 320, & l = 0 \text{ or } l = 7 \\ 288, & l \neq 0 \text{ or } l \neq 7 \end{cases} \quad (8.14)$$

The total number of samples per symbol:

$$N_{NR,l}^{0} = \begin{cases} 320 + 4096, & l = 0 \text{ or } l = 7 \\ 288 + 4096, & l \neq 0 \text{ or } l \neq 7 \end{cases} \quad (8.15)$$

The number of symbols in 1ms is 14, and 2 of them ($l = 0$ and $l = 7$) are $4,096 + 320 = 4,416$ long, while the remaining 12 symbols are $4,096 + 288 = 4,384$ long, so the total number of symbols over 1-ms subframe is

$$2 \times 4,416 + 12 \times 4,384 = 61,440$$

The number of cyclic prefix samples for different μ values is illustrated in Fig. 8.3.

Alternatively, if we set the κ value to 1/2, the number of CP samples is $N_{CP,l}^{0} = 72$ for $l \neq 0$ or $l \neq 7$, which is aligned with the window length specified in TS 38.104 [2, Table B.5.2-1].

8.1.2.1 5G Frame Structure

The time domain frame structure in 5G is similar to the time domain frame structure of 4G. This is by design, as it allows for dynamic spectrum sharing, where part of resources are allocated to 4G and the rest of resources is allocated to 5G, which means the frame structures need to be compatible. However, instead of 7 symbols

8.1 Time Domain Numerology

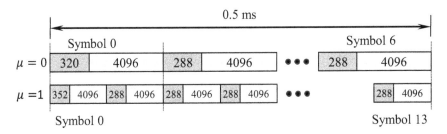

Fig. 8.3 Illustration of different number of cyclic prefix samples over different symbols, for $\mu = 0$ and $\mu = 1$ subcarrier spacing selection

Table 8.2 5G supported number slots per frame and subframe

μ	Number of symbols/slot	Number of slots/frame	Number of slots/subframe
0	14	10	1
1	14	20	2
2	14	40	4
3	14	80	8
4	14	160	16
5	14	320	32
6	14	640	64

per slot, in 5G the number of symbols per slot is 14. For example, at a subcarrier spacing of 15 kHz, one NR slot is 1 ms while an LTE slot is 0.5 ms. Also, because of the variable subcarrier spacing, the number of slots per frame and subframe changes. Different number of slots per frame and subframe is listed in Table 8.2. The number of symbols per slot is defined as follows.

On a larger timescale, frames are assigned system frame numbers (SFN) in the range from 0 to 1023. The periodicity of the SFN is 10.24 seconds, and this applies both to 4G (LTE) and 5G (NR).

8.1.3 4G Frequency Domain Numerology

In the frequency domain, similarly to the idea we have introduced in Chap. 6, not all the subcarriers are utilized for data transmission. Some of them are reserved for pilot signals, as shown below, and some of them are not used at all to ensure a guardband (Fig. 8.4).

Three types of subcarriers are defined in 4G:

1. Data subcarriers: Data and control channels (to carry data traffic)
2. Pilot subcarriers: Channel estimation (coherent demodulation), channel tracking, and synchronization

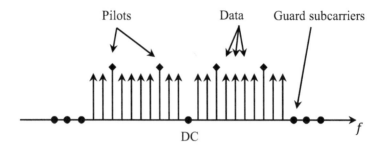

Fig. 8.4 Illustration of pilot, data, and guardband subcarriers in 4G

Table 8.3 LTE channel bandwidths and the number of subcarriers corresponding to each bandwidth

Channel bandwidth (MHz)	1.4	3	5	10	15	20
Total data and pilot subcarriers	72	180	300	600	900	1200

3. Guardband subcarriers: Not transmitted, used for adjacent channel interference protection

Additionally, the DC subcarrier is not transmitted, to avoid DC signal degradation due to LO leakage into the receiver.

Different numbers of data subcarriers are used to support different channel bandwidths. LTE supports six different channel bandwidths, and the number of data subcarriers (including pilots) for a given LTE bandwidth is listed in Table 8.3.

8.1.3.1 LTE Resource Grid Definitions

LTE transmission is based on a discrete set of resources, where the smallest unit of transmission is called a resource block (at least in frequency domain, as will be clarified). A resource block (RB) in LTE consists of one slot in time duration (7 symbols) and 12 subcarriers in frequency domain, as shown in Fig. 8.5. The smallest unit in the time/frequency grid is called a resource element (RE). An RB therefore consists of $12 \times 7 = 84$ REs.

The sizes of the resource block, expressed in terms of the time and frequency dimensions, are 0.5 ms and 180 kHz. Given that a TTI is 1ms, and consists of two slots, at least two resource blocks are needed as the smallest scheduling unit in time domain, while one RB may be sufficient as the smallest scheduling unit in frequency domain. A different number of resource blocks can be selected to support different channel bandwidths. LTE supports six such choices [3], as shown in the Table 8.4, where the transmission bandwidth is calculated as a product of the number of resource blocks N_{RB} and 180 kHz.

8.1 Time Domain Numerology

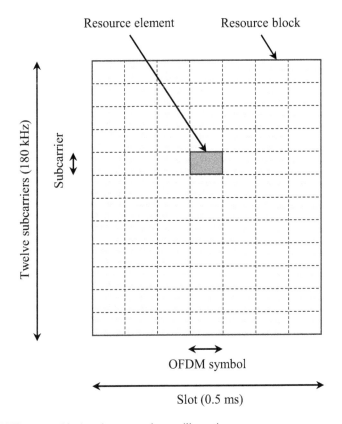

Fig. 8.5 LTE resource block and resource element illustration

Table 8.4 LTE channel bandwidths and corresponding transmission bandwidths

Channel bandwidth (MHz)	Number of RBs (N_{RB})	Transmission bandwidth (MHz)
1.4	6	1.08
3	15	2.7
5	25	4.5
10	50	9
15	75	13.5
20	100	18

By comparing the transmission bandwidth with the channel bandwidth, it is evident that there is 10% of guardband left to account for slow decay of sinc pulse energy (except for the 1.4-MHz channel bandwidth, where the guardband is around 22.86 %).

Table 8.5 LTE channel bandwidths and parameters associated with them

Channel bandwidth (MHz)	1.4	3	5	10	15	20
FFT size (N_{FFT})	128	256	512	1024	2048	2048
Number of RBs (N_{RBx})	6	15	25	50	75	100

Example 8.2 Given the LTE channel bandwidth table (Table 8.5) and the associated FFT sizes, determine:

- Sampling frequency for each channel bandwidth
- Number of data and pilot subcarriers
- Transmission bandwidth

Solution

- The sampling frequency is calculated as $f_{samp} = \Delta f \times N_{FFT}$. For example, for $N_{FFT} = 128$, the sampling frequency is $f_{samp} = 1.92$ Msps, etc.
- The number of data and pilot subcarriers can be calculated from the provided number of resource blocks. For example, for $N_{RB} = 6$, the number of subcarriers is $N_{SUB} = 6 \times 12 = 72$, and so on.
- The LTE channel bandwidth is typically calculated as a product of the number of subcarriers for a given channel bandwidth and the subcarrier spacing: $W_{transmission} = N_{SUB} \times \Delta f$. For the channel bandwidth $W = 1.4$ MHz, the transmission bandwidth is $72 \times 15 \times 10^3 = 1.08$ MHz. Note that this bandwidth is not the actual 99% bandwidth but rather a measure of how many subcarriers are occupied with data or pilots.

After calculating all the values, the LTE channel bandwidth table and its associated parameters are provided in Table 8.6.

8.1.4 5G Frequency Domain Numerology

In 5G, that is, in NR technology, an RB consists of 12 subcarriers in frequency domain (same as in 4G), but in time domain there is a difference as the 5G RB is only 1 symbol long. The RE is the same as in 4G and consists of one subcarrier by one symbol. Given that 5G can use different subcarrier spacings, the size of a RB changes along the time and frequency domains, as illustrated in Fig. 8.6.

The average symbol duration, by accounting for subcarrier spacing, is

8.1 Time Domain Numerology

Table 8.6 LTE channel bandwidths and parameters associated with them

Channel bandwidth (MHz)	1.4	3	5	10	15	20
FFT size (N_{FFT})	128	256	512	1024	2048	2048
Number of RBs (N_{RBx})	6	15	25	50	75	100
Sampling frequency (Msamples/s)	1.92	3.84	7.68	15.36	30.72	30.72
Data and pilot subcarriers (N_{SUB})	72	180	300	600	900	1200
Occupied/transmission bandwidth (MHz)	1.08	2.7	4.5	9	13.5	18

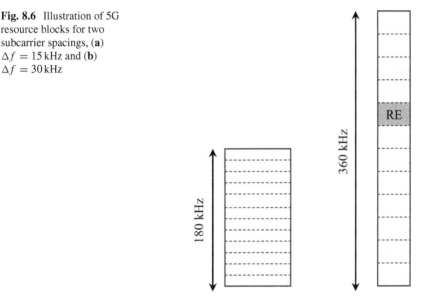

Fig. 8.6 Illustration of 5G resource blocks for two subcarrier spacings, (**a**) $\Delta f = 15$ kHz and (**b**) $\Delta f = 30$ kHz

$$T_{OFDM,avg} = \frac{1 \text{ ms}}{14 \times 2^\mu} = \begin{cases} 71.43\,\mu s, & \mu = 0 \\ 35.71\,\mu s, & \mu = 1 \\ 17.86\,\mu s, & \mu = 2 \end{cases} \quad (8.16)$$

Technical specification 38.101-1 [4] specifies all the potential channel bandwidths a UT can support in 5G within the FR1 range. Table 8.7 lists a subset of transmission bandwidths, indicating the number of RBs for different channel bandwidths and subcarrier spacings.

A similar table for the FR2 range is provided in TS 38.101-2 [5]. The transmission bandwidth is calculated as

$$W_{transmission} = N \times \Delta f = 12 \times N_{RB} \times \Delta f, \quad (8.17)$$

Table 8.7 Maximum transmission bandwidth configuration in 5G

SCS (kHz)	10 MHz N_{RB}	20 MHz N_{RB}	40 MHz N_{RB}	50 MHz N_{RB}	80 MHz N_{RB}	100 MHz N_{RB}
15	52	106	216	270	N/A	N/A
30	24	51	106	133	217	**273**
60	11	24	51	65	107	135

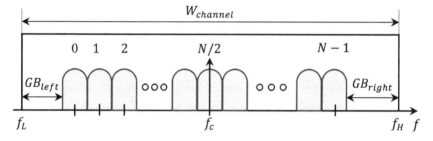

Fig. 8.7 Subcarrier locations and guardbands in 5G (numbers above subcarriers are relative indices, not absolute frequencies). The channel bandwidth $W_{channel} = f_H - f_L$, where f_L and f_H are bottom and the top edges of the channel, respectively

where $N = 12 \times N_{RB}$ is the total number of data and pilot subcarriers. What is evident from Table 8.7 is that for $\Delta f = 15$ kHz ($\mu = 0$), the largest supported channel bandwidth is 50 MHz. Similarly, for $\Delta f = 30$ kHz ($\mu = 1$), the widest channel bandwidth is 100 MHz, but within FR1 range the widest channel bandwidth does not go up to 200 MHz although the number of subcarriers with $\Delta f = 60$ kHz could fit within 4096 points (e.g., 3300 subcarriers would have a transmission bandwidth of about 198 MHz).

5G subcarriers, as illustrated in Fig. 8.7, are not symmetric around the carrier frequency. Namely, the carrier frequency is aligned with the subcarrier index $N/2$ (the first subcarrier index starts at 0), so that there are $N/2$ subcarriers to the left of the subcarrier at the center, and $N/2 - 1$ subcarriers to the right.

The left, right, and total guardbands are

$$GB_{left} = f_c - f_L - \frac{N\Delta f}{2} - \frac{\Delta f}{2} \tag{8.18}$$

$$GB_{right} = f_H - f_c - \frac{N\Delta f}{2} + \frac{\Delta f}{2}$$

$$GB_{total} = GB_{left} + GB_{right} = W_{channel} - N\Delta f.$$

If the carrier frequency f_c is exactly at the center of the channel

$$f_c = \frac{f_L + f_H}{2}$$

8.1 Time Domain Numerology

Table 8.8 Minimum guardband for each UE channel bandwidth and SCS (kHz) in 5G

SCS (kHz)	10 MHz	20 MHz	40 MHz	50 MHz	80 MHz	100 MHz
15	312.5	452.5	552.5	692.5	N/A	N/A
30	665	805	905	1045	925	**845**
60	1010	1330	1610	1570	1450	1370

the left guardband is the minimum guardband, and its value in kHz is calculated as [4]

$$GB_{min} = \frac{W_{channel}[\text{MHz}] \times 1000 - N_{RB} \times \Delta f \times 12}{2} - \frac{\Delta f}{2} \quad (8.19)$$

and selected values from the TS 38.101-1, Table 5.3.3-1, are reproduced in Table 8.8.

Once notable difference compared to 4G is that for wider channels in 5G, the guardband is much smaller as a percentage of the total bandwidth. For example, if we look at the 100-MHz channel, with $\Delta f = 30\,\text{kHz}$, the transmission bandwidth is 98.28 MHz, thus leaving only $(100-98.28) \times 1000 = 1720\,\text{kHz}$ for the guardband, distributed between left and right sides of the transmission bandwidth.

8.1.5 5G Deployment Scenarios Based on Subcarrier Spacing

Subcarrier spacing selection in 5G is mainly based on the carrier frequency, and also on the duplexing mode (FDD vs TDD). Ideally, the delay spread should be less than the cyclic prefix duration, which can be achieved by selecting the appropriate subcarrier spacing. However, in practical deployments, the channel bandwidth also plays a role in the subcarrier spacing selection. Namely, FDD bands below 2700 MHz do not have channel bandwidths that exceed 50 MHz, and for these bands the subcarrier spacing of 15 kHz is sufficient. For TDD bands above 2500 MHz (i.e., starting with the NR band n41), the channel bandwidths can go up to 100 MHz, and for these bands the subcarrier spacing of 30 kHz is used, as illustrated in Fig. 8.8. While the delay spread in some deployment scenarios can exceed 20 μs (e.g., in suburban and hilly rural areas), within the urban core it is typically less than 3 μs [6], and the small amount of the ISI caused by shorter cyclic prefix duration associated with the 30 kHz subcarrier spacing has negligible impact on the system performance.

The 60-kHz subcarrier spacing, indicated with a dashed rectangle in Fig. 8.8, is not used in practical deployments within the FR1 range, but it is defined in 3GPP technical specifications.

It is worth nothing that the channel delay spread reduces with the cell size decrease, and so does the required cyclic-prefix duration, meaning that wider subcarrier spacing values (having shorter cyclic prefix) would be more suitable for

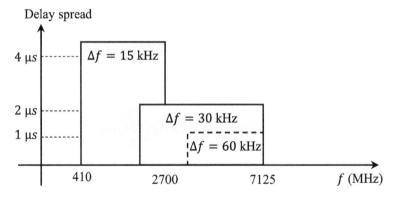

Fig. 8.8 Subcarrier spacing selection based on the frequency range and delay spread in the FR1 range

deployments with smaller cell size. However, within the FR1 range, small cells are designed with the same parameter as the macro stations, so the FDD bands below 2700 MHz are typically deployed with 15-kHz subcarrier spacing regardless of the cell size. On the other hand, in the FR2 range, the cell sizes are typically much smaller, resulting in much lower delay spread, and the 120-kHz subcarrier spacing used in practical deployments is suitable.

8.2 5G DL and UL Frame Alignment and Slot Structures

Both TDD and FDD modes in 5G share a common frame structure, where UL and DL frames are offset in time, as defined in 3GPP TS 38.211 and illustrated in Fig. 8.9.

For reasons that will be explained, the UL frame starts T_{TA} seconds before the corresponding DL frame, where

$$T_{TA} = (N_{TA} + N_{TA,offset}) T_c \qquad (8.20)$$

and T_c has been defined earlier.

The $N_{TA,offset}$ is a fixed offset and is defined as

$$N_{TA,offset} = \begin{cases} 25600, & \text{FDD/TDD, FR1 range} \\ 0, & \text{FDD with DSS in FR1 range} \\ 39936, & \text{TDD with DSS in FR1} \end{cases} \qquad (8.21)$$

For example, if we focus on standalone (SA) 5G deployments in the FR1 frequency range, the fixed offset is $T_{offset} = N_{TA,offset} \times T_c = 25560 T_c = 13\,\mu\text{s}$. The remaining

8.2 5G DL and UL Frame Alignment and Slot Structures

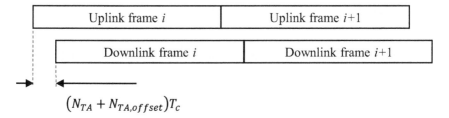

Fig. 8.9 5G frame structure and DL and UL alignment [1]

part of the offset is variable and changes over time. During the initial access, it is estimated based on the delay from the random access channel, and is calculated as

$$N_{TA} = T_A \frac{16 \times 64}{2^\mu}, \tag{8.22}$$

where $T_A = 0, 1, \ldots, 3846$, as defined in TS 38.213 [7]. After the initial access, further updates are iterative:

$$N_{TA,new} = N_{TA,old} + (T_A - 1)\frac{16 \times 64}{2^\mu} \tag{8.23}$$

where $T_A = 0, 1, \ldots, 63$.

Let us first determine why the timing adjustments of the UL transmission are needed.

As illustrated in Fig. 8.10, a reference DL frame is received by UT_1 with some delay τ_1, and the variable part of the time advancement command accounts for round-trip delay, thus aligning the received UL frame with a fixed amount of time T_{offset}, which is necessary for proper TDD operation. The range of timing adjustments needs to be sufficient to account for larger delays (e.g., τ_2 propagation delay associated with UT_2).

In TDD systems, DL and UL transmissions cannot happen simultaneously. Therefore, we need guardtimes between the UL and DL transmission. An example of offset DL and UL slots is illustrated in Fig. 8.11.

The guardtime is imposed by not using some of the symbols that are marked as flexible (F) symbols within a slot. The number of symbols required to achieve the required τ_{guard} is calculated as the following. Let us assume that a UT is located at the distance $d = 10$ km away from the basestation. The guardtime needs to be

$$\tau_{guard} > 2\tau_d + T_{offset} \tag{8.24}$$

where

$$\tau_d = \frac{d}{c} = \frac{10 \times 10^3}{3 \times 10^8} = 33.33 \,\mu s \tag{8.25}$$

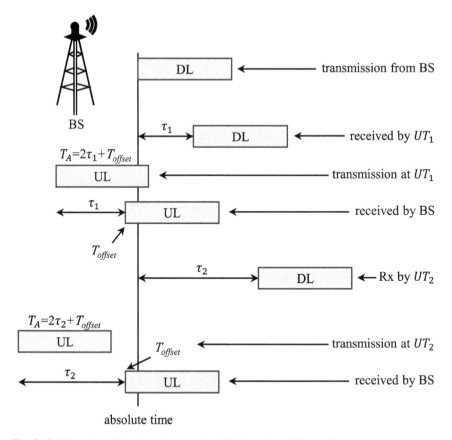

Fig. 8.10 Illustration of timing adjustment for UTs located at different distances

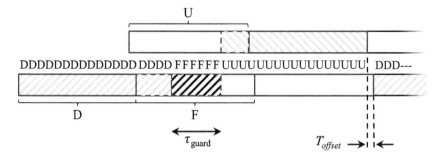

Fig. 8.11 Illustration of DL and UL slots with a flexible slot

and for $T_{offset} = 13\,\mu s$, the required guardtime is

$$\tau_{guard} > 66.67 + 13 = 79.67\,\mu s. \qquad (8.26)$$

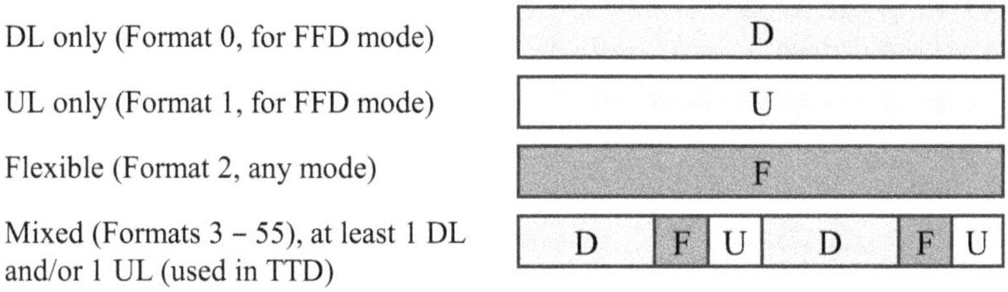

8.3 5G DL PHY Processing Chain

DL only (Format 0, for FFD mode)	D
UL only (Format 1, for FFD mode)	U
Flexible (Format 2, any mode)	F
Mixed (Formats 3 – 55), at least 1 DL and/or 1 UL (used in TTD)	D | F | U | D | F | U

Fig. 8.12 Slot formats for normal cyclic prefix in 5G [7]

The number of flexible symbols that are not used and serve as guard symbols depends on the subcarrier spacing. For example, we would need 2 symbols for $\mu = 0$ to meet the $\tau_{guard} > 79.67\,\mu s$ requirement, 3 symbols for $\mu = 1$, etc.

The $T_{offset} = 13\,\mu s$, between the UL and DL frames, is necessary to allow for the basestation to switch from receive to transmit mode. For larger cells, a larger guardtime is required. Note that in FDD mode, there is no need for guardtime, but for the purpose of maintaining a common frame structure between TDD and FDD, the same T_{offset} is applied, unless the system operates in dynamic spectrum sharing mode, in which case no such offset exists (to align with the 4G frame structure where no such offset is imposed).

This flexibility to configure symbols as DL, UL, or flexible within a slot is allowed by 3GPP specs (TS 38.211). As of Rel-18, there are 56 different slot configurations specified in [7]. These configurations can be divided into four main categories: DL only (Format 0), UL only (Format 1), Flexible (Format 2), and mixed (Formats 3–55), as illustrated in Fig. 8.12.

The slot information is sent to UT via scheduling control dynamically, or can be configured semi-statically via radio resource control signaling.

8.3 5G DL PHY Processing Chain

The DL physical layer processing, at the high level, is shown in the following block diagram (Fig. 8.13).

8.3.1 Cyclic Redundancy Check (CRC) Insertion and Forward Error Control (FEC) Coding

The first step in DL physical layer processing is the insertion of a 24-bit cyclic redundancy check (CRC) parity check for transport blocks with a transport block size (TBS) larger than 3824 bits, and 16-bit CRC if the TBS $\leq$ 3824. An n-bit CRC, applied to a transport (or code) block, detects error bursts up to n-bits in

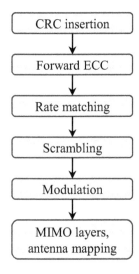

Fig. 8.13 DL physical layer processing block diagram

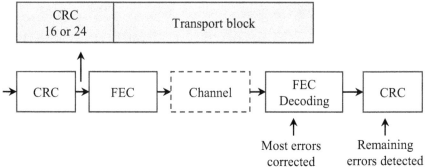

Fig. 8.14 The CRC and FEC block diagram

length. The purpose of having a CRC, in addition to having forward error correction (FEC) mechanisms, is to detect residual errors left over from the FEC block. When errors are detected, the receiver can request a retransmission of the block through a process called Automatic Repeat reQuest (ARQ). The ARQ process in 5G utilizes acknowledgments (ACKs) and negative acknowledgments (NACKs) to eliminate residual errors. This combination of ARQ and FEC is referred to as Hybrid ARQ (HARQ). Chapter 9 provides more details on how HARQ functions. The CRC and FEC block diagram is shown in Fig. 8.14.

The FEC code is low-density parity check (LDPC) based, which is a powerful family of error control codes that are used in many other standards like IEEE 802.11ac, IEEE 802.11ad, IEEE 802.11a/b/g/n, IEEE 802.15.3c, IEEE 802.1e, DVB C2/S2/T2, etc. [8]. The LDPC codes used in 5G consist of two base graphs:

- Base graph 1, which supports a maximum input code block size of 8448 bits and code rates from 1/3 to 8/9, without any puncturing or repetition

8.3 5G DL PHY Processing Chain

- Base graph 2, which supports a maximum input code block size of 3840 bits and code rates from 1/5 to 2/3, without any puncturing or repetition.

In the context of LDPC coding, a base graph represents a carefully designed short LDPC code with desirable properties. By repeating the base graph and reconnecting the nodes across repeated graphs, a longer LDPC code, also with desirable properties, can be constructed. This process is called lifting. Details of 5G-NR base graphs and lifting process are available, for example, in [9].

If the transport block, after the CRC addition, is longer than the max number of bits supported by either of the two base graphs, the transport block to code block segmentation is performed so that the code blocks fit within the FEC code size. Otherwise, no segmentation is required. The algorithm for the code block segmentation is as follows: for a transport block with TBS bits, the total number of bits B, after the CRC addition to the transport block, is

$$B = \begin{cases} \text{TBS} + 24, & \text{if TBS} > 3824, \\ \text{TBS} + 16, & \text{if TBS} \leq 3824, \end{cases} \quad (8.27)$$

and the maximum number of bits per code block, K_{cb}, is calculated as

$$K_{cb} = \begin{cases} 8448, & \text{if base grap 1 is used} \\ 3840, & \text{if base graph 2 is used.} \end{cases} \quad (8.28)$$

Therefore, the number of code blocks C is calculated as

$$C = \left\lceil \frac{B}{K_{cb} - 24} \right\rceil. \quad (8.29)$$

The code block segmentation is illustrated in Fig. 8.15.

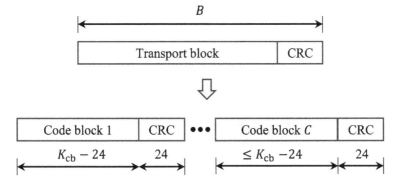

Fig. 8.15 Code block segmentation in 5G

Table 8.9 List of code rates for the DL for 1024-QAM modulation scheme [10]

Modulation order Q_m	Target code rate R	Spectral efficiency
2	120/1024	0.2344
2	193/1024	0.3770
2	449/1024	0.8770
4	378/1024	1.4766
4	490/1024	1.9141
...	...	...
10	853/1024	8.3301
10	900.5/1024	8.7939
10	948/1024	9.2578

Each new code block after the segmentation contains a code block level 24-bit-long CRC, regardless of which base graph is used. These segments are used for retransmission should the error occur, which is more efficient than retransmission of the entire transport block. The details of the code block segmentation are specified in 3GPP TS 38.212.

8.3.2 Rate Matching

The LDPC codes used in 5G increase the number of transmitted bits by adding redundancy bits. The number of redundancy bits can easily exceed the number of information (systematic) bits, especially for base graph 2, where the code rate can be as low as 1/5, and even lower with repetition. Rate matching is employed to reduce the coding overhead in order to match the number of bits to the number of available resources for the DL PDSCH or UL PUSCH transmission. Code rates in the rate matching process are selected from a discrete set of values specified by 3GPP. An example of possible code rates for the DL PDSCH for 1024-QAM modulation scheme, based on Table 5.1.3.1-4 in [10], is shown in Table 8.9.

8.3.3 Scrambling

Neighboring cells use different scrambling sequences which are XOR-ed with the input sequence, so that the interfering signals, after descrambling at the receiver, are randomized. Scrambling sequence c is a function of cell ID, which is selected from a set $\{0, 1, \ldots, 1023\}$ and individual user identity known as the Cell Radio Network Temporary Identifier (C-RNTI). The C-RNTI is used to identify different UTs (phone numbers are not used for this purpose). Details of the pseudorandom scrambling sequence generation and scrambling process are provided in TS 38.211.

8.3 5G DL PHY Processing Chain

Table 8.10 Modulation schemes for different DL physical channels in 5G

Physical channel	Modulation	# of MIMO layers
PDSCH	QPSK, 16QAM, 64QAM, 256QAM, 1024QAM	1–8
PBCH	QPSK	1
PDCCH	QPSK	1

8.3.4 Modulation

Digital modulation order depends on the type of the physical channel (physical channels are covered in the next section). An example of modulation schemes for different DL physical channels is provided in Table 8.10.

The modulation scheme for the data channel (PDSCH) is selected based on the channel quality, and bits to symbols mapping is specified in TS 38.211. For example, QPSK modulation mapping for a pair of bits $b(2i)$ and $b(2i+1)$ is expressed as

$$d(i) = \frac{1}{\sqrt{2}} [(1 - 2b(2i)) + j(1 - 2b(2i+1))]. \tag{8.30}$$

The 16-QAM mapping of the quadruplets of bits $b(4i), b(4i+1), b(4i+2), b(4i+3)$ is expressed as a complex valued symbol $d(i)$:

$$d(i) = \frac{1}{\sqrt{10}} \{(1 - 2b(4i))[2 - (1 - 2b(4i+2))] + \tag{8.31}$$
$$j(1 - 2(b(4i+1))[2 - (1 - 2b(4i+3))]\}$$

and so on.

8.3.5 MIMO Layer Mapping

Following the process of modulation, the MIMO layer mapping is performed. The number of MIMO layers is determined based on the channel rank, reported SNR, and the number of antennas at the basestation and at the UT. 5G supports up to eight MIMO layers in the DL.

Data associated with a single transport blocks are called a code word, and in 3GPP specifications they are indexed using a superscript notation $()^{(q)}$, where $q \in \{0, 1\}$. The complex valued modulated data stream from Sect. 8.3.4 is then denoted as $d^{(q)}(i)$. If $q = 0$, only one transport block/code word is transmitted, and if both values for q are used, two transport blocks/code words are transmitted. One code word can be mapped into up to four MIMO layers, while two code words can be mapped into up to eight MIMO layers, following the rules given in Table 8.11 [1].

Table 8.11 Code words to MIMO layer mapping [1]

Code words	Number of MIMO layers N_L	Mapping
1	1	$x^{(0)}(i) = d^{(0)}(i)$
1	2	$x^{(0)}(i) = d^{(0)}(2i)$ $x^{(1)}(i) = d^{(0)}(2i+1)$
1	3	$x^{(0)}(i) = d^{(0)}(3i)$ $x^{(1)}(i) = d^{(0)}(3i+1)$ $x^{(2)}(i) = d^{(0)}(3i+2)$
1	4	$x^{(0)}(i) = d^{(0)}(4i)$ $x^{(1)}(i) = d^{(0)}(4i+1)$ $x^{(2)}(i) = d^{(0)}(4i+2)$ $x^{(3)}(i) = d^{(0)}(4i+3)$
2	5	$x^{(0)}(i) = d^{(0)}(2i)$ $x^{(1)}(i) = d^{(0)}(2i+1)$ $x^{(2)}(i) = d^{(1)}(3i)$ $x^{(3)}(i) = d^{(1)}(3i+1)$ $x^{(4)}(i) = d^{(1)}(3i+2)$
...	...	...
2	8	$x^{(0)}(i) = d^{(0)}(4i)$ $x^{(1)}(i) = d^{(0)}(4i+1)$ $x^{(2)}(i) = d^{(0)}(4i+2)$ $x^{(3)}(i) = d^{(0)}(4i+3)$ $x^{(4)}(i) = d^{(1)}(4i)$ $x^{(5)}(i) = d^{(1)}(4i+1)$ $x^{(6)}(i) = d^{(1)}(4i+2)$ $x^{(7)}(i) = d^{(1)}(4i+3)$

The mapped output $x^{(l)}(i)$, where $l = 0, 1, \ldots, N_L - 1$ and N_L is the number of MIMO layers, is then mapped to antenna port following the rules outlined in 3GPP TS 38.212.

8.4 5G DL Physical Signals and Channels

The Medium Access Control (MAC) layer, described in Chap. 9, provides transport channels that are mapped onto physical channels. These physical channels originate from higher layers (Fig. 8.16).

Downlink physical signals, which do not originate from higher layers, consist of Primary Synchronization Signal and Secondary Synchronization Signal (PSS/SSS), Demodulation Reference Signal (DM-RS), Phase-Tracking Reference Signal (PS-RS), and Channel State Information Reference Signal (CSI-RS). PSS and SSS signals are used to establish and maintain time and frequency synchronization

8.4 5G DL Physical Signals and Channels

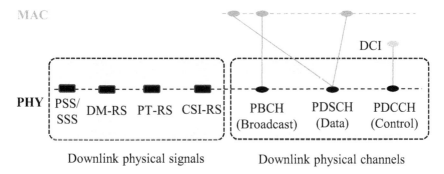

Fig. 8.16 5G DL physical signals and channels based on Rel-15 technical specifications (additional signals are added in subsequent releases)

Fig. 8.17 SSB block in 5G

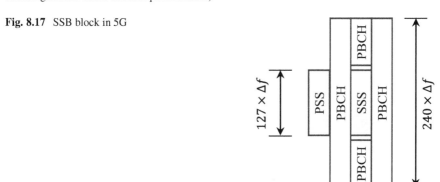

between the basestation and a receiving device. They are transmitted together with a Physical Broadcast Channel (PBCH) over a set of resource elements grouped together into an SSB block. The partition of resources between the PSS, SSS, and PBCH is illustrated above (Fig. 8.17).

The PSS signal provides a cell ID index out of 3 possible values, whereas the SSS provides a cell ID group from a set of 336 possible values, thus providing 1008 possible cell ID values. The SSB block is transmitted periodically, with a configurable period between 5 ms and 160 ms, although a typical value used is 20 ms.

Demodulation Reference Signals (DM-RS) are transmitted together with the physical data shared channel (PDSCH) and are used for channel estimation during the PDSCH demodulation. For faster decoding, channel estimation can be done first if DM-RS signals are at the start of a slot, which is known as front loaded DM-RS. Otherwise, DM-RS can be mapped to other resources, and additional DM-RS can be configured for better channel estimation (which comes with higher overhead).

Phase-Tracking Reference Signal (PT-RS) is a special pilot signal that helps with tracking the phase to reduce phase noise and is intended primarily for higher bands. In the frequency domain, it is added relatively sparsely (every 2 or 4 RBs), and in the time domain, it is more frequent (every 1, 2, or 4 symbols). Specific details can be found in TS 38.211.

Channel State Information Reference Signal (CSI-RS) is intended for channel quality measurement and is associated with individual beams in multiple antenna systems. It is also used for fine time-frequency tracking, radio link monitoring, and beam failure detection.

Physical Broadcast Channel (PBCH) is a part of the SSB block and is used to carry the Master Information Block (MIB). The MIB is a minimum dataset required by UTs for initial access and to acquire other system information blocks, and consists of only 24 bits. It is transmitted periodically, every 80 ms, and within a period, it is repeated several times (e.g., four times if the SSB block is transmitted every 20 ms).

Physical downlink control channel (PDCCH) is the DL control channel. It carries the DL Control Information (DCI) that provides UTs with the information about resource assignment, error control code parameters, and modulation order and also provides UL scheduling grants and information about the UL resources and transport format.

Finally, the physical data shared channel (PDSCH) carries user data. It occupies time and frequency resources within slots and resource blocks, respectively.

8.5 5G UL Physical Signals and Channels

Similarly to the DL, the UL physical layer provides physical channels for the transport channels coming from the MAC layer and adds uplink physical signals, as shown in Fig. 8.18.

We will start by describing the physical random access channel as it plays an important role during the initial access.

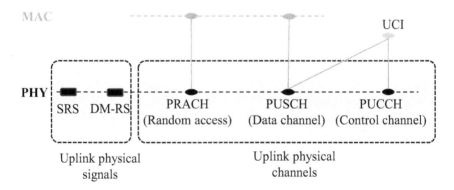

Fig. 8.18 5G UL physical signals and channels

8.5 5G UL Physical Signals and Channels

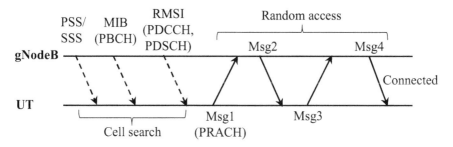

Fig. 8.19 Initial access with cell search and random access phases. The remaining minimum system information (RMSI) is obtained via the PDCCH and PDSCH channels

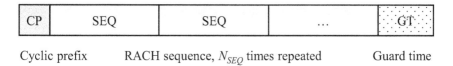

Fig. 8.20 PRACH symbol, structured as OFDM symbol with cyclic prefix, but with a time guard gap at the end

8.5.1 Random Access and PRACH

A UT that is not attached to a network first searches for the SSB block. Once found, the UT can determine time and frequency domain parameters as well as other system parameters broadcast within the PBCH. Once these parameters are determined, the UT initiates a connection request by transmitting a random access preamble, a special data sequence that will be described below. The initial access process is illustrated in Fig. 8.19. More details about the random access procedure are provided in Chap. 9.

The illustration above shows a four-step random access process. Since Rel-16, this process has been shortened into a two-step process.

The PRACH symbol consists of three parts, cyclic prefix, sequence component, and guardtime, which is shown in Fig. 8.20. PRACH preambles are based on Zadoff-Chu sequences and have several possible lengths:

1. Long sequences have a sequence duration $L_{RA} = 839$ and are intended for regular or large cell sizes
2. Short sequences with $L_{RA} = 139$ are intended for small cells
3. Since Rel-16, $L_{RA} = 1151$ and $L_{RA} = 571$ have been added to support NR-U because for the same PRACH power, spread over wider bandwidth we have lower PSD, which is necessary in unlicenced bands that are power limited.

8.5.1.1 PRACH Sequence Generation and Zadoff-Chu Sequences

Zadoff-Chu (ZC) sequences are defined as the following:

$$x_u(n) = \begin{cases} e^{-j\frac{\pi u n^2}{L_{RA}}}, & L_{RA} \text{ is even} \\ e^{-j\frac{\pi u n(n+1)}{L_{RA}}}, & L_{RA} \text{ is odd,} \end{cases} \quad (8.32)$$

where u is called the root sequence index such that $0 < u < L_{RA}$ and $\gcd(u, L_{RA}) = 1$, where $\gcd(\cdot)$ denotes the greatest common divisor. If L_{RA} is odd (which is the design choice in 5G), the sequence is periodic:

$$x_u(u + L_{RA}) = x_u(n). \quad (8.33)$$

Another important property of ZC sequences is that if L_{RA} is a prime, the DFT of a ZC sequence is another ZC sequence. This simplifies the implementation because instead of performing a DFT over a ZC sequence, a ZC output can be directly read from a table. Also, the ZC signal has a constant envelope, which results in low PAPR.

The autocorrelation function of the ZC sequence is defined as

$$R_u(m) = \frac{1}{L_{RA}} \sum_{n=0}^{L_{RA}-1} x_u(n) x_u^* ((n+m) \bmod L_{RA}) = \delta(m) \quad (8.34)$$

where $m = 0, \pm 1, \ldots, \pm(L_{RA} - 1)$. The autocorrelation function $R_u(m)$ is periodic. Similarly, the cross-correlation function is defined as

$$|R_{u_1, u_2}(m)| = \frac{1}{L_{RA}} \left| \sum_{n=0}^{L_{RA}-1} x_{u_1}(n) x_{u_2}^* ((n+m) \bmod L_{RA}) \right| = \frac{1}{\sqrt{L_{RA}}}. \quad (8.35)$$

Ideal autocorrelation and very low cross-correlation, combined with constant signal envelope, offers several advantages over other potential choices, and Zadoff-Chu sequences have been adopted both for 4G and 5G RACH sequences. The UL Sounding Reference Signal (SRS) is also based on a ZC sequence.

Let us examine the long preamble and its formats. The 3GPP specified TS 38.211 lists four formats for the long preamble, as shown in Table 8.12.

Table 8.12 Long RACH preamble format

Format	L_{RA}	Δf^{RA}	N_u	N_{CP}^{RA}
0	839	1.25 kHz	$1 \times 24576\kappa$	3168κ
1	839	1.25 kHz	$2 \times 24576\kappa$	21024κ
2	839	1.25 kHz	$4 \times 24576\kappa$	4688κ
3	839	5 kHz	$4 \times 6144\kappa$	3168κ

8.5 5G UL Physical Signals and Channels

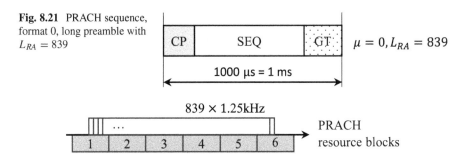

Fig. 8.21 PRACH sequence, format 0, long preamble with $L_{RA} = 839$

Fig. 8.22 PRACH bandwidth compared to six resource blocks when $\Delta f = 15\,\text{kHz}$

In format 0, the $N_{SEQ} = 1$, so only one preamble is used (no repetition), and the whole sequence is contained within a 1-ms processing window, as shown above (Fig. 8.21).

If the sampling time is $T_{samp} = 1/(15\,\text{kHz} \times 4096) = 16.28\,\text{ns}$, and using $\kappa = 2$, individual sequence components have the following duration:

$$T_{CP} = 3168 \times 2 \times T_{samp} = 103.13\,\mu\text{s}$$

$$T_{SEQ} = 24576 \times 2 \times T_{samp} = 800\,\mu\text{s}$$

$$T_{GT} = 1000\,\mu\text{s} - (T_{CP} + T_{SEQ}) = 96.87\,\mu\text{s}$$

We will examine the role of this guardtime, but let us first look at the frequency domain numerology. The RACH sequence, in the frequency domain, has the subcarrier spacing:

$$\Delta f^{RA} = \frac{1}{T_{SEQ}} = 1.25\,\text{kHz}.$$

which is in alignment with the value in Table 8.12. Given that there are 839 subcarriers, the RACH bandwidth is not aligned with a multiple integers of RBs (Fig. 8.22):

$$839 \times \Delta f^{RA} = 1.04877\,\text{MHz} < 6 \times 180\,\text{kHz} = 1.08\,\text{MHz}.$$

Similar calculations can be made for other formats. For example, for format 3, the sequence duration is $T_{SEQ} = 6144\kappa T_{samp} = 200\,\mu\text{s}$, and due to repetition, the entire sequence is $4 \times T_{SEQ} = 800\,\mu\text{s}$ long. For this format, the subcarrier spacing is wider which is more suitable for higher speeds (less subject do Doppler shifts).

The processing window durations, used to calculate guardtime lengths, are:

- format 0 uses a 1-ms processing window
- format 1 uses a 3-ms processing window

Fig. 8.23 The role of guardtime in initial access

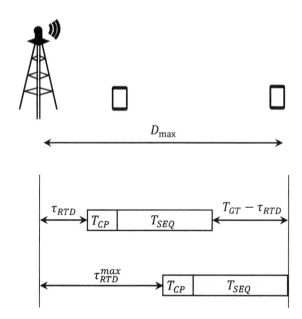

- format 2 uses a 3.5-ms processing window
- format 3 uses a 1-ms processing window (same as format 0)

The guardtime duration is determined by subtracting the repeated sequence from the processing window duration.

8.5.1.2 The Role of Guardtime

During the initial access, the timing offset for the UL transmission, $T_{TA} = N_{TA,offset}T_c$, as the N_{TA} component of timing advance is still unknown. As a result, the UL PRACH sequence will be delayed by the round-trip delay. Let us look at the scenario where two UTs are connected to the basestation (gNodeB). The closer UT will receive the DL frame with delay τ_1, and the further away UT will receive the DL frame with a larger delay τ_2. In both cases, The DL frame is the transmission reference, and the uplink transmission occurs with the delay τ_i, where $i = 1, 2$ (Fig. 8.23).

These uplink transmissions are received by the basestation after $2\tau_i$, which is the delay from start of the DL transmission, and is known as the round-trip delay (RTD). The guardtime duration needs to be sufficiently large to ensure that the entire T_{SEQ} is received at the basestation within a processing window. Therefore:

$$T_{GT} \geq \tau_{RTD} = 2\frac{D_{max}}{c} \Rightarrow D_{max} \leq \frac{1}{2}c \times T_{GT} \qquad (8.36)$$

8.5 5G UL Physical Signals and Channels

where c is the speed of EM waves. For example, for format 0, the $T_{GT} = 97\,\mu s$, and the maximal UT distance from the basestation is $D_{\max} = 14.55$ km. Sometimes, the channel delay spread τ_{DS} is accounted for in maximal cell distance, so the maximal cell radius is

$$D_{\max} = \frac{1}{2}c\left(T_{GT} - \tau_{DS}\right). \tag{8.37}$$

8.5.1.3 ZC Cyclic Shift

From the autocorrelation and cross-correlation properties of ZC sequences, we know that better detection of the PRACH sequence happens if we rely on autocorrelation. However, having only one sequence for the entire system would not work as there would collisions when multiple users try the initial access process. The question is: "how do we design multiple ZC sequences to be used in individual cells?" To answer that question, let us rewrite the odd-length ZC sequence:

$$x_u(n) = e^{-j\frac{\pi u n(n+1)}{L_{RA}}}$$

and define a cyclically shifted version of the same sequence as

$$x_{u,v}(n) = x_u\left((u+v) \bmod L_{RA}\right).$$

The autocorrelation between two cyclically shifted sequences is

$$R_u(v_1, v_2) = \frac{1}{L_{RA}} \sum_{n=0}^{L_{RA}-1} x_{u,v_1}(n) x^*_{u,v_2}(n) \tag{8.38}$$

$$= \begin{cases} 1, & v_1 = v_2 \bmod L_{RA} \\ 0, & v_1 \neq v_2 \bmod L_{RA} \end{cases} \tag{8.39}$$

Therefore, cyclically shifted versions of the same ZC sequence have the same properties as single sequences in terms of autocorrelation, which translates to improved ability to detect the signal while allowing for multiple sequences based on the same root value u.

The next question we need to answer is how long the cyclic shift, $v = N_{CS}$, needs to be. Let us recall that the channel delay has the same impact as the cyclic shift used for CP. Therefore, within the processing window at the receiver, the version of the signal that is delayed by m samples, is $x_{u,m}(n) = x\left((n+m) \bmod L_{RA}\right)$. As long as $m < N_{CS}$, the delayed version of the PRACH sequence will have zero autocorrelation with the cyclically shifted version of the ZC sequence:

Fig. 8.24 Processing window in the presence of the delayed signal component

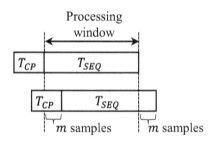

$$R_u(m, N_{CS}) = 0$$

The delay duration associated with the cyclic shift, T_{CS}, relative to the sequence duration T_{SEQ} can be calculated from the following equation:

$$\frac{T_{CS}}{T_{SEQ}} = \frac{N_{CS}}{L_{RA}} \qquad (8.40)$$

The round-trip delay needs to be smaller than the cyclic shift delay:

$$\tau_{RTD} \leq T_{CS} = \frac{N_{CS}}{L_{RA}} T_{SEQ} \qquad (8.41)$$

$$2\frac{D_{max}}{c} \leq \frac{N_{CS}}{L_{RA}} T_{SEQ} \qquad (8.42)$$

which allows us to calculate the cyclic shift value

$$N_{CS} = \left\lceil \frac{L_{RA}}{T_{SEQ}} \left(2\frac{D_{max}}{c} + \tau_{DS} \right) \right\rceil \qquad (8.43)$$

after we account for the delay spread.

Example 8.3 The random access preamble is generated using Zadoff-Chu sequences of length $L_{RA} = 839$. Using preamble format 0, determine the cyclic shift value that guarantees the cell radius to be ≤ 16 km assuming a delay spread of $\tau_{DS} = 6\,\mu s$.

Solution From (8.43), we calculate that $N_{CS} = 119$. Therefore, we can use sequences $x_u(n), x_{u,119}(n), x_{u,238}(n), \ldots x_{u,714}(n)$ to generate different ZC sequences. Note that $839 - 714 > 119$. The seventh cyclic shift would not work as $7 \times 119 = 833$ would be too *close* to unshifted version $x_u(n)$.

8.5 5G UL Physical Signals and Channels

Example 8.4 Using the result from the previous example, determine how many root sequences do we need to generate 64 preambles.

Solution Using a single root sequence, for the given cyclic shift value, we can generate seven different sequences: $x_u(n), x_{u,119}(n), \ldots, x_{u,6 \times 119}(n)$. Therefore, relying on cyclic-shift and autocorrelation properties is not sufficient to generate 64 PRACH sequences (this is the number of sequences adopted by 4G and 5G). To generate all 64 sequences, we need 10 different root sequences $u_1, u_2, \ldots, u_{10}$. This means that the sequences generated with different root values will have to rely on low cross-correlation.

8.5.2 Other UL Physical Channels and Signals

The Physical Uplink Control Channel (PUCCH) carries UL control information (UCI), which consists of:

- HARQ acknowledgments (HARQ-ACQ) for downlink transmissions
- scheduling request (SR) when a UT has something to transmit in the UL and is not otherwise scheduled
- channel state information (CSI) about the downlink channel

PUCCH timing and resources are scheduled by the DCI, although it is possible to semi-statically configure periodic PUCCH resources.

Physical Uplink Shared Channel (PUSCH) carries user data in the UL. Additionally, it can also carry UCI.

UL Demodulation Reference Signals (DM-RS) serve the same purpose as in the downlink—they are used for data demodulation.

Sounding Reference Signals (SRS) are the uplink equivalents of CSI-RS. These *pilot* (reference) signals provide a wideband channel view of the UL channel to the basestation. The SRS has a comblike structure to cover the whole band (unlike DM-RS signals that only go with the resources used for data transmission).

For more details on the UL physical channels and signals, the reader is referred to [11].

Problems

1. For the NR radio access technology, determine the following:

    a. Number of normal cyclic prefix samples for NR subcarrier spacing such that $\mu = 0$ and $\mu = 1$.

b. Number of useful symbol samples for both $\mu = 0$ and $\mu = 1$.
c. Total number of samples over the duration of 1 ms using the results from (a) and (b).

If the ratio κ is

$$\kappa = 2^{\mu+1},$$

show that the total number of samples calculated in (c) matches the number of samples over 1 ms calculated using the sampling rate expression $f_s = \Delta f \times N_{FFT}$.

2. A subset of Release 18 3GPP technical specification TS 38.101-1 lists the following supported NR bandwidths in FR1 range, along with the subcarrier spacings and the number of resource blocks.

BW	60 MHz	70 MHz	80 MHz	90 MHz	100 MHz
SCS	N_{RB}	N_{RB}	N_{RB}	N_{RB}	N_{RB}
30	162	189	217	245	273
60	79	93	107	121	135

Determine the maximal transmission bandwidth for given configurations.

3. A subset of Release 18 3GPP technical specification 38.101-2 lists the following supported NR bandwidths in the FR2 range, along with the subcarrier spacings and the number of resource blocks.

BW	50 MHz	100 MHz	200 MHz	400 MHz
SCS	N_{RB}	N_{RB}	N_{RB}	N_{RB}
60	66	132	264	N/A
120	32	66	132	264

For the entries that have numerical values different from "N/A," determine the maximal transmission bandwidth. Using these results, determine the minimal guardband for each of the bandwidth and subcarrier spacing configuration.

4. Calculate the missing value and enter them in the table below.

FR	μ	SCS [kHz]	$N_{symbols/slot}$	$N_{slots/frame}$	$N_{slots/subfr}$	T_{slot} μs	T_{symbol} μs
1	0		14			1000	symb 0,7: 71.875 else: 71.354
1	1		14				symb 0,14: else:
1,2	2		14				slots 0, 2: symb 0,28: else else:
2	3		14				slots 0, 4: symb 0,56: else else:
2	4		14				slots 0, 8: symb 0,112: else else:

8.5 5G UL Physical Signals and Channels

5. A timing advance command during the initial access random access response is based on an approximate UT distance from the basestation.

   a. If the $T_A = 3846$ has been selected for the numerologies listed in the table below, calculate the missing values:

FR	SCS	N_{TA}	Time (ms)	Distance (km)
FR1	15 kHz			
FR1	30 kHz			
FR1	60 kHz			
FR2	60 kHz			
FR2	120 kHz			

   For calculations, the random access-based time alignment value is $N_{TA} = T_A \times 16 \times 64/2^\mu$ (note that $N_{TA,Offset}$ is not part of the initial access time adjustment). Distance between the UT and the basestation (in meters) is related to the time advance as $d = \frac{1}{2}\text{Time(s)} \times c$, where $c = 3 \times 10^8$ m/s.

   b. Determine the adjustment step size in time units (i.e., in ns) for each of the SCS values.
   c. Assuming the distance is known, calculate the T_A values by rounding to the closest integer.

FR	SCS	T_A	Distance (km)
FR1	15 kHz		10
FR1	30 kHz		8
FR2	60 kHz		6

6. A Zadoff-Chu sequence of odd length L_{RA} is defined as

$$x_u(n) = e^{-j\pi \frac{un(n+1)}{L_{RA}}}$$

   Show that it is periodic.

7. A Zadoff-Chu sequence of odd length L_{RA} is defined as

$$x_u(n) = e^{-j\pi un(n+1)/L_{RA}}$$

   Show that its autocorrelation function:

$$R_u(m) = \frac{1}{L_{RA}} \sum_{n=0}^{L_{RA}-1} x_u(n) x_u^* ((n+m) \bmod L_{RA}) = \delta(m) \qquad (8.44)$$

   where $m = 0, \pm 1, \ldots, \pm(L_{RA} - 1)$.

8. PRACH signal in 5G can be generated using four different Zadoff-Chu sequence lengths L_{RA}, which are further mapped to different subcarrier spacing values Δf_{RA} for the PRACH signal.

L_{RA}	Δf_{RA} for PRACH	N_U (sample count)	Δf for PUSCH	$N_{RB^{RA}}$, PUSCH	PRACH BW (MHz)
839	1.25		15		
839	1.25		30		
839	1.25		60		
839	5		15		
839	5		30		
839	5		60		
139	15		15		
139	15		30		
139	15		60		
139	30		15		
139	30		30		
139	30		60		
139	60		60		
139	60		120		
139	120		60		
139	120		120		
571	30		15		
571	30		30		
571	30		60		
1151	15		15		
1151	15		30		
1151	15		60		

Populate the missing values in the table above by showing all the steps.

9. For preamble formats given in the table below, calculate the guardtime and the maximum cell radius. Assume $\kappa = 2$ and $T_{samp} = 1/(4096 \times 15 \times 10^3)$.

Format	L_{RA}	Δf^{RA}	N_U	N_{CP}^{RA}	T_{GT}	Max cell radius (km)
0	839	1.25 kHz	$1 \times 24576\kappa$	3169κ		
1	839	1.25 kHz	$2 \times 24576\kappa$	21024κ		
2	839	1.25 kHz	$4 \times 24576\kappa$	4688κ		
3	839	5 kHz	$4 \times 24576\kappa$	3169κ		

10. Short preamble formats have been specified for 5G. One of them, format B, is illustrated below (lengths are not to scale; only two of the four formats B

are shown). This format has been designed to support small to medium cell sizes. Format B RACH signals are aligned with the regular symbol edges and consist of the following: (i) cyclic prefix, (ii) N_{seq} times repeated sequence of the duration T_{seq}, and (iii) the time gap T_{GT}. The time gap constrains the max cell radius.

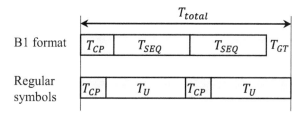

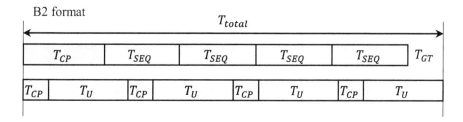

Calculate the missing values in the table below assuming $T_{seq} = T_U = 4096 \times T_{sam}$, where $T_{sam} = 1/61.44 \times 10^{-6}$ seconds. For regular symbols (non-RACH symbols), $T_{CP} = 288 \times T_{sam}$, and for RACH signals, the T_{CP} values are provided in the table. For the bandwidth calculation, round the determined bandwidth value to the nearest integer multiple of resource blocks.

Format	L_{RA}	Δf_{RACH}	BW (MHz)	T_{seq} (μs)	T_{CP} (μs)	N_{seq}	T_{GT} (μs)	T_{total} (μs)	Max cell radius
B1	139			$2048\kappa 2^{-\mu}$	$216\kappa 2^{-\mu}$	2			
B2	139			$2048\kappa 2^{-\mu}$	$360\kappa 2^{-\mu}$	4			
B3	139			$2048\kappa 2^{-\mu}$	$504\kappa 2^{-\mu}$	6			
B4	139			$2048\kappa 2^{-\mu}$	$936\kappa 2^{-\mu}$	12			

11. Consider a TDD system operating in FR1 range that has a channel bandwidth of 80 MHz and is configured with a frame pattern DDFUUDDDDD (verify, D indicates the DL slot, F indicates flexible slot and U indicates the UL slot). Determine the maximal cell radius that can be achieved based on RACH channel.

12. For the short RACH preamble formats A, listed below:

Format	L_{RA}	Δf_{RACH}	N_U	N_{CP}
A1	139	$15 \times 2^\mu$ kHz	$2 \times 2048\kappa 2^{-\mu}$	$288\kappa 2^{-\mu}$
A2	139	$15 \times 2^\mu$ kHz	$4 \times 2048\kappa 2^{-\mu}$	$576\kappa 2^{-\mu}$
A3	139	$15 \times 2^\mu$ kHz	$6 \times 2048\kappa 2^{-\mu}$	$864\kappa 2^{-\mu}$

verify the calculations provided below and add the missing values.

Format	L_{RA}	Δf_{RACH}	Time domain duration	Occupied bandwidth	Cell radius	Typical scenario
A1	139	$15 \times 2^\mu$ kHz	$0.14/2^\mu$ ms	$2.16 \times 2^\mu$ MHz		Small cells
A2	139	$15 \times 2^\mu$ kHz	$0.29/2^\mu$ ms	$2.16 \times 2^\mu$ MHz		Normal cells
A3	139	$15 \times 2^\mu$ kHz	$0.43/2^\mu$ ms	$2.16 \times 2^\mu$ MHz		Normal cells

References

1. 3GPP TS 38.211, NR; Physical channels and modulation, V18.2.0 (2024). https://www.3gpp.org/ftp/Specs/archive/38_series/38.211/38211-i20.zip. Cited May 19, 2024
2. 3GPP TS 38.104, NR; Base Station (BS) radio transmission and reception, V18.5.0 (2024). https://www.3gpp.org/ftp/Specs/archive/38_series/38.104/38104-i50.zip. Cited May 19, 2024
3. 3GPP TS 36.101, User Equipment (UE) radio transmission and reception, V85.5.0 (2024). https://www.3gpp.org/ftp/Specs/archive/36_series/36.101/36101-i50.zip. Cited May 19, 2024
4. 3GPP TS 38.101-1, NR; User Equipment (UE) radio transmission and reception; Part 1: Range 1 Standalone, V18.5.0 (2024). https://www.3gpp.org/ftp/Specs/archive/38_series/38.101-1/38101-1-i50.zip. Cited May 19, 2024
5. 3GPP TS 38.101-2, NR; User Equipment (UE) radio transmission and reception; Part 2: Range 2 Standalone, V18.5.0 (2024). https://www.3gpp.org/ftp/Specs/archive/38_series/38.101-2/38101-2-i50.zip. Cited May 19, 2024
6. T.S. Rappaport, S.Y. Seidel, R. Singh, 900-MHz multipath propagation measurements for US digital cellular radiotelephone. IEEE Trans. Veh. Technol. **39**, 132–139 (1990)
7. 3GPP TS 38.213, NR; Physical layer procedures for control, V18.2.0 (2024). https://www.3gpp.org/ftp/Specs/archive/38_series/38.213/38213-i20.zip. Cited May 19, 2024
8. H. Holma, A. Toskala, T. Nakamura *5G Technology: 3GPP New Radio* (Wiley, New York, 2020)
9. J.H. Bae, A. Abotabi, H.-P. Lin, K.-B. Song, J. Lee, An overview of channel coding for 5G NR cellular communication. APSIPA Trans. Signal Inf. Process. **8**(1), e17 (2019)
10. 3GPP TS 38.214, NR; Physical layer procedures for data, V18.2.0 (2024). https://www.3gpp.org/ftp/Specs/archive/38_series/38.214/38214-i20.zip. Cited May 20, 2024
11. E. Dahlman, S. Parkvall, J. Sköld, *5G NR: The Next Generation Wireless Access Technology*, 2nd edn. (Academic Press, New York, 2021)

Chapter 9
3GPP-Based Air Interface: Layer 2

9.1 Protocol Architecture

Layer 2 (L2) of the 5G or New Radio (NR) protocol is divided into several sublayers, each serving a specific purpose. These sublayers include the Medium Access Control (MAC), Radio Link Control (RLC), Packet Data Convergence Protocol (PDCP), and Service Data Adaptation Protocol (SDAP). In the downlink and uplink directions, the Layer 2 architecture is represented by Figs. 9.1 and 9.2, respectively. Interactions between L2 sublayers are broken down as follows:

- The physical layer provides transport channels to the MAC sublayer. Transport channels describe how and with what characteristics the data is transferred over the air interface, including channels like the Broadcast Channel (BCH) and the Downlink Shared Channel (DL-SCH).
- The MAC sublayer provides logical channels to the RLC sublayer. Logical channels define the type of information being transmitted, categorizing it into control and traffic channels such as the Dedicated Control Channel (DCCH) and Dedicated Traffic Channel (DTCH).
- The RLC sublayer provides channels to communicate with the PDCP sublayer.
- The PDCP sublayer provides channels to offer its services to the SDAP sublayer.
- The SDAP sublayer maps radio bearers to 5GC QoS flows.

Supplementary Information The online version contains supplementary material available at (https://doi.org/10.1007/978-3-031-76455-4_9).

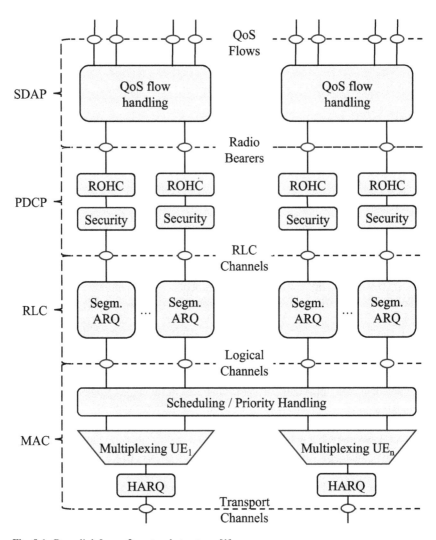

Fig. 9.1 Downlink Layer 2 protocol structure [1]

9.2 The Medium Access Control Sublayer

The Medium Access Control (MAC) sublayer performs various essential services and functions [2]. These include:

- Mapping between logical channels and transport channels, ensuring proper data flow.
- Multiplexing/Demultiplexing: The MAC sublayer efficiently combines or separates MAC service data units (SDUs) from different logical channels into/from

9.2 The Medium Access Control Sublayer

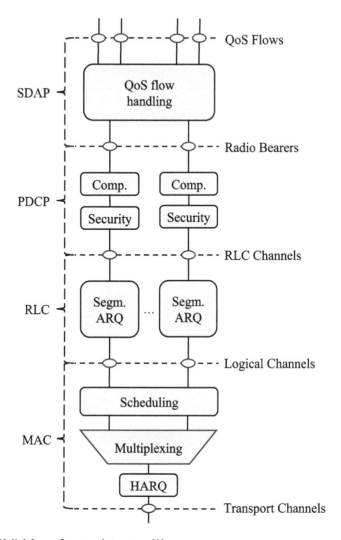

Fig. 9.2 Uplink Layer 2 protocol structure [1]

transport blocks (TB). These TBs are then delivered to/from the physical layer using transport channels.
- Scheduling Information Reporting, including scheduling requests and buffer status reports.
- Hybrid Automatic Repeat Request (HARQ): Error correction is achieved through HARQ, where each cell in the case of carrier aggregation (CA) has a dedicated HARQ entity to handle retransmissions and ensure reliable data transmission.

- Dynamic Scheduling: Prioritization among UEs is managed by dynamic scheduling, which allocates resources based on varying priorities to enhance overall system performance.
- Logical Channel Prioritization (LCP): Within a UE, logical channels are assigned priorities to determine their order of resource allocation and access to the available transmission resources.
- Handling Overlapping Resources: The MAC sublayer efficiently manages overlapping resources within a single UE, including prioritization between dynamic grants, configured grants, and SR resources, thus ensuring optimal utilization.
- Padding: When necessary, the MAC sublayer adds padding bits to ensure proper alignment with the uplink grant size and the number of bits provided to the physical layer.

A single MAC entity has the capability to support multiple numerologies, transmission timings, and cells. LCP mapping restrictions defined in logical channel prioritization control determine which grant characteristics map to logical channels, where characteristics include numerologies, cells, or transmission durations a logical channel can map to.

Radio bearers are categorized into two groups: data radio bearers (DRB) for user plane data and signaling radio bearers (SRB) for control plane data.

9.2.1 Mapping Between Logical Channels and Transport Channels

The MAC sublayer offers different types of data transfer services, each catering to specific information requirements. These services can be categorized into two groups: control channels and data traffic channels. Control channels focus on the transmission of control plane information. The following control channels are handled in the MAC layer:

- Broadcast Control Channel (BCCH): This downlink channel is responsible for broadcasting system control information to all UEs.
- Paging Control Channel (PCCH): Operating in the downlink direction, the PCCH carries paging messages for specific UEs. Paging is explained in Chap. 10.
- Common Control Channel (CCCH): The CCCH facilitates the transmission of control information between UEs and the network. It serves UEs that do not have a Radio Resource Control (RRC) connection established with the network.
- Dedicated Control Channel (DCCH): As a bidirectional point-to-point channel, the DCCH is dedicated to transmitting control information between a UE and the network. It is utilized by UEs that have an active RRC connection.

Traffic channels handle user plane information transfer only:

- Dedicated Traffic Channel (DTCH): The DTCH is a point-to-point channel dedicated to a single UE, facilitating the transfer of user information. It can exist in both the uplink and downlink directions.

In the downlink direction, the following mappings exist between logical channels and transport channels:

- BCCH can be mapped to Broadcast Channel (BCH).
- BCCH can also be mapped to Downlink Shared Channel (DL-SCH).
- PCCH can be mapped to Paging Channel (PCH).
- CCCH can be mapped to DL-SCH.
- DCCH can be mapped to DL-SCH.
- DTCH can be mapped to DL-SCH.

In the uplink direction, the following mappings exist between logical channels and transport channels:

- CCCH can be mapped to Uplink Shared Channel (UL-SCH).
- DCCH can be mapped to UL-SCH.
- DTCH can be mapped to UL-SCH.

These connections between logical channels and transport channels enable the efficient transfer of information within the MAC sublayer and contribute to the overall functionality of the system.

9.2.2 Random Access Procedure

In IDLE and Inactive modes, a random access (RA) procedure is used to access a cell during the initial access procedure to transition to connected mode. In connected mode, random access is initiated during handover at the target cell, during beam failure recovery, after reception of a PDCCH order by the network to renew the UE's timing advance for uplink synchronization, to request an uplink grant if the UE does not have a valid PUCCH resource to transmit a scheduling request, or after the detection of a consistent listen before talk (LBT) failure on an unlicensed primary cell.

In NR, there are two types of random access procedures: four-step random access and two-step random access. A two-step procedure is advantageous to use to facilitate faster access to the cell, as it combines the preamble and data payload in a single transmission in the first step. However, it only works well in case the uplink timing advance is already known by the UE or not required (e.g., for small cells), as the timing synchronization is required to successfully decode the PUSCH payload.

Both types of RA procedure support contention-based random access (CBRA) and contention-free random access (CFRA). During the initiation of the random access procedure, the UE determines the random access type based on the network configuration. If CFRA resources are not configured, the UE uses a reference signal received power (RSRP) threshold to choose between the two-step RA type and four-step RA type. The RSRP threshold acts a mean to ensure that the UE accessing the cell is close to its cell center region, where uplink timing advance errors are not prevalent or timing advance values are limited. If CFRA resources are configured for the four-step RA type, the UE performs random access using the four-step RA type. If CFRA resources are configured for the two-step RA type, the UE performs random access using the two-step RA type.

In a four-step contention-based RA procedure, the UE follows the following steps:

1. the UE transmits a preamble (Msg1);
2. the network responds by transmitting a random-access response (RAR/Msg2) indicating reception of the preamble and providing a time-alignment command, a temporary UE identity (T-C-RNTI), and an uplink grant for the transmission of Msg3. The RAR is received as downlink PDCCH/PDSCH transmission with the corresponding PDCCH transmitted within the common search space, where the PDCCH is scrambled by a specified RA-RNTI which is determined by the UE as a function of the RACH occasion selected for Msg1 transmissions. The RA-RNTI thus serves as an echo to the PRACH resource selected by the UE.
3. the UE transmits the PUSCH payload (Msg3) on the provided grant, and it includes its temporary identity—if not already assigned—and potentially a buffer status report if the grant is big enough.
4. the network transmits a contention resolution message (Msg4) if there is no collision with another UE that has selected the same preamble and RA occasion. This message is scheduled addressed to the UE identity included in Msg3, either the T-C-RNTI or the C-RNTI if already assigned. Upon successful reception of Msg4, the procedure is considered successfully complete, and the UE considers the T-C-RNTI as its C-RNTI. If the UE does not receive a Msg4 within a configured contention resolution timer, the UE goes back to step 1 and retransmits a preamble (Fig. 9.3).

Preambles selected within a RACH occasion are selected by the UE at random. In the event that two UEs select the same RACH occasion and the same preamble, this is referred to as a preamble collision or a contention, which can happen at higher network access loads. In such an event, both UEs transmit Msg3 simultaneously creating contention, as both UEs will use the same grant indicated in the RAR and the RAR is addressed to a common RA-RNTI related to the selected preamble and selected RACH occasion. Neither or only one of the two UEs may get a contention resolution message (Msg4), depending on whether the network was able to decode the PUSCH payload for at least one UE. The colliding UEs can thus restart the procedure using a different RACH occasion and a different preamble selected at random, or the network can reschedule a Msg3 retransmission grant for the colliding UE.

After the random-access procedure is successfully completed, the device transitions to connected state, if not already there, enabling communication between the network and the UE through dedicated transmission channels.

9.2 The Medium Access Control Sublayer

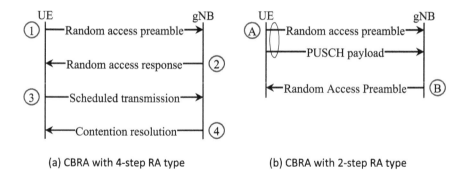

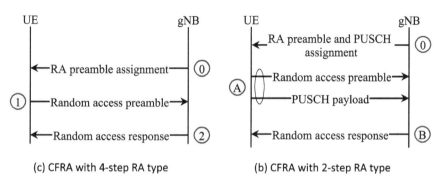

Fig. 9.3 Random access procedures [1]

The preamble transmission may involve retransmitting the preamble multiple times, gradually increasing the transmit power between each transmission due to uncertain transmit power requirements and pathloss estimate to the base station. The initial transmit power is determined based on estimated path loss and a target received preamble power set by the network. The path loss estimation relies on the received power of the acquired synchronization signal block (SSB); the structure of which is described in Chap. 8. The best measured SSB also determines the RACH resource on which the preamble transmission is made. It assumes that if the preamble is received through beamforming, the corresponding SS block was transmitted with a matching transmit beam.

When the cell has multiple SSBs configured with downlink beamforming, the random-access response should align with the beamforming used for the acquired SS block during the initial cell search. This ensures that the UE, which uses receive-side beamforming, can properly orient its receiver beam. By transmitting the random-access response using the same beam as the SS block, the device can use the same receiver beam identified during the cell search. To enable this, the UE is configured with an association between SSBs and PRACH resources, such that network is able to know which SSB is most preferred from the UE's perspective by mere reception of a preamble. From the network perspective, the selected preamble and RA occasion can thus be an indication of the UE's preferred beam, e.g., for the best direction to reach the UE with an acceptable signal strength.

If no random-access response is received within a predetermined timeframe, a configured RAR window, the UE retransmits the preamble with an increased transmit power, offset by a configurable value. This power ramping process continues until a random-access response is received, a maximum number of retransmissions is reached, or the maximum preamble transmit power is reached. Once the UE reaches the maximum number of preamble retransmissions, the random-access attempt is considered a failure, and the UE initiates a radio link failure procedure and a corresponding RRC re-establishment procedure.

For contention-free random access, a dedicated preamble serves as a dedicated resource for the UE thus eliminating contention. Additionally, the UE has already been assigned a unique identity in a C-RNTI which is used to address Msg2 to the UE.

For the two-step RA type, the transmission in the first step (MSGA) consists of a preamble on PRACH and a payload on PUSCH. After transmitting the MSGA, the UE waits for a response from the network within a configured timeframe, known as the MsgB window.

For two-step CBRA, if contention resolution is successful upon receiving the network response, the UE concludes the random access procedure. However, if a fallback indication is received in MSGB, the UE falls back to a four-step procedure and proceeds with MSG3 transmission using the UL grant specified in the fallback indication and monitors contention resolution (Fig. 9.4). If contention resolution is not successful after MSG3 transmission or retransmission, the UE returns to MSGA transmission. In addition, if the random access procedure with the two-step RA type is not completed after a certain number of MSGA transmissions, the UE can be configured to switch to CBRA with the four-step RA type.

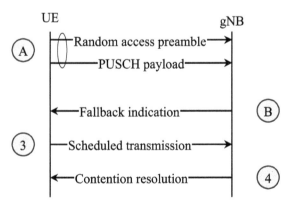

Fig. 9.4 Fallback for CBRA with two-step RA type

9.2 The Medium Access Control Sublayer

When carrier aggregation is activated, when using the four-step RA type for the random access procedure, the first three steps of contention-based random access (CBRA) always occur on the primary cell (PCell), while the fourth step, which is contention resolution, can be cross-scheduled by the PCell. The three steps of contention-free random access (CFRA) initiated on the PCell remain confined to the PCell. On the other hand, CFRA on the secondary cell (SCell) can only be initiated by the gNB to establish timing advance for a secondary timing advance group (TAG). The CFRA procedure initiated by the gNB includes the following steps: PDCCH order (step 0) transmitted on an activated SCell of the secondary TAG, preamble transmission (step 1) taking place on the SCell, and random access response (step 2) occurring on the PCell.

When performing random access in a cell configured with Supplementary Uplink (SUL), the network has the option to explicitly indicate which carrier to use, whether it is the Uplink (UL) or SUL carrier. In the absence of explicit signaling, the UE decides to employ the SUL carrier only if the measured quality of the downlink RSRP falls below a predefined broadcast threshold. Carrier selection takes place prior to choosing between the two-step and four-step RA type. Once initiated, all uplink transmissions of the random access procedure remain confined to the selected carrier.

9.2.3 Scheduling

The operation of the network scheduler involves the allocation of radio resources to UEs based on considerations such as UE buffer status and QoS requirements. The scheduler takes into account the specific requirements of each UE and its associated radio bearers when assigning resources. In addition, the scheduler may consider the radio conditions at the UE, which can be determined through measurements conducted at the gNB or reported by the UE itself. These measurements aid in making informed decisions about resource assignment.

Resource assignment primarily involves allocating radio resources in the form of resource blocks. These resource blocks are the fundamental units of radio frequency and time that are allocated to UEs to support their communication needs.

In LTE and in NR, the network grants uplink radio resources to the UE for a transmission on the UL shared channel (UL-SCH). The UE may receive such a resource allocation either in a grant received on the PDCCH or in a configured resource (a semi-persistently scheduled UL grant). Conceptually, the MAC layer provides the HARQ entity with the necessary information for the UL transmission. This information may consist of one or more of the following:

- a new data indication (NDI), controlling whether or not the uplink transmission should be a new transmission or a retransmission;
- a transmission unit, i.e., a transport block (TB) size, indicating the number of bits available for the uplink transmission;

- a redundancy version (RV);
- a transmission duration (transmit time intervals or TTI);

In the absence of carrier aggregation and spatial multiplexing, there is at most a single TB of a given transmission duration at any given time.

The HARQ entity typically identifies the HARQ process for which the transmission should take place. The HARQ entity also routes HARQ feedback and the modulation and coding scheme (MCS) to the HARQ process.

The values of the NDI, TB size, RV, TTI, and MCS are controlled by the network and selected to meet the QoS requirements such as the Packet Delay Budget (PDB), the Packet Error Loss Rate (PLER), and corresponding radio Block Error Rate (BLER) target of the different radio bearers established for the UE, based on, e.g., Buffer Status Reporting (BSR) information, reported Channel Quality Indications (CQI) and HARQ feedback received from the UE.

To facilitate the efficient operation of the scheduler, several measurements are used in 5G networks. These measurements provide valuable information to optimize packet scheduling based on specific considerations. Measurements, such as uplink buffer status reports and power headroom reports, support the scheduler's decision-making process, ensuring efficient resource allocation and enhancing overall network performance.

9.2.4 Uplink Scheduling

For uplink data transmission of a cellular network in connected mode, the gNB can either dynamically allocate resources to UEs by dynamic PDCCH scheduling addressed to the UE's unique C-RNTI or configure semi-static periodic uplink resources, known as configured grants. For dynamic grants, the UE constantly monitors the PDCCH to identify potential dynamic grants for transmitting uplink data.

For dynamically scheduling low-latency transmissions for certain UEs configured with stringent QoS services, the gNB can interrupt ongoing PUSCH or SRS transmissions of lower priority for the same UE or a different UE. The gNB configures other UEs to monitor indications of these canceled transmissions by using a Cell-specific cancelation indication CI-RNTI on the PDCCH. If a UE receives the cancelation indication, it is required to terminate the ongoing PUSCH transmission, starting from the earliest symbol that overlaps with the affected PUSCH resource allocation or the SRS transmission overlapping with the indicated resource allocation. This allows the gNB to perform inter-UE cancelation, in cases where the available bandwidth is not sufficient to complete a low-latency transmission in time.

For configured grants, the gNB can allocate uplink resources for both initial HARQ transmissions and HARQ retransmissions for a given UE. Retransmissions are allocated using dynamic grants scheduled by the PDCCH or can be UE autonomous, whereby the UE can retransmit the transport block after the expiry of a configured retransmission timer, if configured. Two distinct types of configured uplink grants have been specified:

- Type 1: Configuration by the radio resource control (RRC) layer, defined in Chap. 10, directly provides the configured uplink grant to the UE, including the periodicity at which it is granted. This grant type can only be (de)-activated using RRC (re)-configuration.
- Type 2: Configuration by RRC defines the periodicity of the configured uplink grant. Subsequently, PDCCH addressed to the UE's configured scheduling RNTI (CS-RNTI) can either signal and activate the configured uplink grant or deactivate it. The CS-RNTI can also be used by the gNB scheduler to schedule a dynamic grant for a retransmission of a transport block initially transmitted on a configured grant.

Withing a serving cell, the UE can be assigned a maximum of 12 active configured uplink grants in a given bandwidth part, defined in Section 9.2.14. In cases where multiple configured uplink grants are allocated, the network can activate multiple configured grants simultaneously, which can be useful when the UE supports services of different periodicities or to retransmit a TB that failed LBT when another configured grant is on a different LBT sub-band, for example.

For Type 2 configured uplink grants, the activation and deactivation of grants are independent among the serving cells. When multiple Type 2 configured grants are configured, each grant is activated individually using a DCI command. Similarly, deactivation of Type 2 configured grants is achieved using a DCI command, which can deactivate either a single configured grant configuration or multiple configured grant configurations collectively.

The NR ConfiguredGrantTimer, configured per HARQ process, serves as a mean to control and govern HARQ processes of configured grants. While the timer is running, the UE stores the transmitted transport block in the associated HARQ buffer until the timer expires. Upon the expiry of the timer, the UE flushes the HARQ buffer. This timer serves both as means for the network to control the time during which it can schedule a dynamic retransmission (e.g., by a dynamic grant scheduled to the UE's CS-RNTI), but also a means to avoid sending explicit HARQ ACK signaling just to clear the UE HARQ buffer. The UE starts this timer once a TB is transmitted on the associated HARQ process.

In NR systems, a HARQ process ID for a new transmission on a configured grant is determined by a formula that considers the time occasion in which the transmission is made

$$\text{HARQ Process ID} = \left\lfloor \frac{\text{CURRENT\_symbol}}{periodicity} \right\rfloor \mod nforHARQ - Process \tag{9.1}$$

where *periodicity* is the CG periodicity in symbols, i.e., the number of symbols between consecutive CG occasions and

$$\text{CURRENT\_symbol} = \text{SFN} \times numberOfSlotsPerFrame \times numberOfSymbolsPerSlot +$$
$$\text{slot number in the frame} \times numberOfSymbolsPerSlot +$$
$$\text{symbol number in the slot}$$

This HARQ formula results in uniformly distributing the HARQ process IDs in the time domain while maintaining a common understanding between the network and the UE on the HARQ process ID used (i.e., there is no ambiguity on which process ID was selected as it is a function of synchronization and the timing of the starting symbol).

Example 9.1 (Max Uplink Data Rate Determination Using a Configured Grant)
In an NR-FDD system using a subcarrier spacing of 15 kHz and a 1 ms slot duration, a device capable of four HARQ processes is configured with a configured grant with the following parameters: a periodicity of two slots, a grant size of 512 bits, and a configured grant timer of 16 ms. Assuming that the device has enough data (full buffer) to transmit at each configured grant occasion, determine the following:

1. the time a given HARQ process repeats itself;
2. the minimum time until a HARQ process can be reused to transmit new data;
3. Over a period of 28 ms, compute the maximum uplink data rate using the available configured grant, assuming no retransmissions are required.
4. Over a period of 28 ms, compute the maximum uplink data rate using the available configured grant, if one retransmission grant for HARQ process 1 is received 11 ms after the initial transmission on this HARQ process. Assume HARQ processes are identified by 0, 1, 2, and 3.

Solution

1. For simplicity, we can determine the HARQ process allocation in SFN 0 as a reference and work within a given frame. Further, since the periodicity of the configured grant is unit of slots, we can simplify CURRENT_symbol to be

$$\text{HARQ Process ID} = \left\lfloor \frac{\text{CURRENT\_symbol}}{periodicity} \right\rfloor \mod nforHARQ - Process$$
$$= \left\lfloor \frac{\text{CURRENT\_symbol}}{2} \right\rfloor \mod 4$$

Therefore, a HARQ process repeats itself every $4 \times 2 = 8$ slots. More clearly, a CG occasion occurs for each HARQ process at the following slots:
HARQ process 0: 0, 8, 16, 24.

9.2 The Medium Access Control Sublayer

HARQ process 1: 2, 10, 18, 26;
HARQ process 2: 4, 12, 20, 28;
HARQ process 3: 6, 14, 22, 30;

2. The UE cannot transmit new data until the configured grant timer expires (which allows time for the gNB to schedule a retransmission during such period using a dynamic grant if needed). The minimum time until a HARQ process can be reused to transmit new data is thus 16 ms.
3. In the first 16 ms of the 28 ms evaluation period, all four HARQ process can be used to transmit four new TBs; thus $512 \times 4 = 2048$ bits are transmitted. Assuming there are no retransmissions required, all HARQ processes are flushed 16 ms after the initial transmission is made. In the next 8 ms, all HARQ process start to get flushed as the configured grant timer expires given there are no retransmission grants received, and thus the UE can use the configured grant HARQ processes to transmit new data. Four additional new TBs are thus transmitted, totaling another 2048 bits. In the final 4 ms of the 28 ms period, no HARQ process can be used as the configured grant timer is still running. The overall uplink data rate is therefore 2048×2 bits/28×10^{-3} seconds = 146.29 kbps.
4. In the first 16 ms, 2048 new bits are transmitted, per the explanation in (c). For HARQ process 1, the configured grant timer is started at slot 2, when the first initial TB is transmitted, then restarted again at slot 13 when the retransmission grant is received. In the next 8 ms, HARQ processes 0, 2, and 3 can be reused as the configured grant timer is not running; thus an additional $512 \times 3 = 1536$ new bits are transmitted. For HARQ processes 1, the process will be flushed again at slot 29 if no other retransmission grant is received, which is after the evaluation period. The overall uplink data rate is therefore $(2048 + 1536)$ bits/28×10^{-3} seconds = 128 kbps. ∎

One limitation of this HARQ process ID allocation formula is the fact that uniform time domain distribution of HARQ processes does not consider slots in which uplink cannot be transmitted (e.g., slots part of the downlink portion of a TDD frame, slots preempted by DL transmission or dynamically indicated as downlink). This is illustrated in the following example.

Example 9.2 (HARQ Process Allocation in TDD)
Using the legacy HARQ PID formula, Fig. 9.5 plots how the UE determines the HARQ process ID for a configured grant with the following configurations:

- a configured grant periodicity of one slot
- a subcarrier spacing of 15 kHz
- a maximum of eight HARQ processes
- over a period of five frames (50 ms)
- a TDD frame split of [D, D, D, U, U, D, D, D, U, U]

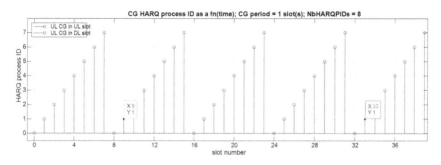

Fig. 9.5 CG HARQ PID selection over a period of 50 ms, using the legacy formula

Solution As illustrated in Fig. 9.5, for this given TDD frame split, HARQ process ID #1 repeats itself only once every 24 ms. This is a considerable delay for the purpose of retransmitting or repeating the TB (e.g., when CG repetition is configured for coverage enhancement). The delay can be even larger if there are fewer UL slots within the TDD frame. Furthermore, Fig. 9.5 illustrates the inefficiency of uniformly spreading HARQ process IDs over time when some slots are downlink only or not valid for CG transmission. As illustrated, it takes 24 ms to use all available HARQ processes, though 10 uplink slots have elapsed already. In slot 19, the existing formula yields selecting a HARQ process ID that has already occurred (in slot 3), though HARQ process 7 has not been used so far and is sitting IDLE. This can result in a reduction in the overall TDD uplink data rate, especially in certain TDD frame split configurations. ∎

In both dynamic grant and configured grant scenarios, multiple repetitions of a transport block can be accommodated within a single slot or spread across slot boundaries in consecutive available slots, with each repetition contained within one slot. Additionally, for both dynamic grant and configured grant Type 2 situations, the number of repetitions can be dynamically indicated by Layer 1 signaling. In such cases, the dynamically indicated number of repetitions takes precedence over the number of repetitions configured by RRC, if both are present.

9.2.5 UL TB Construction and Logical Channel Prioritization

Logical channel prioritization (LCP) is a mechanism used to associate uplink buffered data available for transmission and UL-SCH resources available for uplink transmissions. Multiplexing of data with different QoS requirements within the same transport block can be supported as long as such multiplexing neither introduces negative impacts to the service with the most stringent QoS requirement nor introduces unnecessary waste of system resources.

9.2 The Medium Access Control Sublayer

When the UE assembles the PDU for transmission, it multiplexes one or more MAC SDUs from one or different logical channels (LCH) onto the TB to be delivered to the physical layer on the proper transport channel. In NR, such multiplexing may further include considerations for mapping restrictions between data from a LCH and a given TB, based on one or more characteristics of the transmission of the TB. Such characteristics include the OFDM subcarrier spacing, the maximum PUSCH transmission duration, the type of configured grant (e.g., Type 1, Type 2), the serving cell(s) allowed for transmission of the data for a LCH, the priority index associated with the UL grant, or the HARQ feedback suppression mode associated with the UL grant.

When assembling a transport block for UL transmission, the UE typically allocates data to the uplink grant from one or more LCH(s) using the following procedure: the UE performs LCP using three rounds:

Round 1 (or step 0): the UE selects the logical channels that can be applicable for transmission on the associated uplink grant, whereby such selection depends on the configured logical channel selection restrictions configured by RRC. These are the logical channels than can contend for resource allocation in the following rounds of the LCP procedure.

Round 2 (or equivalently, steps 1, 2): data from logical channels is taken up to Prioritized Bit Rate (PBR) in decreasing LCH priority order. For each LCH, data can be allocated to the grant in this step up to a bucket "Bj" that fills up according to the configured PBR. Data allocation can exceed the allotted amount of data allocation for the LCH for transmission in a given TTI, i.e., the "token bucket", typically to avoid unnecessary RLC segmentation. Round 3 (or equivalently, steps 3): remaining buffered data from logical channels beyond what has been allocated in the second round is allocated in strict decreasing LCH priority order to fill the remaining resources in the uplink grant.

To perform this LCP procedure, the UE is configured by RRC with the following LCP parameters per logical channel:

- A logical channel priority;
- A prioritized bit rate (PBR), which is the bit rate which the token bucket fills up for the amount of data to be served in Round 2.
- The bucket size duration (BSD). This parameter sets the maximum token bucket size, which is computed as PBR x BSD. The bucket size represents the maximum amount of data that can be served for the LCH in Round 2.
- Logical channel selection restrictions: whether the logical channel is applicable for LCP according to at least one of the following traits:
    - allowedSCS-List which sets the allowed subcarrier spacing(s) for transmission; this configuration can be useful to restrict data to only a subset of OFDM subcarrier spacings that are capable of supporting the latency and reliability requirements of the data. If present, data from this logical channel can only be mapped to the indicated numerology. Otherwise, data can be mapped to any configured numerology.

- maxPUSCH-Duration which sets the maximum PUSCH duration allowed for transmission; this configuration can be useful to restrict data to be multiplexed only on grants capable of meeting the latency requirements of the associated data.

  If present, data from this logical channel can only be transmitted using uplink grants that result in a PUSCH duration shorter than or equal to the duration indicated by this parameter. Otherwise, if the LCH mapping restriction is not met, data can be transmitted using an uplink grant resulting in any PUSCH duration, and the data remains the UE data buffer until it is transmitted.
- configuredGrantType1Allowed which sets whether a Configured Grant Type 1 can be used for transmission for data;

  If present, UL data from this logical channel can be transmitted on a configured grant Type 1. Otherwise, data cannot be transmitted on a configured grant Type 1.
- allowedServingCells which sets the allowed cell(s) for transmission; this configuration can be useful to restrict data to be multiplexed on a subset of cells, especially for LCHs of DRBs with CA-based PDCP duplication enabled.

  If present, data from this logical channel can only be mapped to the serving cells indicated in this list. Otherwise, data can be mapped to any configured serving cell.
- allowedCG-List which sets the allowed configured grant(s) for transmission, This restriction applies only when the UL grant is a configured grant. If present, data from this logical channel can only be mapped to the indicated configured grant configuration.
- allowedPHY-PriorityIndex which sets the allowed PHY priority index (indices) of a dynamic grant for transmission. This restriction applies only when the UL grant is a dynamic grant. This restriction is beneficial to allow a multitude of scheduling strategies, whereby the gNB scheduler may configure the priority index associated with a certain scheduling strategy to meet a certain QoS requirement. If present and the dynamic grant has a PHY-priority index, data from this logical channel can only be mapped to the dynamic grants indicating a PHY-priority index equal to the values configured by this field.
- allowedHARQ-mode which sets the allowed UL HARQ mode for transmission. Indicates the allowed HARQ mode of a HARQ process mapped to this logical channel. A HARQ mode corresponds to whether HARQ feedback is expected to be suppressed for the uplink transmission or not, which can be applicable to NTN systems, for example, due to the propagation delay. If the parameter is absent, there is no restriction for HARQ mode for the mapping.

9.2 The Medium Access Control Sublayer

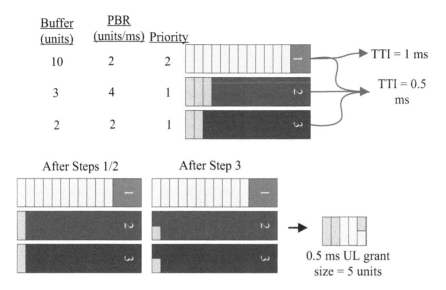

Fig. 9.6 Illustration of legacy LCP procedure used on LCHs mapping to multiple TTI durations

> **Example 9.3 (LCP Procedure and Max PUSCH Duration)**
> During Step 1 of the LCP procedure, the UE attempts to serve all logical channels mapped to the TB up to their Bj values in decreasing priority order. In an example where the UE is configured with 3 LCHs, LCHs 2 and 3 are configured a higher priority and with a max PUSCH duration of 0.5 ms, as they support low-latency data transmissions. LCH 1 is not configured with a max PUSCH duration restriction, so it can contend on any grant of any PUSCH duration. Each LCH will occupy some resources in step 1 as a first step of LCP, as illustrated in Fig. 9.6, which illustrates the UE buffer after each step of LCP. In step 3, remaining resources in the grants are allocated in priority order.
>
> If the grant was issued with a PUSCH duration of 1 ms, only data from LCH1 can be multiplexed. Remaining data from non-mapped LCHs remains at the UE data buffer. ∎

> **Example 9.4 (LCP Procedure with Priority-Based Scheduling)**
> Priority indices can be used to configure a LCH selection restriction in LCP based on the physical layer traits of the resource and whether it can support

(continued)

Example 9.4 (continued)
the QoS requirement associated with the data. From the perspective of the network, a priority index can correspond to a scheduling strategy associated to the transmission of a transport block. From the UE MAC's perspective, MAC would use a priority index indicated by the physical layer for an UL grant to determine which LCH(s) to consider when filling the transport block, considering the priority indices configured for each LCH by RRC. The mapping procedure would then be performed by the UE without explicit knowledge of the physical layer traits of the grant or the scheduling strategy from the gNB. In this example, assuming that the scheduler considers LCH A for the transport of an SRB, LCH B for the transport of eMBB data, and LCH C for transport of URLLC data, the UE may receive the following LCH configurations for priority index and LCH priority:

LCH C: priority index = (0), LCH priority = 1
e.g., URLLC data will contends only on grants with indication 0 with highest priority;

LCH B: priority index = (0, 1), LCH priority = 3
e.g., eMBB data contends only on grants with indication 0, 1 with lowest priority;

LCH A: priority index = (0), LCH priority = 2
e.g., SRB data will contends only on grants with indication 0 with highest priority; ∎

9.2.6 Buffer Status Reporting

Uplink buffer status reports play a crucial role in supporting QoS-aware packet scheduling. These reports provide insight into the amount of data currently buffered in the UE's logical channel queues at a given point in time. By considering the buffer status, the scheduler can make informed decisions to meet the QoS requirements of different services and applications.

BSR reports provide valuable information about the data buffered in logical channel groups (LCGs) within the UE. In NR and LTE, four reporting formats are used for transmitting uplink buffer status reports, allowing for flexible and efficient communication of this essential information. The formats are as follows:

- Short BSR Format: This format enables the reporting of a single BSR pertaining to one LCG.
- Long BSR Format: With the flexible long format, multiple LCGs can be reported, allowing for the transmission of up to all eight LCGs' buffer status.
- Extended Short Format: The extended short format is used to report a single BSR of one LCG.

9.2 The Medium Access Control Sublayer   313

- Extended Long Format: The extended long format facilitates the reporting of multiple BSRs, accommodating up to all 256 LCGs. The extended versions of the BSR formats are exclusively applicable to integrated access and backhaul (IAB) nodes.
- Padding and truncated BSR: a padding BSR is multiplexed when the number of bits remaining in the uplink grant after all data has been allocated is large enough to fit a BSR MAC control element (MAC CE). A short BSR is multiplexed if the number of padding bits does not allow a long BSR MAC CE; a long BSR MAC CE is multiplexed otherwise. A padding BSR can be truncated when a full BSR cannot fit, by reducing the number of LCGs reported, whereby the UE reports by order of LCG priority.

Uplink buffer status reports are transmitted using MAC control elements. When a BSR is triggered, typically when new data arrives in the UE's transmission buffers, the UE triggers the transmission of a scheduling request (SR) if no resources are available for transmitting the BSR or when the available resources cannot be used to transmit buffered data. This mechanism ensures that the network is aware of the buffer status even when immediate resources for BSR transmission are not available.

A BSR is triggered if any of the following events occur:

- New UL data becomes available at the mobile's buffer; and this UL data belongs to a logical channel with higher priority than the priority of any logical channel with available UL data (new data of higher priority) or none of the logical channels contain any available UL data (new data arrival on empty buffer). This BSR type is referred to as "Regular BSR".
- Uplink resources are allocated, and there is remaining space in the grant after applying LCP and filling the grant with data. The number of padding bits must be fit the size of the BSR MAC CE. This BSR type is referred to as "Padding BSR".
- Retransmission BSR of a BSR that has not been yet acknowledged by the network. Upon expiry of BSR retransmission timer, at least one logical channel contains UL data. This BSR type is referred below to as "Regular BSR".
- Periodic reporting of BSR. Upon expiry of a periodic BSR timer, the UE reports a 'Periodic BSR'.

With these reporting formats and the associated MAC signaling, the scheduler can effectively utilize uplink buffer status reports to make informed decisions regarding QoS-aware packet scheduling. This enables the network to optimize resource allocation based on the buffer status of different logical channel groups, ensuring efficient, reliable, and QoS aware resource allocation.

9.2.7 Scheduling Requests

Dynamic scheduling relies on the UE requesting uplink resources when the UE has buffered data to transmit but no grants available to transmit the data. The scheduler thus needs to know about data arrival at the UE's buffer. This information can be conveyed by the BSR if the UE has a grant available; a scheduling request is otherwise transmitted by the UE when it does not have a grant available for uplink transmission.

A scheduling request (SR) is a 1 bit indication transmitted by the UE to request uplink resources from the gNB scheduler when the UE does not have resources available to transmit buffered data. The scheduling request is transmitted on preconfigured PUCCH resources or can be multiplexed on uplink control information (UCI) on PUSCH resources in the physical layer. The network is able to identify which UE is requesting resources from the dedicated PUCCH resources configured from each UE.

A scheduling request can be triggered by one of multiple triggers. A typical SR is triggered when a BSR is triggered and the UE has no uplink resources to transmit the BSR MAC CE. In NR, an SR can also be triggered if the UE has newly arrived buffered data that cannot be transmitted on an available PUSCH grant, due to configured LCH LCP restrictions restricting such newly arrived data from being multiplexed on the available PUSCH grant. An SR can also be triggered in NR to report a channel failure, including a beam failure detected on a secondary cell or a consistent LBT failure in the uplink direction.

NR supports multiple SR configurations per UE to support reporting of data arrival from different types of data LCHs. An LCH can be configured to map to at most one SR configuration. Therefore, when data arrives for a LCH and an SR is triggered, the network scheduler is able to know the type of newly arrived data that triggered the SR and thus can allocate an appropriate grant to support the QoS associated with such data.

Each SR configuration is configured with an SR resource id, which includes a periodicity, configuration of frequency/time resources, and a PUCCH format used. Each SR configuration can thus be configured to map to one or more LCHs. The gNB scheduler can then infer the numerology, PUSCH duration, and other physical layer PUSCH traits required for the first PUSCH transmission based on the specific SR configuration used by the UE.

For each SR configuration, at most one SR can be triggered. Once an SR is transmitted, it is considered pending until canceled. As long as the SR is pending, the SR is transmitted on each applicable PUCCH occasion configured for the associated SR configuration. To control the retransmission frequency of pending SRs, the network can configure the UE with an SR prohibit timer, which (re-)starts upon each SR transmission. While the SR prohibit timer is running, the UE refrains from retransmitting a pending SR.

An SR is canceled when the BSR MAC CE is multiplexed eventually on a PUSCH resource received after the reception of a grant, as at such point the network is aware of the UE buffer status and can issue more grants as needed.

When multiple LCHs have newly arrived data that trigger a BSR and a corresponding SR, the highest priority logical channel is the one considered for SR triggering and selecting an associated SR configuration.

When an SR is triggered, and the UE does not have any valid PUCCH resources or SR configuration configured for the LCH that triggered the SR, the UE triggers a contention-based random access procedure. In such a random access procedure, the UE can report a BSR MAC CE part of the msg3 or MsgA PUSCH payload. Such an RA-SR procedure can be more appropriate for reporting data arrival from lower priority logical channels or when there are a large number of UEs served in the cell.

The network also configures each SR configuration with a maximum number of SR retransmission attempts, after which the UE initiates a random access procedure and cancel all pending SRs. The UE also releases configured PUCCH and any configured downlink assignments and uplink grants at this point.

9.2.8 Power Headroom Reporting

Power headroom reports (PHR) are useful for power-aware packet scheduling. Power headroom is a measure of the difference between the UE's nominal maximum transmit power and the estimated power required for uplink transmission. Given the amount of power needed to transmit uplink data at a constant signal-to-noise ratio is scaled by the number of PRBs assigned for uplink transmission, the scheduler typically avoids over-allocating uplink resources for the PUSCH if the UE would reach its maximum power. By analyzing power headroom reports, the scheduler can effectively manage power resources while ensuring reliable and efficient communication.

PHR provide vital information to facilitate efficient power management during transmission. The gNB can infer at least some of the following information from a PHR report:

- The gap between the UE's actual or hypothetical transmit power and the UE's maximum power.
- The maximum power reduction (MPR) reduction implemented by the UE for each cell. This information allows the gNB to learn, over time, the specific MPR reduction taken by the UE for different grant sizes.
- Assessing whether the UE has exceeded its maximum power limit on any cell. The gNB can use such info to determine whether it is feasible to increase or reduce scheduling activities on a particular cell.
- Whether the UE has surpassed its overall maximum power limit, leading to scaled channels. Understanding the presence of scaled channels assists the gNB in evaluating the combined scheduling across multiple cells.

- Determining the current value of the combination of pathloss and transmission power control (TPC) accumulator terms in the power control equation for each cell. This enables the gNB to adjust more accurate power control calculations for future scheduling decisions.

PHR reports are transmitted using MAC control elements for added reliability. Typically, the UE triggers power headroom reporting when it is scheduled to transmit uplink data or about to transmit uplink data. In NR, the UE can trigger the reporting of PHR upon satisfying one of the following triggers:

- When the measured pathloss changes by more than a configured threshold. A PHR is triggered when the difference between the current power headroom and the last reported power headroom is larger than a configured threshold.
- Upon expiry of a periodic reporting timer.
- When there is remaining space in the uplink grant, the UE adds a padding PHR reporting—similar to padding BSR.
- The value of the power backoff value due to power management (P-MPR) applied to meet Frequency Range 2 (FR2) maximum permissible exposure (MPE) requirement above a configured threshold or has changed by more than a configured threshold since the last PHR transmission due MPE P-MPR. This triggers a "MPE P-MPR PHR" PHR report type.
- PUCCH resources are available on a carrier, and the required power backoff due to power management (as allowed by P-MPR) for this cell has changed more than a configured threshold since the last transmission of a PHR.

Three PHR types of reporting are supported based on different transmission scenarios:

- Type 1: reports power headroom assuming PUSCH-only transmission on the carrier. It is valid for a single carrier. This PHR type of reporting is used for power control scheduling when transmitting PUSCH data. This PHR type can also be reported when there is no actual PUSCH transmission, as the power headroom is computed assuming a default transmission configuration corresponding to the minimum possible resource assignment.
- Type 2: similar to Type 1, but the transmit power computation includes an assumption that both PUSCH and PUCCH transmissions occur simultaneously. This PHR type is specifically designed for dual connectivity deployment scenarios where both PUSCH and PUCCH transmissions occur within a cell group.
- Type 3: this PHR type is designed to aid with SRS switching on secondary cells (SCells) configured with SRS only transmission and is reported when the mobile device is transmitting sounding reference signals but no PUSCH or PUCCH. This type of reporting allows the gNB to evaluate the uplink channel quality of alternative SCells and thus instruct the mobile device to activate such carriers for uplink transmission instead, if deemed better.
- MPE P-MPR PHR: the power backoff to meet the maximum permissible exposure (MPE) requirements for a serving cell operating on FR2 frequencies.

The power headroom $PH_c(t)$ associated with a particular serving cell "c" during subframe t can be expressed using a straightforward formula. In simple form, $PH_c(t)$ is obtained by subtracting the transmit power of serving cell c in subframe t, before any adjustments to prevent exceeding the maximum power limit. The calculation is as follows:

$$PH_c(t) = P_{cmax,c}(t) - P_{T,c}(t) \qquad (9.2)$$

where $P_{T,c}(t)$ is the transmit power of serving cell c in subframe t. The value of $P_{cmax,c}(t)$ represents the actual maximum power per carrier configured for the UE within the serving cell during subframe t. This value is configured by the network and is thus known to the scheduler. This value takes into consideration the MPR allowed by the UE. It is important to note that the MPR value can be equal to or less than the maximum allowed value. The transmit power value used in the formula is not the actual uplink transmit power, but rather the transmit power that would have been used assuming that there is no upper limit on the transmit power cap. The reported PHR can therefore be a negative value, thus indicating to the network that the UE would reach its maximum transmit power on the carrier for a given resource allocation, and thus the mobile device cannot transmit the scheduled amount of data without power reduction. The scheduler can thus reduce the selected MCS and schedule less PUSCH data to sustain enough energy per bit without having the UE exceed its maximum transmit power.

In carrier aggregation scenarios, a reference power is employed to generate a virtual report when no transmission takes place on an activated SCell. This virtual report assists in maintaining a comprehensive view of the power headroom situation across the network.

To ensure compliance with the maximum permissible exposure (MPE) exposure regulations for FR2, which limits RF exposure on the human body, PHR reports may include Power Management Maximum Power Reduction (P-MPR) information. This information is utilized by UE to ensure adherence to the specified MPE requirements.

9.2.9 Downlink Scheduling

Downlink resource allocation is achieved through the use of the UE's unique connected mode identity, the C-RNTI (see previous chapter), which scrambles UE-specific PDCCH signaling. The UE continuously monitors the PDCCH in order to detect any possible resource allocation. This monitoring activity is typically governed by the DRX mechanism to conserve battery life. In cases where multiple carriers are activated for a given UE, the same C-RNTI is used for all of them.

In some situations, the gNB may need to interrupt an ongoing data transmission to one UE in order to send time-critical data to another UE. To facilitate this, the gNB can configure UEs to monitor interrupted transmission indications (ITIs) using a unique code, the INT-RNTI, on the PDCCH. If a UE receives an interrupted transmission indication, it assumes that no useful information was carried by the resource elements included in the indication, even if some of those resource elements were previously scheduled for that UE. Overall, this enables efficient and flexible resource allocation while ensuring that time-critical data transmissions can be prioritized when necessary.

Semi-persistent scheduling (SPS) is used in NR and LTE networks for the allocation of semi-static downlink resources for initial transmissions to UEs. SPS allows the gNB to allocate periodic resources to the UE in a predetermined manner, providing an efficient means of scheduling resources for the transmission of data. For SPS, the RRC configuration defines the periodicity of the downlink assignments, which is typically determined based on the specific service requirements and the network configuration. Once the periodicity is set, the gNB can allocate downlink resources to the UE in a semi-persistent manner, meaning that the UE can expect to receive resources at regular intervals. SPS thus allows the network to ensure that certain services with periodic traffic have resources already provided and monitored by the UE, including periodic VOIP or time-sensitive networking services.

To activate or deactivate the configured downlink assignment, the gNB uses PDCCH DCI signaling addressed to the UE's unique configured scheduling identifier (CS-RNTI), which is used by the network to schedule the UE with dynamic assignments or grants for dynamic retransmissions of TBs initially transmitted on DL or UL configured grants. The DCI can either signal and activate the configured downlink assignment or deactivate it, depending on the specific requirements of the network. A PDCCH addressed to CS-RNTI can indicate that the downlink assignment can be implicitly reused according to the periodicity defined by RRC until it is explicitly deactivated. This provides an efficient means of scheduling resources for the transmission of data while also ensuring that the resources are used in a manner that is consistent with the requirements of the network. When required, retransmissions are explicitly scheduled on PDCCH.

If the dynamic allocation overlaps with the previously configured downlink assignment in the time domain, the dynamically allocated downlink reception takes precedence over the configured downlink assignment. However, if the dynamic allocation and the configured downlink assignment do not overlap in time, the UE assumes downlink reception according to the configured downlink assignment, provided that it is activated. This mechanism allows the system to prioritize dynamically allocated downlink resources over previously configured assignments when they overlap while still allowing for configured assignments to be used when they are activated and do not overlap with the dynamically allocated resources.

9.2.10 Discontinuous Reception (DRX) and UE Power Savings

Discontinuous reception in Connected mode (C-DRX) is used to give the UE an opportunity to sleep briefly in connected mode and save power consumption while

9.2 The Medium Access Control Sublayer

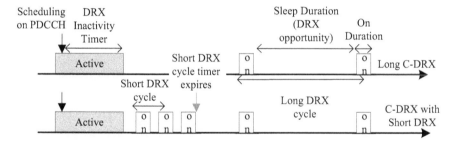

Fig. 9.7 Illustration of the C-DRX procedure in NR and LTE

maintaining all DRBs and QoS parameters in Connected mode. C-DRX is structured as a configurable sleep cycle with a brief "On Duration" period, where the UE wakes up to read the PDCCH in anticipation of further downlink data arrivals. DRX enables a UE in connected mode to discontinuously monitor PDCCH during inactivity periods under the control of the eNB based on the DRX configuration and scheduling activity.

Following any new data scheduled for the UE on the PDCCH, the UE resets an inactivity timer, which is used a trigger to start sleeping upon timer expiration. The brief sleep opportunity period is made possible due to traffic characteristics of the service, which often either intermittent (e.g., web traffic) or patterned (e.g., VoIP traffic).

The gNB can tailor the DRX configuration based on service requirements and/or based on cell load (e.g., using DRX start offset). DRX thus provides the minimal PDCCH decoding requirements in time for specific RNTIs. A UE may then save power by turning off at least parts of the radio outside of its PDCCH active time.

C-DRX operation allows for two C-DRX cycles, long and short C-DRX, where the usage of short C-DRX is optional. The UE falls back to the long DRX cycle upon the expiration of a configured short DRX cycle timer. This is illustrated in Fig. 9.7. The short cycle is useful for the scheduler to keep the UE monitoring PDCCH at reduced periodicities, which allows the scheduling of subsequent data and retransmissions without having to take the UE out of C-DRX.

From the UE perspective, in order to decode the PDCCH when waking up upon the "On Duration", the UE must be time synchronized and aware of the serving cell.

9.2.11 Beam Failure Detection and Recovery

Beam failure detection and recovery procedures are implemented to ensure reliable and uninterrupted communication between the UE and the base station. The first step in beam failure detection is the monitoring of beam-specific reference signals. The UE continuously receives these signals from the base station, allowing it to track the beamforming performance. If the quality of the received signals deteriorates below a certain threshold, indicating a potential beam failure, the UE triggers a beam failure event. Once a beam failure event is detected, the UE initiates the beam failure recovery procedure. The UE first tries to restore the connection with the base station by using the last known beamforming configuration. If this attempt fails, the UE searches for alternative beams and performs beam sweeping, where it scans different beamforming configurations in search of a stronger and more reliable beam.

In a beamformed NR system, the UE can be configured to maintain one or multiple beam pairs. The UE monitors certain periodic CSI-RS on a serving DL beam to assess its quality and computes a corresponding quality metric. If the beam's quality in a given RS period is below a configured threshold, the UE's PHY entity reports a beam failure instance (BFI) to the MAC sub-layer.

In order to re-establish lost beam pair(s) in a faster manner compared to the RLM/RLF procedure, the UE's MAC layer employs a BFR procedure in which a beam failure recovery request is reported to the network upon detecting a beam failure. BFR can be configured for beam maintenance on the PCell and/or SCell.

The MAC entity maintains a beam failure instance counter (BFI_counter) for the purpose of beam failure detection. The MAC entity counts the number of beam failure instance indications received from the PHY entity. If the BFI counter exceeds a certain maximum number of BFIs, a BFR request is triggered to notify the serving gNB that a beam failure has been detected.

The MAC entity resets the BFI counter only after a beam failure detection timer (BFD_timer) has expired. This can help provide some hysteresis in the detection function. In this case, the UE resets the BFD timer each time a BFI is indicated by the PHY layer. For example, the MAC entity may only reset the BFI counter after observing no BFI indications from PHY for three consecutive CSI-RS periods if the BFD timer is configured to three CSI-RS periods.

To report a BFR request, the UE initiates a random access procedure with certain parameter values (e.g., PreambleTransMax, power ramping step, and the target received preamble power). Such a random access procedure may be used for beam re-establishment, as the UE may select an appropriate PRACH preamble and/or PRACH resource dependent on the best measured downlink beam (or DL SSB). The UE may have the means to re-establish a beam pair when it can determine an association between DL beams and UL preambles and/or PRACH occasions, whereby the downlink beam selected by the UE is tested by receiving the random access response (RAR) on it. This re-establishment RA procedure can be made faster if the gNB configures a certain set of contention-free PRACH preambles/resources, which can be prioritized for selection by the UE upon initiating the random access procedure. Upon completion of the random access procedure, beam failure recovery for PCell is considered complete.

The beam failure and detection procedure also address secondary cells. Once beam failure is detected on a SCell, the UE triggers beam failure recovery by transmitting a BFR MAC CE specifically for that SCell. The UE also selects a suitable beam for the affected SCell if there is an alternative available and includes this information, along with details about the beam failure, in the BFR MAC CE. Upon reception of a PDCCH indicating an uplink grant for a new transmission for the HARQ process used for the transmission of the BFR MAC CE, beam failure recovery for this SCell is considered complete.

In multi-TRP operation, beam failure detection involves configuring the UE with two sets of beam failure detection reference signals. The UE identifies beam failure for a specific TRP/BFD-RS set when the number of beam failure instance indications from the corresponding set of reference signals reaches a configured threshold within a set timer duration.

9.2.12 Timing Advance

The UE may be in one of three RRC states; Idle, Inactive, or Connected states. These states are explained in Chap. 10. Within the RRC_CONNECTED state, maintaining synchronization between the gNB and the L1 layer is the responsibility of the gNB. This synchronization is achieved through the management of timing advance, as explained in Chap. 8. To ensure synchronization across multiple serving cells that share the same timing advance and reference cell for uplink transmissions, these cells are grouped into a TAG (timing advance group). RRC configuration is responsible for assigning each serving cell to a specific TAG.

In the primary TAG, the PCell serves as the timing reference for the UE. However, in certain cases where shared spectrum channel access is utilized, an SCell can also be employed as a timing reference alongside the PCell. In a secondary TAG, the UE has the flexibility to use any of the activated SCells within that TAG as a timing reference cell. However, the UE should avoid changing the timing reference cell unless it becomes necessary.

Updates to the timing advance are communicated from the gNB to the UE using MAC CE commands. These commands trigger the restart of a timer specific to each TAG. The state of this timer indicates the synchronization status of the L1 layer. When the timer is running, it signifies that the L1 layer is considered synchronized. Conversely, when the timer is not running, the L1 layer is regarded as non-synchronized. In such cases, uplink transmissions can only occur through MSG1/MSGA procedures.

By managing the timing advance and synchronization status through TAGs and timers, the gNB ensures effective coordination with the UE for seamless uplink transmissions in the RRC_CONNECTED state.

9.2.13 HARQ

The hybrid automatic repeat request (HARQ) functionality plays a crucial role in ensuring reliable data delivery between peer entities. The behavior of HARQ processes depends on the configuration of the physical layer regarding downlink/uplink spatial multiplexing.

Without spatial multiplexing, a single HARQ process is responsible for supporting one transport block (TB). It focuses on the transmission and acknowledgment of a single TB, ensuring its successful delivery. With spatial multiplexing, a single HARQ process has the capability to support not only one but multiple TBs. It efficiently handles the transmission and acknowledgment of these multiple TBs, ensuring their reliable and successful delivery.

The higher the number of HARQ processes the UE supports, the larger the data rate as the UE and the network are then able to schedule multiple transport blocks simultaneously on different HARQ processes without having to wait for a HARQ process to be acknowledged and flushed before scheduling another transport block. However, additional HARQ processes come at a device cost in terms of memory requirements and also complexity.

Example 9.5 (HARQ Process Dimensioning)
Given the following device capabilities:

- a high capability smart phone with a HARQ memory of 4 Mbits
- a reduced capability UE with a HARQ memory of 2 Mbits
- an NTN device with a HARQ memory of 500 Kbits
- an IoT device with a HARQ memory of 100 Kbits

Assuming a configured grant configured with a periodicity of 1 slot, a grant size of 128 K bits, and a configured grant timer of 10 ms, determine the following:

- The minimum number of HARQ processes required in the UE needed to meet a guaranteed bit uplink data rate of 100 Mbps using this configured grant.
- Which of the types of devices can meet a given data rate requirement, given the HARQ memory constraints.

Solution In the duration during which the configured timer is running (10ms), the following needs to be maintained at a minimum to achieve the guaranteed minimum data rate:

$$\frac{128 \times 10^3 \times nrofHARQ - Processes}{10 \times 10^{-3}} = 100 \text{ Mbps} \tag{9.3}$$

nrofHARQ-Processes therefore needs to be at least $\lceil \frac{100 \times 10}{128} \rceil = 8$ HARQ processes.

Given each HARQ process needs to accommodate up to 128 kb, the required HARQ memory size is $8 \times 128 = 1024$ kb. Therefore, the high capability smartphone and the reduced capability device will be able to meet this guaranteed bit rate, while the NTN and IoT device will not be able to. ∎

9.2.14 Wideband Carrier Operation

The UE can be configured with one or more bandwidth parts (BWP) allocated on a specific component carrier, although only one can be active at any given time. The active bandwidth part determines the UE's operational bandwidth within the overall bandwidth of the cell. During the initial access procedure, before receiving the UE's configuration in a particular cell, the initial bandwidth part detected from system information is used.

In connected mode, bandwidth adaptation can be used, where the receive and transmit bandwidth of the UE can be dynamically adjusted, allowing it to adapt up to the cell's total bandwidth. This adjustment encompasses changes in width, such as shrinking the bandwidth during periods of low activity to conserve power, modifying the frequency domain allocation to enhance scheduling flexibility, and ordering changes in the subcarrier spacing to accommodate diverse services. A portion of the cell's total bandwidth is designated as a bandwidth part, and bandwidth adaptation is achieved by configuring the UE with one or more BWPs and explicitly indicating which of the configured BWPs is presently active (Fig. 9.8).

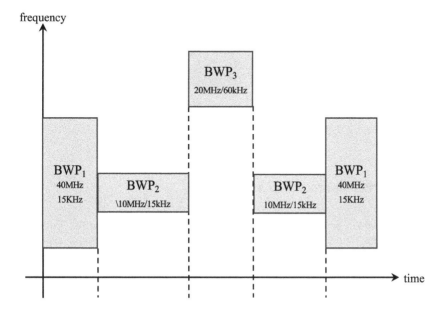

Fig. 9.8 An example of different configured bandwidth parts in an NR carrier [1]

9.3 The Radio Link Control Sublayer

The Radio Link Control (RLC) sublayer offers support for three distinct transmission modes: transparent mode (TM), unacknowledged mode (UM), and acknowledged mode (AM). The selection of the transmission mode is based on the specific requirements of the logical channel, and it is independent of numerologies and transmission durations.

In TM, the RLC sublayer does not provide any error correction or retransmission mechanism for faster processing of latency critical transmissions, making it suitable for transmitting essential SRB0 data (defined in Chap. 10), paging, and broadcast system information. UM operates without explicit acknowledgment, allowing for reduced latency but with the possibility of missing or corrupted data. AM, on the other hand, provides reliable data delivery through error correction and ARQ (automatic repeat request), ensuring accurate and complete transmission of RLC SDUs, making it suitable for most SRBs and DRBs in the system.

TM mode is used for SRB0, which handles paging and broadcast system information. However, for other SRBs, AM mode is employed. As for DRBs, both UM and AM modes can be used.

The RLC sublayer's services and functions vary depending on the transmission mode and include:

- Transfer of upper layer PDUs.
- Sequence numbering, independent of the sequence numbering in the PDCP sublayer (UM and AM).
- Error correction through automatic repeat request (ARQ) mechanism (AM only).
- Segmentation (AM and UM) and re-segmentation (AM only) of RLC service data units (SDUs).
- Reassembly of SDUs (AM and UM).
- Duplicate detection (AM only).
- RLC SDU discard (AM and UM).
- RLC sublayer re-establishment.
- Protocol error detection (AM only).

9.3.1 Error Correction Through ARQ

The automatic repeat request (ARQ) mechanism, nestled within the Radio Link Control (RLC) sublayer, plays a pivotal role in ensuring the reliable delivery of RLC service data units (SDUs) or segments. This functionality operates by initiating retransmissions based on RLC status reports, which convey crucial information about the state of the transmission. To obtain these reports, the RLC sublayer utilizes a polling mechanism when deemed necessary. Furthermore, the RLC receiver is equipped with the capability to autonomously generate RLC status reports upon identifying the absence of an expected RLC SDU or segment. This comprehensive approach to ARQ functionality facilitates efficient error recovery and enhances the overall reliability of data transmission within the RLC sublayer.

These features of the RLC sublayer and ARQ mechanism ensure reliable and efficient data transfer between the RLC sublayer and upper layers, with appropriate segmentation, reassembly, error correction, and status reporting.

9.4 Packet Data Convergence Protocol

The Packet Data Convergence Protocol (PDCP), specified in 3GPP TS 38.323, takes on several essential tasks within the communication system. It is responsible for performing vital operations such as IP header compression, ciphering, and integrity protection, ensuring the robustness and security of data. Additionally, PDCP facilitates retransmissions, maintaining the correct order of data delivery, and eliminating duplicates in handover scenarios. In the context of dual connectivity with split bearers, PDCP demonstrates its capability in offering routing and duplication functionality. Each radio bearer configured for a device corresponds to a dedicated PDCP entity, ensuring efficient and reliable data processing at the bearer level.

- Header compression and decompression (ROHC): Header compression and decompression is performed at the PDCP layer to optimize bandwidth utilization and reduce transmission overhead. This process involves compressing the headers of IP packets before transmission and decompressing them upon reception. By removing redundant information, such as repetitive field values and unnecessary headers, ROHC effectively reduces the size of the transmitted packets. This compression technique ensures efficient data transfer, especially in bandwidth-constrained environments, while maintaining the integrity and reliability of the transmitted information. Upon reception, the decompression process restores the original headers, allowing for seamless and accurate packet processing by the receiving end.
- Reordering and duplication detection: reordering at PDCP ensures that packets are delivered in the correct order, which is particularly important in scenarios such as handovers or multi-path transmission. Duplication detection helps eliminate redundant or duplicated packets, preventing unnecessary processing and improving overall network efficiency.

9.5 Service Data Adaptation Protocol

The Service Data Adaptation Protocol (SDAP) plays a pivotal role in ensuring effective mapping and coordination between QoS flows from the 5G core network and the corresponding data radio bearers. SDAP is responsible for establishing the link between these QoS flows and the associated radio bearers while also performing the essential function of marking the quality-of-service flow identifier (QFI) within both uplink and downlink packets. A single protocol entity of SDAP is configured for each individual PDU session.

The introduction of SDAP in NR is motivated by the enhanced QoS handling capabilities compared to LTE when interconnected with the 5G core network. Consequently, SDAP serves as the primary mechanism for establishing the mapping relationship between QoS flows and radio bearers, facilitating efficient and seamless data transmission in the 5G ecosystem.

The main function of the SDAP is to:

- Map a QoS flow to a data radio bearer,
- Provide marking for QoS flow ID (QFI) in both DL and UL packets.

In 3GPP, more details about the SDAP sublayer can be found in TS 37.324.

Problems

1. In an NR-FDD system using a subcarrier spacing of 15 kHz and a 1 ms slot duration, a device capable of 8 HARQ processes is configured with a configured grant with the following parameters: a periodicity of four slots, a grant size of 2048 bits, and a configured grant timer of 16 ms. Assuming that the device has enough data (full buffer) to transmit at each configured grant occasion, determine the following:

    a. the time a given HARQ process repeats itself;
    b. the minimum time until a HARQ process can be reused to transmit new data;
    c. Over a period of 56 ms, compute the maximum uplink data rate using the available configured grant, assuming no retransmissions are required.
    d. Over a period of 56 ms, compute the maximum uplink data rate using the available configured grant, if one retransmission grant for HARQ process 1 is received 11 ms after the initial transmission on this HARQ process. Assume HARQ processes are identified by 0, 1, 2, and 3, 4, etc.

2. Given the following device capabilities:
    - a high capability smart phone with a HARQ memory of 16 M bits
    - a reduced capability UE with a HARQ memory of 4 M bits
    - an NTN device with a HARQ memory of 2048 K bits
    - an IoT device with a HARQ memory of 1024 K bits

    Assuming a configured grant configured with a periodicity of one slot, a grant size of 512 K bits, and a configured grant timer of 10 ms, determine the following:

9.5 Service Data Adaptation Protocol

- The minimum number of HARQ processes required in the UE needed to meet a guaranteed bit uplink data rate of 100 Mbps using this configured grant.
- Which of the types of devices can meet a given data rate requirement, given the HARQ memory constraints.

3. Describe the functionality of user plane protocol layers: SDAP, PDCP, RLC, and MAC.
4. For a smart phone configured with 4 LCHs, LCHs 3 and 4 are configured a higher priority and with a max PUSCH duration of 0.5 ms, as they support low latency data transmissions. LCHs 1 and 2 are not configured with a max PUSCH duration restriction, as they carry eMBB best effort data. Assuming the following is applicable or configured for each LCH:

LCH	Amount of buffered data (Kbits)	PBR (Kbps)	LCH priority	LCP LCH selection restrictions
1	8	0	3	None configured
2	10	20	3	None configured
3	6	400	1	Max PUSCH duration = 0.5 ms
4	4	200	2	Max PUSCH duration = 0.5 ms

For an uplink grant A of PUSCH duration 0.5 ms and size 10 Kbits:

a. Compute the number of bits allocated to each LCH after step 1 of LCP.
b. Compute the number of remaining buffered bits for each LCH after step 1 of LCP.
c. Compute the number of bits allocated to each LCH after step 3 of LCP.
d. Compute the number of remaining buffered bits for each LCH after step 3 of LCP.
e. For an uplink grant B of duration of 1 ms and size 10 Kbits received after a transport block is transmitted on grant A, repeat steps (a) to (d) for grant B.

5. Explain the benefit of having multiple SR configurations in NR (as opposed to having a single SR PUCCH configuration in LTE).
6. Explain the benefit of C-DRX short and long cycles. Provide an example where the short C-DRX can be useful to have.
7. A narrowband adaptive multi-rate codec (AMR) supports a range of code rates, as shown in the table below, where each packet is assembled over a 20 ms time interval.

Bandwidth	Bit rate (kbps)
Narrowband (NB)	1.80, 4.75, 5.15, 5.90, 6.70, 7.40, 7.95, 10.20, 12.20

Adaptive switching between the three top code rates (7.95, 10.20, and 12.20) has been enabled. Assuming the following:

- IPv4 header is 40 bytes long
- 12 bits are added to every AMR packet for RTP formatting; additional filler bits may be added if the total number of bits is not a multiple of 8
- ROHC reduces the IP/UDP/RTP header down to 4 bytes
- PDCP header is 2 bytes long
- RLC header is 1 byte long
- MAC header is 4 bytes long

Determine:

a. Average MAC throughput if a user is assigned 7.95 kbps code rate 40% of the time, 10.20 kbps code rate 30% of the time, and 12.2 kbps code rate 30% of the time, and no robust header compression is enabled.
b. Assuming the highest AMR rate of 12.2 kbs, what is the instantaneous data rate per user if each 20 ms packet is transmitted over one slot duration if the subcarrier spacing $\Delta f = 30$ kHz.
c. For the highest AMR rate, how many simultaneous users can be supported if the error control rate is $1/2$ and QPSK modulation is used for a 20 MHz system bandwidth? Assume that the number of REs (N_{RE}) for the voice transmission within the slot is $N_{RE} = 144$ (over one PRB), single-layer MIMO transmission is used, and not all the users need to be scheduled over the same slot.
d. Repeat (a), (b), and (c) assuming ROHC is enabled.

References

1. NR; NR and NG-RAN Overall Description; Stage 2. https://www.3gpp.org/ftp/Specs/archive/38_series/38.300/38300-h90.zip. Cited 21 Jul 2024
2. NR; Medium Access Control (MAC) protocol specification. https://www.3gpp.org/ftp/Specs/archive/38_series/38.321/38321-h90.zip. Cited 21 Jul 2024

Chapter 10
3GPP Based Air Interface, Layer 3

10.1 The Radio Resource Control

Layer 3 (L3) of the 4G and 5G radio access network encompasses the radio resource control (RRC) sublayer and the non-access stratum (NAS) sublayer; the L3 structure is illustrated in Fig. 10.1. The RRC sublayer plays a critical role in the Uu interface's operational efficiency and robustness. One of its principal functions is the broadcast of system information pertaining to access stratum (AS) and NAS. This sublayer is responsible for paging, which can be initiated by either the 5th Generation Core (5GC) or the Next-Generation Radio Access Network (NG-RAN). The RRC sublayer also manages the establishment, maintenance, and termination of the RRC connection between the UE and NG-RAN. This encompasses various functionalities, ranging from carrier aggregation (covered in Chap. 12) management to establishment Dual Connectivity in NR or between the LTE and NR.

On the security front, the RRC sublayer is responsible for the configuration of the security context of the UE, which includes security keys and security algorithms to be used for the encryption and integrity protection of signaling radio bearers (SRBs) and data ratio bearers (DRBs) (integrity protection of DRBs is optional).

RRC also ensures the robust establishment, configuration, and management of both signaling radio bearers (SRBs) and data radio bearers (DRBs). RRC further specifies crucial mobility functions and operations, managing tasks such as handovers, context transfers, UE cell selection and reselection in IDLE mode, and the overarching control of these selections. The RRC sublayer is also responsible for handling inter-RAT mobility and guaranteeing a seamless transition between different radio access technologies.

Supplementary Information The online version contains supplementary material available at (https://doi.org/10.1007/978-3-031-76455-4_10).

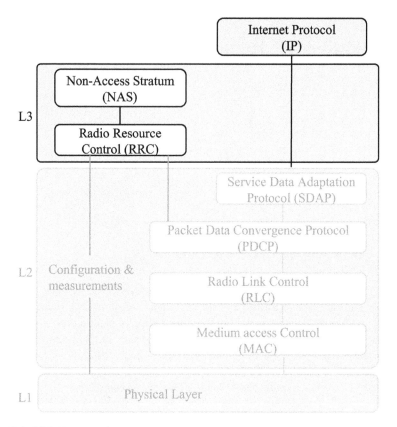

Fig. 10.1 5G RAN protocol stack

QoS management functions configured by the RRC ensure the prioritization of traffic as per its significance and requirement. RRC is used to configure the radio bearers with the proper parameters that is aligned with the QoS requirements of the bearers (e.g., priorities, bit rates, delay budget, etc.). The network can ensure the UE gets the needed resources for the transmission of UL data and reception of DL data at RRC level via the configuration of semi-persistence scheduling resources (SPS), and configured grants (CGs), respectively. Additionally, RRC sublayer specifies UE measurement reporting, controlling its frequency and consistency, as well as ensuring rapid detection and recovery from any radio link failures. The RRC acts as a conduit for the transfer of NAS messages between NAS and UE, maintaining bidirectional communication integrity.

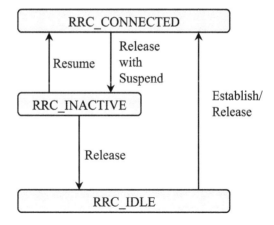

Fig. 10.2 UE RRC state machine and state transitions within NR

10.1.1 RRC States

In LTE and NR networks, the UE can be in one of the three connection states (or modes):

- RRC_CONNECTED
- RRC_INACTIVE
- RRC_IDLE

The UE maintains a state machine to transition between the different states. Upon being turned on, the UE starts with RRC IDLE then transitions to connected mode. From connected mode, the UE can transition to either RRC INACTIVE or RRC IDLE. From RRC INACTIVE, the UE can transition to either RRC Connected or RRC IDLE. An illustration of UE states in NR is shown in Fig. 10.2 (from TS 38.331 [1, Figure 4.2.1–1]).

- **RRC IDLE:**
  A UE can be in RRC_IDLE state when it is not actively transmitting or receiving data (mainly for power saving reasons) or just before connection setup as in the case of the UE being turned on or upon recovering from radio link failure (RLF). In RRC_IDLE state, several critical operations ensure the system's optimal functioning and efficient resource usage. The process begins with a public land mobile network (PLMN) selection, which aids in determining the most appropriate network for the UE to connect to. Concurrently, the network broadcasts vital system information, allowing UE to understand and adapt to the network environment. Mobility within this state is managed through cell reselection, ensuring that the UE consistently maintains an association with the optimal cell based on factors like the cell's signal strength and quality.
  An essential aspect of RRC_IDLE is the initiation of paging for mobile terminated data by the 5GC, which ensures timely data delivery without a persistent connection. In IDLE mode, CN paging during DRX on period is configured by

NAS and used to enhance UE battery life, allowing periodic wake-up of the UE to check for paging, rather than maintaining a continuous active state. There is no AS context stored by the UE in RRC_IDLE.

If the network expects the UE to become inactive for a long duration, it can send the UE to RRC_IDLE state. While in RRC_IDLE, the UE camps at the best cell (the cell with the best signal level at the highest priority RAT and highest priority frequency within that RAT) that will facilitate the UE establishing the connection via that cell if a need arises for the UE to transition back to the connected state. The UE also monitors the downlink paging channel to detect for DL data arrival. The UE will initiate the connection setup/establishment procedure if it detects a paging from the network indicating an arrival of a DL data or if the UE needs to send an UL data. During connection setup or resume, the UE has first to perform a random-access procedure before sending the RRCSetupRequest or the RRCResumeRequest message.

- **RRC INACTIVE:**

  Though the RRC_IDLE state helps the UE in power saving, it is inefficient if the UE has to transition back and forth between RRC_IDLE and RRC_CONNECTED, as the connection establishment from IDLE to CONNECTED is a lengthy process that involves not only the base station (gNB) but also the core network (CN). To solve this issue, 3GPP introduced the RRC_INACTIVE state in Rel-15, wherein the UE behaves like in RRC_IDLE state in most aspects, but it maintains the AS context and as such transitions back and forth between INACTIVE and CONNECTED state that can be performed without involving the CN.

  RRC_INACTIVE is a state where a UE remains in CM-CONNECTED and can move within an area configured by NG-RAN (the RNA) without notifying NG-RAN. The RRC_INACTIVE state is a compromise between the ACTIVE/CONNECTED state and the IDLE state, where the UE has almost the same power saving benefits of the IDLE state, but it can be brought back quickly to connected sate when the need arises (e.g., when UL/DL data concerning that UE arrives).

  When a UE is suspended to an INACTIVE state, it is allocated an I-RNTI by the last serving gNB. The I-RNTI includes a UE specific part (e.g., CRNTI used in the cell where the UE got suspended) and a network specific part (e.g., ID of the gNB that was hosting the cell where the UE got suspended). The last serving gNB node keeps the UE context and the UE-associated NG connection with the serving AMF and UPF.

  In the RRC_INACTIVE state, paging is driven by NG-RAN, referred to as RAN paging, ensuring timely communication without a continuous active connection. A RAN-based notification area (RNA) is adeptly managed by NG-RAN, marking the UE's location in relation to the network. This state ensures that the NG-RAN remains aware of the RNA to which the UE is affiliated, optimizing paging, handover, and other mobility operations. From a core network perspective, a UE in RRC Inactive remains in CN Connected mode.

If the last serving gNB receives DL data from the UPF or DL UE-associated signaling from the AMF the UE is in RRC_INACTIVE, it pages in the cells corresponding to the RNA and may send XnAP RAN Paging to neighbor gNB(s) if the RNA includes cells of neighbor gNB(s). When the UE resumes (e.g., in a cell/gNB different from where it was suspended), it will provide the I-RNTI in the resume request and the target gNB will be able to identify the source gNB from the network specific part of the C-RNTI and send the context fetch request to the source node.

A UE in the RRC_INACTIVE state can be configured by the last serving gNB with an RNA, where the RNA can cover a single or multiple cells and shall be contained within the CN registration area; a RAN-based notification area update (RNAU) is periodically sent by the UE and is also sent when the cell reselection procedure of the UE selects a cell that does not belong to the configured RNA. The RNAU is done via a two-step resume procedure (i.e., UE sending an RRC resume request with a cause value indicating RANU and the network responding immediately with an RRC release)

To further enhance resource utilization, DRX for RAN paging is configured by the NG-RAN. One of the critical aspects of RRC_INACTIVE is the establishment of a 5GC-NG-RAN connection on both the C- and U-planes for the UE, ensuring robust connectivity while still saving UE battery. The UE Inactive AS context is retained in both the NG-RAN and the UE, facilitating swift transitions between RRC states.

The AS context includes parameters essential for maintaining communication between the UE and the network, including security parameters, encryption, and integrity protection keys, which safeguard data transmission between the UE and the network. It also holds the UE's capabilities, such as supported radio access technologies and frequency bands, which ensure compatibility and optimal connection with the network infrastructure. The AS context includes the UE's radio bearer settings, which define the flow of data streams, quality of service attributes for maintaining data traffic priorities, and the measurement configuration for monitoring signal quality and neighboring cells to assist in mobility management and handover procedures.

Transfer of unicast data and signaling between the UE and the network over radio bearers specifically configured for small data transmission (SDT), maintaining data integrity and efficient signaling. Transitioning from inactive into connected mode just to send a small amount of data creates increased signaling overhead in the network and increased battery consumption. For devices supporting eMBB services, applications can have frequent background small data (e.g., app refresh data, notifications etc.), which may be periodic or aperiodic. Further, sensors and IoT devices may have considerable amount of signaling and small data (e.g., periodic heartbeat or stay-alive signals, surveillance updates, periodic video stream, non-periodic video based on motion sensing, etc.). Requiring the UE to move to connected mode for such small data or signaling thus affects power consumption considerably, especially for power- or battery-limited sensor/IoT devices or for eMBB mobile devices aiming to reduce battery consumption. Such data can instead be exchanged using the SDT framework in Inactive state.

- **RRC CONNECTED:**
  In the RRC_CONNECTED, robust connection between the 5GC and NG-RAN is ensured on both the C and U-planes for the UE. This state ensures that the UE's AS context is stored in both NG-RAN and the UE, preserving critical communication parameters. The NG-RAN knows the specific cell the UE is associated with, e.g., for the purpose of scheduling, mobility, and data transfer. Efficient transfer of unicast data between the UE and the network is performed in RRC CONNECTED. Further, mobility operations within this state are predominantly controlled by the network, encompassing tasks such as measurement reporting and evaluating mobility conditions.

  In connected mode, the UE is configured and maintains several SRBs and DRBs. SRBs are responsible for carrying control plane information, which includes all the necessary signaling messages between the UE and the network that establish and maintain the connection, handle mobility, and enforce security measures. SRBs ensure that the UE remains reliably connected and managed by the network. DRBs carry user plane data, which encompasses the actual user-generated traffic, such as voice, video, or Internet data. DRBs are designed to handle the diverse QoS requirements of different data streams, providing the necessary bandwidth and latency specifications to meet the service demands.

  In RRC_CONNECTED, the UE is actively connected to the network, with signaling and data radio bearers established (SRB and DRBs), and able to receive downlink data from the network in a unicast fashion and also send uplink data to the network. The mobility of the UE from one cell/node to another is controlled by the network. Network may configure the UE to send measurement reports periodically or when certain conditions are fulfilled (e.g., a neighbor cell becomes better than a serving cell by more than a certain threshold) and based on these reports may send the UE a handover command to move the UE to another cell/node. The network may also send the UE a HO command without receiving any measurement report (e.g., based on implementation, such as the determination of current location).

  Keeping the UE in connected mode is power intensive for the UE, as the UE needs to continuously monitor the PDCCH of the serving cell for determining the arrival of DL data, for UL data scheduling, etc. As gNB is able to accommodate a certain number of UEs in connected mode (e.g., due to resource limitations), when there is no activity in the UL or DL for a certain duration (e.g., based on an inactivity timer kept at the network), the network may send the UE to the RRC_INACTIVE or RRC_IDLE state.

Figure 10.3 illustrates the RRC connection establishment/setup and connection resume procedures [1].

10.1 The Radio Resource Control 335

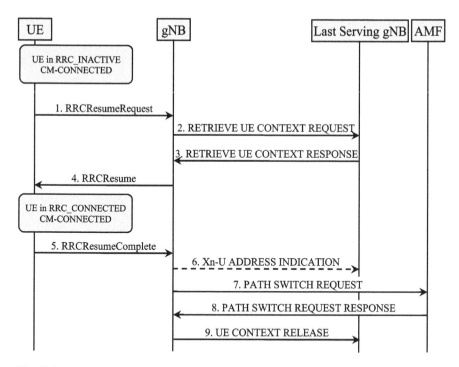

Fig. 10.3 The RRC connection establishment/setup and connection resume procedures [1]

When the UE performs the connection setup/establishment or resume procedure, it includes (in the RRCSetupRequest or RRCResumeRequest) the establishment or resume cause. Currently, the following causes are defined:

```
EstablishmentCause ::= ENUMERATED {
    emergency, highPriorityAccess, mt-Access, mo-Signalling,
    mo-Data, mo-VoiceCall, mo-VideoCall, mo-SMS,
    mps-PriorityAccess, mcs-PriorityAccess,spare6, spare5,
    spare4, spare3, spare2, spare1}

ResumeCause ::= ENUMERATED {
    emergency, highPriorityAccess, mt-Access, mo-Signalling,
    mo-Data, mo-VoiceCall, mo-VideoCall, mo-SMS, rna-Update,
    mps-PriorityAccess,mcs-PriorityAccess,
    spare1, spare2, spare3, spare4, spare5 }
```

For example, if the connection is being setup or resumed due to a voice call or video call originating from the UE, the UE will set the establishment/resume cause to mo-VoiceCall (mobile originated voice call) or mo-VideoCall (mobile originated video call). As another example, if the connection is being setup/resumed due to downlink paging indicating DL data, the UE will set the establishment/resume cause to one of mt-Access (mobile terminated access), highPriorityAccess, mps-PriorityAccess, or mcs-PriorityAccess (depending on the access category of the UE).

10.2 Initial Access Procedure

The initial access procedure occurs when the UE is powered ON and is not yet connected to the network. This procedure consists of a series of steps, where each step unlocks the parameters or triggers signaling required for the next step. Some of the steps that are part of the initial access procedure, such as scanning of the supported frequency to identify the potential channel for connection are not discussed here.

10.2.1 Synchronization Procedure

During an initial access or an IDLE mode camping procedure, the first step is for the UE to attempt to read broadcast synchronization signals. The synchronization signal and PBCH block (SSB), introduced in Chap. 8, consists of primary and secondary synchronization signals (PSS, SSS), each occupying 1 symbol and 127 subcarriers, and PBCH spanning across 3 OFDM symbols and 240 subcarriers, but on one symbol leaving an unused part in the middle for SSS as shown in Fig. 10.4 [2]. The possible time locations of SSBs within a half-frame are determined by subcarrier spacing. The periodicity of the half-frames where SSBs are transmitted is also configured by the network. During a half-frame, different SSBs may be transmitted in different spatial directions (i.e., using different beams, spanning the coverage area of a cell).

Within the frequency span of a carrier, multiple SSBs can be transmitted. The physical cell IDs (PCIs) of SSBs transmitted in different frequency locations do not have to be unique, i.e., different SSBs in the frequency domain can have different PCIs. However, when an SSB is associated with an RMSI, the SSB is referred to as a cell-defining SSB (CD-SSB), and such SSB defines the cell's PCI. A PCell is always associated to a CD-SSB located on the synchronization raster.[1]

The UE assumes a band-specific subcarrier spacing for the SSB unless a network has configured the UE to use a different subcarrier spacing. Several beams may be associated with a given, and multiple SSBs can be transmitted within a given cell on different beams (i.e., beam sweeping).

Multiple SSBs are being transmitted at a certain interval. Each SSB can be identified by a unique number called SSB index. Each SSB is transmitted via a specific beam radiated in a certain direction. The UE measures the signal strength of each SSB it detected for a certain period (a period of one SSB Set). From the measurement result, UE can identify the SSB index with the strongest signal strength.

---

[1] Synchronization raster is not covered in this textbook, but relates to discrete frequency points in frequency where the SSB and/or carrier frequency can be found.

Fig. 10.4 Time-frequency structure of SSB

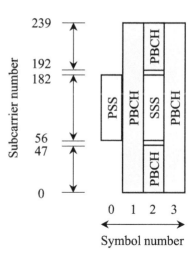

The number of different transmitted beams is determined by the number of transmitted SSBs within a SSB Burst Set (a set of SSBs being transmitted in 5 ms window of SSB transmission). In FR1, the maximum number of SSBs within an SSB set is 4 or 8, while for FR2 it can be up to 64 beams.

10.2.2 System Information

Once the UE is synchronized, the next step in the initial access procedure in NR or LTE is to acquire the system and broadcast information. Broadcast of system information (SI) via the PBCH channel introduced in Chap. 8 is pivotal to cellular networks, as the cell communicates vital parameters and configurations to the UE. The following describes the architectural division and various components of SI.

SI contains the Master Information Block (MIB) and a series of System Information Blocks (SIBs). SI can be classified either as minimum system information (Minimum SI) or additional system information (Other SI). The SI acquisition procedure is illustrated in Fig. 10.5.

When the UE evaluates a particular cell or frequency for the purpose of camping (part of the process of selecting and remaining on a specific cell or frequency), it is not mandated to acquire the exhaustive system information (SI) available for that target. Instead, the UE retrieves only the MSI pertaining to the prospective cell or frequency layer.

Recognizing the recurring mobility patterns and the potential for revisits to previously camped cells, UEs can store SI from cells they've previously camped on. By doing so, when the UE revisits a previously camped cell or frequency, it can forgo the process of re-acquiring SI that it has previously stored. The UE stores value tags associated with a camped cell for ease of retrieval. The UE can thus recall the cached SI, facilitating a faster re-attachment process. Such mechanism not only

Fig. 10.5 System information provisioning [3]

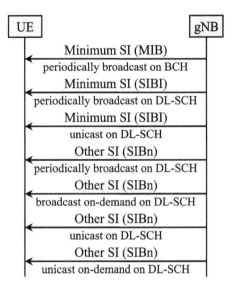

accelerates the UE's transition between cells but also aids in optimizing system performance by reducing redundant SI acquisition.

10.2.3 Minimum System Information

The Minimum system information (MSI) encapsulates rudimentary data requisite for initial access and the acquisition of supplementary SI. MSI includes the Master Information Block (MIB) and SIB1.

- Master Information Block (MIB): The MIB is disseminated periodically via the Broadcast Channel (BCH) via the SSB block for ease of accessibility. The MIB information is small in size; it consists of only 24 bits and encompasses:
  - Status information indicating if a cell is barred.
  - Physical layer metrics of the cell required for the reception of subsequent system information, such as the CORESET#0 configuration. Coreset#0 refers to the primary set of PDCCH control resource sets that defines the default Coreset used by the network to schedule initial downlink control information to the UE. Coreset 0 is used for scheduling SIB1 in IDLE and Inactive states. This is crucial for the initial access and ongoing communication between the base station and the mobile devices.
- System Information Block 1 (SIB1): SIB1 is also recognized as the Remaining Minimum SI (RMSI). SIB1 outlines the scheduling framework for the other SIBs and incorporates data essential for initial access. Depending on the system's state and configuration, SIB1 can either be broadcast periodically through DL-SCH or be relayed by dedicated signaling on DL-SCH to UEs in the RRC_CONNECTED state.

10.2.4 Additional System Information

Additional system information, or Other SI, contains all SIBs excluded from the Minimum SI. These SIBs have various means of broadcast and transmission, including:

- Regular broadcast on the DL-SCH.
- On-demand broadcast on the DL-SCH.
- Dedicated transmission on DL-SCH to UEs in the RRC_CONNECTED state under specific conditions, such as absence of common search space in the active BWP or having different PCI between the serving TRP and the serving cell.

A brief function of each block under Other SI is described:

- SIB2: Contains cell reselection parameters specific to the serving cell.
- SIB3: Contains data regarding serving frequency and intra-frequency neighboring cells critical for cell reselection.
- SIB4: Contains metrics related to other NR frequencies and inter-frequency neighboring cells vital for reselection and measurements in idle/inactive states.
- SIB5: Dedicated to E-UTRA frequencies and associated neighboring cells relevant for reselection.
- SIB6 & SIB7: Reserved for primary and secondary Earthquake and Tsunami Warning System (ETWS) notifications, respectively. ETWS is a standard used in mobile communications, especially in countries prone to earthquakes and tsunamis, to alert users of impending natural disasters. When the system detects signs of a potential earthquake or tsunami, it sends out an immediate alert to users in the affected areas to warn them of the possible danger. This rapid dissemination of information can be crucial in providing people with the time they need to take protective measures or evacuate.
- SIB8: Holds the Commercial Mobile Alert System (CMAS) warning notification. CMAS is a system used to disseminate emergency alerts to mobile devices. CMAS warnings can include alerts about severe weather, amber alerts (for missing children), and other significant or imminent threats to public safety.
- SIB9: Houses information connected to GPS time and Coordinated Universal Time (UTC).
- SIB10: Lists Human-Readable Network Names (HRNN) of nonpublic networks (NPNs) referenced in SIB1.
- SIB11: Dedicated to information pertaining to idle/inactive measurements.
- SIB15: Contains data associated with disaster roaming.
- SIB16: Contains slice-centric cell reselection data.
- SIB17: Contains TRS configuration details for UEs in RRC_IDLE/RRC_INACTIVE states.

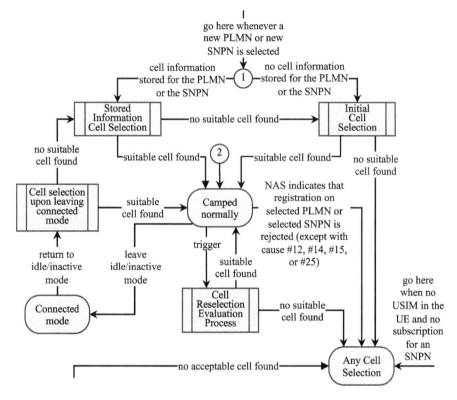

Fig. 10.6 Cell selection and reselection procedures summary in NR [4]

10.2.5 Cell Selection and Reselection

The LTE and NR cell selection/reselection procedure is intricate procedure, as it controls mobility in RRC_IDLE and RRC_INACTIVE states. Figure 10.6 illustrates the NR cell selection/reselection procedure, where aspects related to transitioning between RRC_CONNECTED to RRC_IDLE/RRC_INACTIVE, upon the reception of an RRC Release message or transitory cell selection done during RRC Re-establishment depicted.

10.2.5.1 Cell Selection

A UE in RRC_IDLE or RRC_INACTIVE state performs cell selection to find a best cell to camp on. The UE searches and measures predefined frequency bands, and for each carrier, it identifies the strongest cell as per the broadcast cell-defining SSB. The UE then reads cell system information broadcast to identify its PLMN(s) to find a suitable cell to camp on. A suitable cell is one for which the measured cell

10.2 Initial Access Procedure

attributes satisfy the cell selection criteria; the cell PLMN is the selected PLMN, registered or an equivalent PLMN.

When a subscriber travels outside the coverage area of their home PLMN, they enter the coverage area of another network. If this network allows the subscriber to provide its services, usually based on prior agreements with the home network, it becomes the visiting PLMN (VPLMN) for that subscriber.

Upon transition from RRC_CONNECTED to RRC_INACTIVE or RRC_IDLE, the UE also performs cell selection to camp on a cell. The UE may be configured to perform the cell selection the same way as it is does during powering on, or it may be configured with additional parameters in the RRC release message to perform the cell selection differently (e.g., prioritize a certain radio access network, frequency, etc.).

During cell selection, the UE determines a "suitable" cell to camp on, and this cell has to satisfy a number of criteria including the following: the cell is not being barred to the UE, the cell belongs to the PLMN selected by the NAS layer, and the cell belongs to a tracking area which is not forbidden. Once all cell selection criteria are satisfied, the UE camps on the cell. Once a suitable cell is found, the UE can remain camping in that cell as long as the cell reselection criteria are not fulfilled (described in 10.2.5.2).

If the camped cell fulfills certain criteria, UE doesn't attempt to make neighbor cell measurements to save energy. However, when these camped cell measurement criteria are not met, the UE evaluates additional neighboring cells. During cell selection, the UE evaluates a cell selection criterion per carrier, known as criterion S.

The cell selection criterion is fulfilled when:

$$Srxlev > 0 \text{ AND } Squal > 0 \quad (10.1)$$

where:

$$Srxlev = Qrxlevmeas - (Qrxlevmin + Qrxlevminoffset) - \quad (10.2)$$
$$Pcompensation - Qoffsettemp;$$

$$Squal = Qqualmeas - (Qqualmin + Qqualminoffset) - Qoffsettemp;$$

These parameters are defined in Table 10.1.

The signaled values Qrxlevminoffset and Qqualminoffset are only applied when a cell is evaluated for cell selection as a result of a periodic search for a higher priority PLMN while camped normally in a VPLMN. During this periodic search for higher priority PLMN, the UE checks the S criteria of a cell using parameter values stored from a different cell of this higher priority PLMN.

Table 10.1 NR cell selection parameters

Srxlev	Cell selection received signal strength level value (dBm)
Squal	Cell selection quality value (dB)
Qrxlevmeas	Measured cell received signal strength level value, measured as the RSRP (dBm)
Qqualmeas	Measured cell quality value (RSRQ)
Qrxlevmin	Minimum required received signal strength level in the cell (dBm). Qrxlevmin is obtained from q-RxLevMin in SIB1, SIB2, and SIB4. If Qrxlevminoffsetcell is present in SIB3 and SIB4 for the concerned cell, this cell-specific offset is added to the corresponding Qrxlevmin to achieve the required minimum RX level in the concerned cell.
Qqualmin	Minimum required quality level in the cell (dB). Additionally, if Qqualminoffsetcell is signaled for the concerned cell, this cell-specific offset is added to achieve the required minimum quality level in the concerned cell.
Qrxlevminoffset	Offset to the signaled Qrxlevmin taken into account in the Srxlev evaluation as a result of a periodic search for a higher priority PLMN while camped normally in a VPLMN.
Qqualminoffset	Offset to the signaled Qqualmin taken into account in the Squal evaluation as a result of a periodic search for a higher priority PLMN while camped normally in a VPLMN.
Qoffsettemp	Offset temporarily applied to a cell to bias the selection criterion (dB)
Pcompensation	The parameters relates to the compensation for power variations between UL and DL due to path loss differences between them and thus ensures that the UE's decision to select or reselect a cell considers both DL and UL conditions. For FR1,
	Pcompensation = max($P_{UE\_MAX} - P_{0,PowerClass}$, 0) (dB), where: P_{UE_MAX} is the maximum UL transmit power level of the UE. $P_{0,PowerClass}$ represents the nominal actual transmit power of the UE for the reference uplink channel. For FR2, Pcompensation is set to 0.

10.2.5.2 Cell Reselection

A UE performs cell reselection upon mobility or changing channel conditions in RRC_IDLE and RRC_INACTIVE states. The UE can perform intra-frequency, inter-frequency or inter-RAT cell reselection.

The UE is configured with priorities among different available radio access technologies (e.g., prioritize camping on NR over LTE whenever an NR cell is available) or among frequencies within the same RAT (e.g., frequency band A being the highest priority, frequency B has medium priority, frequency C has lowest priority, etc.). The UE thus attempts to reselect and camp on a cell operating with the highest priority RAT and with the highest priority frequency.

A neighbor cell list (NCL) can be provided to the UE, indicating which neighbor cells (e.g., intra-frequency, inter-frequency, inter-RAT) shall be considered for cell reselection. Allow-lists can be provided to the UE, indicating the only neighboring

10.2 Initial Access Procedure

cells that could be considered for reselection. Exclude-lists can be provided to the UE, indicating the neighboring cells that should not be considered for reselection.

If the serving cell fulfills Srxlev > SIntraSearchP and Squal > SIntraSearchQ, the UE can skip performing intra-frequency measurements to save power. Otherwise, the UE performs intra-frequency measurements for cell reselection as follows:

SIntraSearchP specifies the Srxlev threshold (in dB) for intra-frequency measurements.

SIntraSearchQ specifies the Squal threshold (in dB) for intra-frequency measurements.

If the serving cell fulfills Srxlev > SnonIntraSearchP and Squal > SnonIntraSearchQ, the UE can skip performing measurements of NR inter-frequency cells of equal or lower priority, or inter-RAT frequency cells of lower priority. Otherwise, the UE performs measurements of NR inter-frequency cells of equal or lower priority or inter-RAT frequency cells of lower priority.

SnonIntraSearchP specifies the Srxlev threshold (in dB) for NR inter-frequency and inter-RAT measurements.

SnonIntraSearchQ specifies the Squal threshold (in dB) for NR inter-frequency and inter-RAT measurements.

When the UE performs intra-frequency measurements for cell reselection based on the criteria above, it performs the cell rankings of the concerned cells. Inter-frequency and inter-RAT reselection is based on absolute priorities where a UE tries to camp on the highest priority frequency available. The cell-ranking criterion, referred to us Criteria R (Rs for the serving cell, and Rn for neighboring cells) is defined by

$Rs = Qmeas,s + Qhyst - Qoffsettemp$
$Rn = Qmeas,n - Qoffset - Qoffsettemp$

where the parameters are defined in Table 10.2:

The UE performs ranking of all cells that fulfill the cell selection criterion. Cells are ranked according to the R criteria by deriving Qmeas,n and Qmeas,s and calculating the R values using averaged RSRP results.

Generally, the UE performs cell reselection to the highest ranked cell. If a parameter rangeToBestCell is configured, then the UE performs cell reselection to

Table 10.2 NR cell re- selection parameters

Qmeas	RSRP measurement quantity used in cell reselections.
Qhyst	A hysteresis parameters used to provide a margin for preventing frequent cell reselections due to small variations in signal levels. This margin helps ensure that more stable connections.
Qoffset	For intra-frequency: Equals to Qoffsets,n, if Qoffsets,n is valid, otherwise this equals to zero.
	For inter-frequency: Equals to Qoffsets,n plus Qoffsetfrequency, if Qoffsets,n is valid, otherwise this equals to Qoffsetfrequency.
Qoffsettemp	Offset temporarily applied to a cell as specified in TS 38.331.

the cell with the highest number of beams above the threshold (i.e., absThreshSS-BlocksConsolidation) among the cells whose R value is within rangeToBestCell of the R value of the highest ranked cell. This ensures cell reselection to FR2 cells is more stable, i.e., to avoid frequent cell reselection procedures when beam qualities change rapidly. If there are multiple such cells, the UE performs cell reselection to the highest ranked cell among them.

10.3 Connected Mode Mobility

The goal of mobility in connected mode is to preserve uninterrupted service continuity and quality as the UE moves. To achieve this, the UE persistently measures the radio signal levels of serving and neighbor cells, according to the measurement configuration provided from the network. These measurements are executed on SSBs or on configured CSI-RS.

In standard network-controlled mobility scenarios, the UE evaluates channel measurements and conditions, such as comparing the serving cell's SSB signal strength relative to a neighbor cell, and relays measurement outcomes to the network via RRC signaling. The network, upon evaluation of these reports, instructs the UE to handover to an optimal cell.

A typical handover procedure involves synchronization with a new target cell, by performing a random-access procedure by the UE, due to mismatched uplink transmission timing post-handover. This synchronization is often achieved through contention-free random access, utilizing dedicated resources that do not have collision risks and are solely intended for timing alignment with the new cell. This process is truncated to only the initial two steps of random access—preamble transmission followed by the reception of a random-access response, which provides the UE with the necessary timing adjustment information.

Inter-gNB handover process involves several steps, starting with the source gNB issuing a handover request via the Xn interface. The target gNB then conducts admission control and responds with an acknowledgement to the handover request, which contains the handover command to the UE (i.e., the RRC reconfiguration the UE has to apply on connecting to the target cell). Subsequently, the source gNB conveys the handover command to the UE, which basically relays the RRC configuration prepared by the target to the UE. This message includes the target cell ID and comprehensive access information for the target cell, allowing the UE to bypass reading system information at the target cell. The RRC message can also contain details for both contention-based and contention-free random access resources at the target cell, including beam-specific information when applicable. The process concludes with the UE transitioning the RRC connection to the target gNB following a successful random access procedure at the target cell and dispatching an RRC Reconfiguration Complete message to confirm completion. Figure 10.7 illustrates the inter-cell handover procedure in LTE or NR networks.

10.3 Connected Mode Mobility

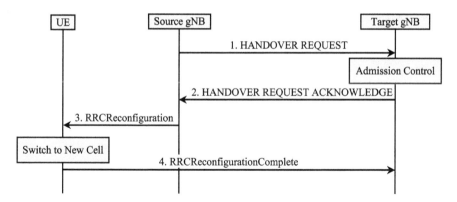

Fig. 10.7 Inter-gNB handover procedures [3]

For connected mode mobility, UE performs several types of measurements, e.g., intra-frequency, inter-frequency, and inter-RAT (e.g., for 3G or 4G cells). Each measurement type is linked to a specific measurement object, which contains a number of parameters like carrier frequencies. These objects are associated with various measurement reporting configurations, dictating the criteria for sending reports associated with the measurements.

To ensure data forwarding, in-sequence delivery, and the prevention of data duplication during handovers, the target gNB uses the same DRB configuration as that of the source gNB. Data forwarding during handovers can follow either a late or early approach. If late data forwarding is applied, the source gNB starts forwarding data to the target gNB node only after confirming the UE's successful access to the target node. With early data forwarding, the source gNB node initiates the data forwarding process to a prospective target node before the UE performs the handover. In Fig. 10.7, this step typically occurs before the labeled messages 3 and 4.

Reporting configurations primarily revolve around three criteria: event-triggered (i.e., measurement reporting triggered where a serving cell or/and neighbor cell's radio conditions fulfill certain relative or absolute thresholds), periodic (e.g., measurement reporting triggered every configured time period), and a combination of the two. The relation between a measurement object and its reporting configuration is defined with a measurement identity. This identity not only links an object to a configuration within the same RAT but also enables flexible associations, allowing multiple configurations per object.

Within 3GPP, TS 38.331 [1] specifies measurement reporting events for 5G NR as listed in Table 10.3:

Handover events, such as A3, A4, and A5, are defined for maintaining call quality and connectivity as a user moves across different cells. Though it is completely up to network implementation on how to use these events, the typical usage of these events is the following:

Table 10.3 Measurement reporting events for connected mode mobility

Intra-RAT events	Inter-RAT events
Event A1 (Serving becomes better than threshold)	Event B1 (Inter-RAT neighbor becomes better than threshold)
Event A2 (Serving becomes worse than threshold)	Event B2 (PCell becomes worse than threshold1 and inter-RAT neighbor becomes better than threshold2)
Event A3 (Neighbor becomes offset better than SpCell)	
Event A4 (Neighbor becomes better than threshold)	
Event A5 (SpCell becomes worse than threshold1 and neighbor becomes better than threshold2)	
Event A6 (Neighbor becomes offset better than SCell)	

- Event A1: To prevent the UE from performing/reporting some measurements (e.g., inter-frequency/RAT measurements, as this is an indication that the serving cell is in very good conditions);
- Event A2: To make the UE to perform/report some measurements as the serving cell is not in good conditions (i.e., inverse of A1);
- Event A3: For handing over the UE to a neighbor cell, as this is an indication that a neighbor cell of better quality than the serving cell is available. The A3 event is triggered when a neighboring cell's signal becomes stronger than the current cell's signal by a predefined threshold, indicating it might be more suitable for the user.
- Event A4: For inter-frequency HO, for adding of secondary cells (SCell) for carrier aggregation or for adding a secondary cell group (SCG) for dual connectivity. The A4 event occurs when the signal strength of a neighboring cell surpasses a specified threshold, but it doesn't necessarily compare it to the current cell signal; this is often used for load balancing or coverage reasons
- Event A5: Similar to A3; the A5 event is a combination of A3 and A4, where it is triggered if the neighboring cell's signal is stronger than the current cell's signal (like A3) and also exceeds a specific threshold (like A4).
- Event A6: For changing of an SCell or an SCG (Fig. 10.8)

Once a handover event is triggered, the UE sends a measurement report to the current serving cell indicating that the criteria for the event have been met. This report includes details of the measured signal strengths of the serving cells and neighboring cells.

Handover measurements can contain other neighbor cells as well, e.g., if the UE is configured to include other neighbor cells in the measurement report. For example, if a handover measurement report is triggered due to a cell A condition, and UE includes measurements of cell A and Cell B, where cell B is slightly worse

10.3 Connected Mode Mobility

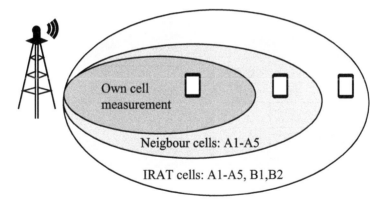

Fig. 10.8 Illustration of UE measurements within a cell

than A. In such case, the network may decide to HO the UE to cell B, if cell A is very loaded while cell B is not, for example.

Mobility scenarios, including handovers, might occur within the same RAT and CN or involve transitions across them. Inter-RAT fallbacks or handover can be triggered when the core network does not support emergency, voice, or specific services or for load balancing cases. In the 5G core network, the Access and Mobility Management Function (AMF) determines the target core network type/RAT based on service support requirements or load balancing needs, instructing the gNB node accordingly. The handover decision is then communicated to the UE via the RRC Release Message.

10.3.1 Conditional Handover and CPC in NR

NR Release-16 introduced the concept of conditional handover (CHO) and conditional PSCell Addition/Change (CPA/CPC) or collectively referred to as CPAC, with the main aim of reducing the likelihood of radio link failures and handover failures.

Regular LTE or NR handover is typically triggered by measurement reports, even though there is nothing preventing the network from sending a HO command to the UE without receiving a measurement report. For example, the UE is configured with an A3 handover event that triggers a measurement report to be sent when the radio signal level/quality (RSRP, RSRQ, etc.) of a neighbor cell becomes better than the PCell or the PSCell, e.g., in the case of dual connectivity. The UE monitors the serving and neighbor cells and will send a measurement report when the conditions get fulfilled. When such a report is received, the network (current serving node/cell) prepares a HO command and sends it to the UE. The UE then executes the HO command immediately, resulting in the UE connecting to the target cell.

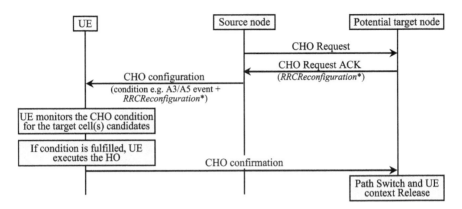

Fig. 10.9 Conditional handover configuration and execution [3]

CHO differs from legacy handover in two main aspects: first, multiple handover targets are prepared in advance for a given UE, compared to only one target in the legacy case. Second, the UE does not immediately execute the CHO as in the case of the legacy handover. Instead, the UE is configured with one or more triggering conditions (i.e., a set of radio conditions), and the UE autonomously executes the handover toward one of the targets only when/if the triggering conditions are fulfilled for that target, without needing to send a measurement report and receive a RRC handover command.

The CHO command can be sent while the radio conditions toward the current serving cell are still favorable, thus reducing the two main points of failure in legacy handovers: the failure to send the measurement report because the link quality to the current serving cell has fallen below acceptable levels and the failure to receive the handover command when the link quality to the current serving cell falls below acceptable levels after the UE has sent the measurement report but before it has received the HO command.

Triggering conditions for CHO could be based on the radio quality of the serving cell and neighbor cells, like the conditions that are used in legacy NR or LTE to trigger measurement reports. For example, the UE could be configured with a CHO condition that has A3-like triggering conditions and associated handover command. The UE monitors the current and serving cells, and when the A3 triggering conditions are fulfilled, it will, instead of sending a measurement report, execute the associated HO command and switches its connection toward the target cell (Fig. 10.9).

Another benefit of CHO is in helping prevent unnecessary re-establishments in case of a radio link failure. For example, assume the UE is configured with multiple CHO targets and the UE experiences an RLF before the triggering conditions with any of the targets gets fulfilled. Legacy operation would have resulted in RRC re-establishment procedure that would have incurred considerable interruption time for the bearers of the UE. However, in the case of CHO, if the UE determines that the cell that it has selected after the RLF is a cell for which it already has an associated CHO configuration (i.e., that cell was a CHO candidate cell), the UE will execute the HO command associated with this target cell directly instead of continuing with the full re-establishment procedure.

When faced with multiple candidates meeting configured CHO conditions, the UE selects the candidate the meets the configured conditions the most (e.g., based on the evaluated signal strength, quality, load balancing, or network priorities).

CPC and CPA are just extensions of CHO, but in dual connectivity scenarios. Dual connectivity and PSCell are described in Chap. 12. A UE could be configured with triggering conditions for PSCell change or addition, and when the triggering conditions are fulfilled, the UE executes the associated PSCell change or PSCell add commands.

10.3.2 Measurement Event Configurations for Handover

The main components of the measurement configuration are:

- Measurement objects
- Reporting configurations
- Measurement ID configurations
- S-measure configuration
- Quantity configuration
- Measurement gap configuration

A measurement object specifies what exactly the UE has to measure and some information regarding how the measurement is to be performed. This includes information such as the RAT, frequency, subcarrier spacing, SSB periodicity, SSB offset, measurement duration, reference signals, signal types to be measured, list of allowed or excluded neighbor cells on the concerned RAT and/or frequency to be measured, measurement gaps, and measurement level offset that can be applied to prioritize/de-prioritize certain cells.

A UE can be configured with multiple measurement objects, and a UE can have measurement configurations that can be related to different frequencies or even different RATs. A UE can be configured with up to 64 measurement objects, and each measurement object is identified by a measurement object ID.

A reporting configuration specifies what is to be reported, including reference signal type (e.g., a CSI-RS or SSB), the beam and cell level quantities to be reported (e.g., RSRP/RSRQ), a maximum number of cells or beams to be reported, and a measurement reporting criteria, upon the fulfillment of which the UE either sends a measurement report or executes an associated HO configuration in the case of CHO. The reporting criteria can be just the expiry of a periodic timer (periodic reporting configuration) or based on some radio conditions of serving and/or neighbor cells. A UE can be configured with up to 64 reporting configurations, and each reporting configuration is identified by a reporting configuration ID.

A measurement object can be associated with one or more reporting configurations. This association is made through a measurement ID. A measurement ID configuration consists of a measurement ID, a measurement object ID, and a reporting configuration ID.

Event A3, A5, and B2 can only be configured for the PCell or PSCell. Events A1, A2, A3, A5, and B2 can be configured for any serving cell. Event A6 can be configured only for SCells in carrier aggregation. Events A4 and B1 are only related to neighbor cell measurements and thus are not related to any serving cell. Each event configuration is associated with a threshold (offset), hysteresis value, and a timeToTrigger (TTT) parameter.

In the case of CHO, instead of sending a measurement report when the reporting conditions are fulfilled, the UE executes a HO command. For CHO, the following event triggered reporting configurations can be defined and configured:

- CondEvent A3 (Neighbor becomes offset better than SpCell)
- CondEvent A4 (Neighbor becomes better than threshold)
- CondEvent A5 (SpCell becomes worse than threshold1 and neighbor becomes better than threshold2)

A CHO configuration contains a conditional reconfiguration ID, a conditional reconfiguration triggering condition, and an RRC reconfiguration to be executed when the conditions are fulfilled, i.e., a handover command.

The triggering condition for CHO is a reference to 1 or 2 measurement IDs. If two measurement IDs are configured, the two must refer to the same measurement object. For example, one measurement IDs associating the measurement object related to the PCell with an A3 event and another measurement IDs associating the same measurement object with an A5 event.

The UE's measurement configuration can contain an s-measure configuration (s-MeasureConfig). When the serving cell's RSRP is above the s-measure threshold, the UE is not required to perform neighbor cell measurements, thereby saving UE battery.

10.3.3 Measurement Gaps

The UE needs measurement gaps to perform measurements when it cannot measure the target carrier frequency while simultaneously transmitting/receiving on the serving frequency. The UE may need measurement gaps to perform inter-frequency and inter-RAT measurements. During measurement gaps, the UE is not able to receive/transmit from/to the serving cell.

A typical measurement gap in LTE is 6 ms which accommodates 5 ms measurement time (PSS and SSS are transmitted once every 5 ms) and RF re-tuning time of 0.5 ms before and after the measurement gap. The measurement gap repeats with a periodicity of either 40 ms or 80 ms.

10.3 Connected Mode Mobility

The need for measurement gap in NR depends on the capability of the UE, the active BWP of the UE, and the current operating frequency. In NR, measurement gaps might be required for intra-frequency, inter-frequency, and inter-RAT measurements. For intra-frequency measurement gaps in NR, a measurement gap may be needed if the intra-frequency measurements are to be done outside of the active BWP.

Depending on the UE capability to support independent FR measurement and network preference, there are two types of measurement gaps defined in NR: per-UE and per-frequency range (FR). In per-FR gap, two independent gap patterns (i.e., FR1 gap and FR2 gap) are defined for FR1 and FR2, respectively. Per-UE gap applies to both FR1 (E-UTRA and NR) and FR2 (NR) frequencies.

The main parameters of a measurement gap configuration are:

- mgrp (Measurement Gap Repetition Period): the periodicity in ms at which measurement gap repeats.
- gapOffset: the gap offset of the gap pattern. The offset values point to the starting subframe within the period; its value range is from 0 to mgrp-1. For example, if the periodicity is 40 ms, the offset ranges from 0 to 39.
- mgl (Measurement Gap Length): the length of measurement gap in ms.
- mgta (Measurement Gap Timing Advance): if configured, the UE starts the measurement mgta ms before the gap subframe occurrence, i.e., the measurement gap starts at time mgta ms advanced to the end of the latest subframe occurring immediately before the measurement gap. The amount of timing advance can be 0.25 ms (FR2) or 0.5 ms (FR1).

Measurement gap lengths of 1.5, 3, 3.5, 4, 5.5, and 6 ms with measurement gap repetition periodicities of 20, 40, 80, and 160 ms are defined in NR. In NR, the RF re-tuning time is 0.5 ms for carrier frequency measurements in FR1 and 0.25 ms for FR2. For example, a gap length of 4 ms for FR1 measurements would allow 3 ms for actual measurements, and a gap length of 3.5 ms for FR2 measurements would allow 3 ms for actual measurements.

During measurement gaps, the measurements are to be performed on SSBs of the neighbor cells. The network provides the timing of neighbor cell SSBs using SS/PBCH Block Measurement Timing Configuration (SMTC). The measurement gap and SMTC duration are configured such that the UE can identify and measure the SSBs within the SMTC window; the SMTC duration should be sufficient to accommodate all SSBs that are being transmitted.

For SSB-based intra-frequency measurements, the network always configures measurement gaps if any of the UE configured BWPs do not contain the frequency domain resources of the SSB associated with the initial DL BWP. For SSB based inter-frequency measurements, the network always configures measurement gap if the UE supports per-FR measurement gaps and if the carrier frequency to be measured is in the same frequency as any of the serving cells or if the UE only supports per-UE measurement gaps.

Inter-RAT measurements in NR are limited to E-UTRA/LTE measurements. For a UE configured with E-UTRA inter-RAT measurements, a measurement gap configuration is always provided when the UE only supports per-UE measurement gaps or when the UE supports per-FR measurement gaps and at least one of the NR serving cells is in FR1.

10.4 Paging

In 5G networks, paging facilitates contacting UEs in RRC_IDLE and RRC_INACTIVE states while allowing the UE to save power in such states. Paging notifies UEs about system information changes, data arrival, and emergency alerts. Paging messages and short messages, both addressed to the UE's P-RNTI on PDCCH, are transmitted on PCCH and PDCCH, respectively. UEs in different states monitor paging channels variably: those in RRC_IDLE for CN-initiated paging, while those in RRC_INACTIVE monitor for both RAN and CN-initiated paging. Paging DRX cycles, defining the frequency of UE monitoring and the UE's sleep opportunity, vary based on the initiation source and are either broadcast in system information or configured via NAS and RRC signaling. The DRX cycle length controls the amount of battery savings in IDLE/Inactive states vs. the frequency at which the network can reach the UE. Configuration of the DRX cycle thus becomes a tradeoff.

Paging is delivered through the paging channel (PCH) from the PCCH logical channel. From a physical layer perspective, paging is transmitted using PDSCH transmissions, similar to system information messages. Paging PDSCH messages are scheduled by a distinct P-RNTI in the DCI. The P-RNTI serves as a common identifier used by multiple UEs for paging purposes, given a single paging transmission might contain messages for multiple UEs. The P-RNTI is transmitted within the DCI and tells the UE to read the corresponding PDSCH where the actual paging message is carried.

The paging early indicator (PEI) in 5G NR enhances paging efficiency by providing UEs advanced notice of incoming paging messages. This preemptive alert allows UEs to align their listening schedules to be ready for the message, improving energy efficiency by reducing the time UEs spend in high-power states. The PEI is particularly beneficial in densely populated network environments, as it helps UEs respond more quickly to paging while also reducing the network's signaling load and optimizing resource management. Further, to conserve UE power, UEs are grouped into subgroups for paging monitoring, based on CN control or UE ID. These subgroups are managed through PEI, which is determined by the UE's support and last used cell information. CN-controlled subgrouping, managed by AMF, assigns subgroup IDs to UEs, ensuring efficient and power-optimized paging across the network.

10.4.1 Paging Occasion and Paging Frames

A paging occasion (PO) is defined as a specific subframe during which a P-RNTI is likely to be transmitted on the PDCCH on which a paging message can be scheduled. A paging frame (PF) represents a radio frame that may include one or several paging occasions. The combination of PF and PO informs a UE about the precise timing to wake up and receive the paging message addressed to it. Within a given paging occasion, there can be several PDCCH monitoring occasions, where multiple can be configured for the purpose of beam sweeping the paging message.

In the simplest case, the formula for determining the paging frame is given as [4]

$PF = SFN \bmod T = (T \text{ div } N) \times (UE\_ID \bmod N)$

where T represents DRX cycle of the UE. T is derived either from the UE-specific DRX value, if provided by upper layers, or from a default DRX value broadcast in system information. Absence of a UE-specific DRX configuration defaults to the system information value. N is defined as min(T, nB), where nB is a parameter configured in SIB2, varies among a set of predefined ratios of T: 4T, 2T, T, T/2, T/4, T/8, T/16, T/32.

10.4.2 Radio Link Monitoring and Failure

In RRC connected mode, the UE performs radio link monitoring (RLM) within the active bandwidth part, relying on reference signals such as SSB and CSI-RS. The network configures these reference signals and corresponding signal quality thresholds. RLM based on SSB typically focuses on the SSB linked to the initial DL BWP, and this monitoring can extend to other DL BWPs that include the SSB from the initial DL BWP. For DL BWPs not containing the initial SSB, RLM relies solely on CSI-RS.

The RLM procedure works by having the UE periodically monitor the SINR of the serving cell (if the UE has more than one cell configured due to carrier aggregation, the UE monitors the RLM of the primary serving cell, PCell).

The UE is configured with thresholds to determine whether the radio link being monitored is good/reliable enough, specifically:

- Qout: the level at which the DL cannot be reliably received and shall correspond to out-of-sync block error rate (BLERout) which is the 10
- Qin: the level at which the DL can be significantly more reliably received than at Qout and shall correspond to in-sync block error rate (BLERin), which is 2
- The UE is also configured with timers and counters that are used to determine the reliability of the link being monitored, specifically:
- n310: the number of consecutive times that an out of sync (OOS) indication is received at the RRC from the lower layers (e.g., PHY) before RRC starts considering the link being monitored as experiencing reliability problems.

- n311: the number of consecutive times that an in-sync (IS) indication is received at the RRC from the lower layers (e.g., PHY) before RRC considers the link being monitored has become reliable again.
- t310: the duration of the timer that is started upon n310 consecutive OOS indications received from lowers and stopped upon n311 consecutive IS indications.

The UE starts the T310 timers upon detecting n310 consecutive OOSs, and if the timer expires before the reception of n311 consecutive in-sync indications from lower layers, RRC will consider the link has failed and declares an RLF.

The UE may use another timer (T312) to detect RLF, which is associated with measurement reporting. A measurement reporting configuration could be associated with t312. When the reporting conditions are fulfilled and a measurement report is to be sent, and if this measurement reporting configuration has been associated with t312, the UE will check if T310 is already running (i.e., RLM has already identified a problem and is waiting for the recovery). If so, the UE starts the T312 timer (which typically has a lower value than T310), and if the problem is not resolved before the timer expires, then the UE also declares an RLF. Basically, T312 is used to detect a late HO (i.e., had the measurement reporting been sent earlier than the radio link problem started, the UE would probably have been handed over to a target cell in time).

Figure 10.10 illustrates the NR RLM and RLF detection mechanisms:

The UE keeps track of the link quality in case the link signal quality deteriorates quickly. If RSRP or RSRQ metrics fall below predefined thresholds, known as an out of synch event, the UE starts the t310 timer. If the signal quality does not improve

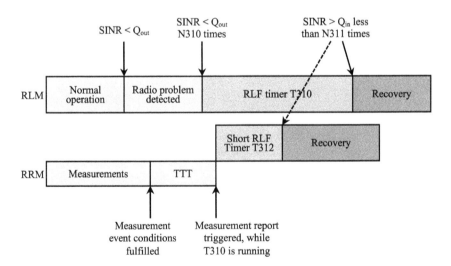

Fig. 10.10 Illustration of the RLM and RLF detection mechanisms

10.4 Paging

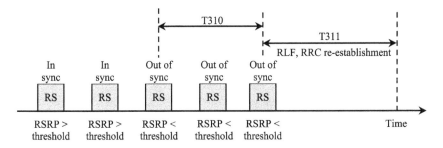

Fig. 10.11 Illustration of the RLM/RLF procedures

and the UE remains out-of-sync with the cell for the entire duration of the t310 timer, the UE then declares RLF.

Once an RLF is declared (post the expiry of t310), the t311 timer is started. This timer governs the duration for which the UE attempts to re-establish a connection with the network through a random access procedure. The length of t311 is also network-configurable and determines how long the UE should try to regain synchronization with a cell before moving to RRC_IDLE, where it performs a more extensive search for a suitable cell. If the UE successfully re-establishes a connection before t311 expires, it resumes normal operation; if not, it transitions to the RRC_IDLE state. The RLM and RLF procedure is illustrated in Fig. 10.11 from a time domain perspective.

In addition to RLM, RLF can be declared by the UE under several other reasons, including the expiry of a measurement report-related timer, while another radio problem timer is active, failure in the random access procedure, a handover failure, an RLC failure, or consistent failures in uplink LBT for shared spectrum channel access. These criteria are critical for maintaining reliable communication and ensuring timely response to potential radio link issues in the RRC_CONNECTED state.

If RLF is declared due to CHO or handover, the UE stays in RRC_CONNECTED state. Subsequent actions taken by the UE are contingent upon cell selection and network configurations. If the UE selects a cell that qualifies as a CHO candidate, the UE performs the execution of CHO toward the selected CHO candidate. If CHO is not applicable or not successful, the UE proceeds with re-establishment procedures.

Following RLF declaration, the UE transitions to RRC IDLE state only if the UE fails to find a suitable cell within a predefined duration after the declaration of RLF. This mechanism is a critical part of the mobility management process in cellular networks, ensuring that the UE maintains connectivity, either by re-establishing the link or by moving to a state where it can search for a new cell to regain network access.

10.5 NAS Message Transfer

For core network operations, a distinction is made between the non-access stratum (NAS) and the access stratum (AS). The NAS refers to the set of functionalities that occur between the core network, specifically the Access and Mobility Management Function (AMF), and the UE. This contrasts with the AS, which encompasses the functionalities between the UE and the radio-access network. The separation of NAS from AS is essential for delineating the different layers of network communication and management in mobile networks.

The NAS functionality operates between the AMF in the core network and the device. NAS functionality includes authentication, IP assignment to the UE, security, and various control-plane procedures such as paging.

NAS messages are delivered over SRBs reliably and in-sequence. During handover events, there's a possibility of message loss or duplication due to PDCP re-establishment. NAS messages are transmitted using transparent containers, and piggybacking of these messages is common in scenarios such as bearer establishment, modification, DL release, as well as during the initial NAS message transfer in connection setup and UL resume procedures.

NAS messages undergo integrity protection and ciphering, both at the NAS level and at the PDCP level. When multiple NAS messages are sent in a single DL RRC message, e.g., during PDU session resource establishment or modification, the sequencing of NAS messages within the RRC message is maintained as per their order in the corresponding Next-Generation Application Protocol (NG-AP) message. This ordering is critical to ensure in-sequence delivery of NAS messages.

In situations where non PDU session-related NAS PDUs from the AMF fail to be delivered, the gNB may initiate the NAS non-delivery indication procedure. This procedure is a crucial step to report any issues in the delivery of these NAS PDUs.

Problems

1. a. Describe one scenario where it is beneficial for the network to keep the UE to be in RRC_IDLE
   b. Describe one scenario where it is beneficial for the UE to be in RRC_Inactive state.
   c. Explain at least three triggers for the UE to transmit an RRC establishment request.
   d. In an initial access procedure, arrange the following procedures in order: system information acquisition, synchronization, random access, and cell selection.
2. a. on which downlink physical channel is the minimum system information transmitted?
   b. on which downlink physical channel is SIB1 transmitted?

c. if other SI is not broadcast by the network, explain how the UE can obtain other S
3. For a UE in connected mode, the following cell measurements are made:
cell A RSRP = −126.5 dBm
cell B RSRP = −130 dBm
cell C RSRP = −120 dBm
Assuming that:

- The UE is served by cell B.
- the UE is configured with an A3 mobility reporting event of 4 dB, an A4 mobility event with threshold of −126 dBm, an A5 mobility event with a serving cell threshold of −127 dBm, and a second threshold neighboring cell threshold of −122 dBm.

determine the following:

a. Which of the configured mobility events are triggered with these measurements.
b. For the triggered mobility events, which cell(s) are viable target cells for handover?
c. Assuming the UE is configured with other cells as CHO candidates and a conditional handover trigger of RSRP with a threshold of −126 dB, which CHO candidate will the UE select for handover execution?
d. Repeat (a) to (c) assuming the UE is served by cell A.

4. Determine the first and second paging frames for a UE camped in a cell, given the formula for determining the Paging Frame is given as:
PF = SFN mod T = (T div N) x (UE_ID mod N);
N = min(T, nB);
UE_ID = IMSI mod 1024
where:

- nB is configured as T/4,
- the UE's IMSI is 1168
- T is 20 ms

References

1. NR, Radio Resource Control (RRC), Protocol specification. https://www.3gpp.org/ftp/Specs/archive/38_series/38.331/38331-i20.zip Cited 21 Jul 2024
2. NR, Physical layer procedures for control. https://www.3gpp.org/ftp/Specs/archive/38_series/38.213/38213-i30.zip Cited 21 Jul 2024
3. NR, NR and NG-RAN Overall Description, Stage 2. https://www.3gpp.org/ftp/Specs/archive/38_series/38.300/38300-h90.zip Cited 21 Jul 2024
4. NR, User Equipment (UE) procedures in Idle mode and in RRC Inactive state. https://www.3gpp.org/ftp/Specs/archive/38_series/38.304/38304-i20.zip Cited 21 Jul 2024

Chapter 11
Modeling and Performance Analysis of Cellular Systems

11.1 Cellular Network Performance Metrics and Modeling

Cellular network analysis involves computing various user key performance indicators (KPIs) or metrics, which could be visualized on geographical locations within a cell's coverage area. Metrics are computed per user location and are important drivers for network performance optimization and/or comparison. The definitions and modeling of the following key cellular network performance metrics are discussed in more detail at a given location on a map that will be analyzed.

11.1.1 Reference Signal Received Power (RSRP)

This KPI provides the reference signal received power (RSRP) expressed in dBm. From the network design point of view, it is used to evaluate the propagation characteristics and the reference signal strength in the coverage area. It is also commonly used for initial network designs based on an RSRP design target (e.g., -105 dBm as a design target for urban areas). Within base station protocol layers, it is used for cell selection, cell reselection, power control calculations, mobility procedures, beam measurements, etc.

RSRP shows the received reference signal power per resource element from the selected best serving cell. The RSRP is measured at the user equipment (UE) location, so the UE antenna gain is accounted for, as well we as any potential penetration loss. For a given cell n

---

Supplementary Information The online version contains supplementary material available at (https://doi.org/10.1007/978-3-031-76455-4_11).

$$RSRP_{n|\text{dBm}} = (RS_{TX\_power})_{n|\text{dBm}} - PL_{Masked|\text{dB}} - L_{penetration|\text{dB}} - G_{UT|\text{dBi}}$$

where the *Masked* pathloss $PL_{Masked|\text{dB}}$ is the pathloss from the base station transmitter to the UE's location that accounts for the base station's antenna gain in the direction of the UE. The average transmitted reference signal power is a reference signal over one resource element (RE):

$$RS_{TX\_power} = \frac{P_{RS/symbol}}{N_{RS \; subcarriers/symbol}}$$

In 5G, RSRP is determined either by a filtered measurement over configured reference signals (e.g., Layer-3 RSRP) or measured in the physical layer (e.g., a Layer-1 RSRP) determined from SSBs or configured CSI-RS. The Layer-1 SS-RSRP, in linear scale, is determined based on the SSB block by performing the linear averaging over the received power contributions of the resource elements that carry secondary synchronization signals (SSS)

$$RSRP = \frac{\sum_{k=1}^{127} P_{Rx,SSS}(k)}{127} \qquad (11.1)$$

where $P_{Rx,SSS}(k)$ is the received linear power of the k-th SSS subcarrier. For that reason, it is referred to as synchronization signal (SS) RSRP or SS-RSRP. The measurement is conducted at the antenna connector of the UE for FR1 range, whereas for FR2 range the SS-RSRP is measured as a combined signal from antenna elements for a given receiver branch. Details of the SS-RSRP measurements are specified in 3GPP TS 38.215 [1].

SS-RSRP calculations based on the physical layer signals referred to as Layer 1 RSRP (L1-RSRP) are part of channel state information that is used when minimal delay is required, such as beam management. Otherwise, the L1-RSRP measurements are averaged over longer time duration (e.g., 40 ms) for radio resource management. This averaging reduces the impact of short-term noise variations and fast fading, and these averages are used for Layer 3 (L3) measurement reports.

The RSRP measurement reports are mapped to the range from -156 dBm to -31 dBm, with the values specified in TS 38.133 [2], and reproduced in Table 11.1.

L3-RSRP, whose reporting is defined from -156 dBm to -31 dBm, has 128 different values, which is represented by 7 bits. L1-RSRP reporting range, on the other hand, is defined for the range from -140 to -44 dBm. The range of RSRP values is quite generous. If, for example, we consider a 15 kHz subcarrier space, thermal noise floor over 15 kHz, in dBm scale, is

$$P_{N|\text{dBm}} = -174 + 10\log_{10}\left(15 \times 10^3\right) = -132.24 \; \text{dBm} \qquad (11.2)$$

11.1 Cellular Network Performance Metrics and Modeling

Table 11.1 RSRP measurement mapping table [2]

Reported value	Measured quantity value (L3 SS-RSRP) and CSI-RSRP	Unit
RSRP_0	SS-RSRP < −156	dBm
RSRP_1	−156 ≤ SS-RSRP < −155	dBm
RSRP_2	−155 ≤ SS-RSRP < −154	dBm
RSRP_3	−154 ≤ SS-RSRP < −153	dBm
...	...	...
RSRP_124	−33 ≤ SS-RSRP < −32	dBm
RSRP_125	−32 ≤ SS-RSRP < −31	dBm
RSRP_126	−31 ≤ SS-RSRP	dBm
RSRP_127	Infinity	dBm

which is higher than −140 dBm. We can also examine the upper limit for the RSRP measurement. The maximum signal input level at the UE is defined in TS 38.101-1 (for FR1 range, [3]) and is equal to −20 dBm. If we assume that this power comes solely from the SSB block, the maximum RSRP would be

$$-20 - 10\log_{10}(N_{RB} \times 12) = -44 \text{ dBm} \quad (11.3)$$

where $N_{RB} = 20$ for the SSB block (equal power per subcarrier is assumed).

For a network designer, a simple approach for pathloss considerations is to compare the known transmit power of the resource elements (RE) carrying SSS signals (typically in the range between 9 dBm and 18 dBm for macro base stations) with the received RSRP level. Assuming equal power between the SSS and PDSCH subcarriers, a simple way to calculate the transmit reference signal RE power is to divide the total conducted power P_T for a fully loaded system with the number of subcarriers:

$$RS_{TX\_power} = \frac{P_T}{N_{RB} \times 12} \quad (11.4)$$

This is not necessarily true as the SSB block may be transmitted over a single polarization, whereas the rest of transmission uses two polarizations, and there is a 3 dB difference in subcarrier powers but is typically used as a first-order approximation. For example, channel with a channel bandwidth of 100 MHz and subcarrier spacing of 30 kHz has N_RB = 273 resource blocks, and the transmit reference signal power is

$$RS_{TX\_power} = \frac{P_T}{273 \times 12} \quad (11.5)$$

In dBm scale, the reference signal becomes

$$RS_{TX\_power}|_{dBm} = P_T|_{dBm} - 35.15 \text{ dBm}. \quad (11.6)$$

This offset between the transmit reference signal power and the conducted power remains constant (on average) at the receive side, which helps us establish the relationship between the received RSRP and the received total power.

11.1.2 Received Signal Strength Indication (RSSI)

This measurement provides the Received Signal Strength Indication (RSSI) for the best carrier. The RSSI is the sum of all the signals received on a symbol carrying the reference signal, including the signals transmitted by the best serving cell, but also interference signals from other cells and noise

$$RSSI = P_S + \sum_k P_I(k) + P_N \quad (11.7)$$

where P_S is the serving cell received power, $P_I(k)$ is the interference power of the k-th interfere, and P_N is the noise power at the receiver. In the absence of interference from other cells, and when the receiver's noise is relatively low, we can establish a relationship with the RSRP and RSSI as[1]

$$RSSI|_{dBm} = P_S|_{dBm} = RSRP|_{dBm} + 10\log_{10}(N_{RB} \times 12) \quad (11.8)$$

However, in practice, it is difficult to find an isolated cell without interference from other cells, and the relationship in (11.8) is no longer valid. For that reason, 3GPP has introduced another measurement that establishes the relationship between the RSSI and RSRP in the presence of interference and/or high noise, which is covered next.

11.1.3 Reference Signal Received Quality (RSRQ)

This KPI provides the Reference Signal Received Quality (RSRQ) for the best carrier. It is typically used for initial network design where design targets are defined as a function of the RSRQ. The RSRQ is function of the RSRP and the RSSI as shown in the following formula:

---

[1] This is valid if the reference signal REs have the same power as the data carrying subcarriers, which is not necessarily true.

11.1 Cellular Network Performance Metrics and Modeling

Table 11.2 RSRQ reported value mapping table [2]

Reported value	Measured quantity value	Unit
SS-RSRQ_0	SS-RSRQ < −43	dB
SS-RSRQ_1	−43 ≤ SS-RSRQ < −42.5	dB
SS-RSRQ_2	−42.5 ≤ SS-RSRQ < −42	dB
SS-RSRQ_3	−42 ≤ SS-RSRQ < −41.5	dB
...	...	...
SS-RSRQ_124	18.5 ≤ SS-RSRQ < 19	dB
SS-RSRQ_125	19 ≤ SS-RSRQ < 19.5	dB
SS-RSRQ_126	19.5 ≤ SS-RSRQ < 20	dBm
SS-RSRQ_127	20 ≤ SS-RSRQ	dBm

$$RSRQ_{|dB} = 10\log_{10}\left(\frac{N_{RB} \times RSRP_{|mw}}{RSRP_{|mw}}\right) \quad (11.9)$$

The reporting range for RSRQ, defined in TS 38.133 [2], is from −43 dBm to 20 dBm, with 0.5 dBm resolution (Table 11.2).

Example 1 Let us consider a base station that transmits only 5% of its total power P_T. The base station is configured with 100 MHz channel with SCS = 30 kHz that operates as an isolated cell, without interference from other cells. Determine the RSRQ value as a function of distance d by ignoring the noise impact.

Solution Received power is only 5% of the maximal possible receive power, and the RSSI can be expressed as

$$RSSI = \frac{0.05 P_T}{PL(d)} \quad (11.10)$$

where $PL(d)$ is the pathloss at distance d and the RSRP level at the same distance is

$$RSRP = \frac{P_T}{273 \times 12} \times \frac{1}{PL(d)} = \frac{P_T}{3276 PL(d)} \quad (11.11)$$

and the RSRQ value is

$$RSRQ = \frac{RSRP \times 373}{RSSI} = \frac{20}{12} = 1.66, \quad (11.12)$$

which is $RSRQ_{|dB} = 2.21$ dB in dB scale.

Example 2 A fully loaded NR system (all resource blocks are utilized) is transmitting $P_T = 40$ watts of power and has an antenna with a gain of $G_T = 15$ dBi. Assuming a pathloss of $PL(d) = 128.1 + 37.6\log_{10}(d)$, where d is distance in km, determine the RSSI value at UE located 1.5 km away from the serving base station. How does the RSSI value change if there is an adjacent cell 4 km away from the serving cell and the UE is in the straight line between the two base station? Determine the RSRP and RSRQ values under both scenarios (one cell and two cells affecting the measurements). Assume a 20 MHz bandwidth and the subcarrier spacing $\Delta f = 15$ kHz.

Repeat the whole calculation assuming each of the NR cells uses only 5% of available resources.

Solution Conducted power $P_T = 40$ watts, converted to dBm scale is $P_{T|\text{dBm}} = 46$ dBm. The EIRP value, obtained by adding the antenna gain to the total conducted power, in dBm is

$$EIRP_{|\text{dBm}} = 46 + G_T = 46 + 15 = 61 \text{ dBm} \tag{11.13}$$

and the pathloss at $d = 1.5$ km is

$$PL(d) = 128.1 + 37.6\log_{10}(1.5) = 134.72 \text{ dB} \tag{11.14}$$

Ignoring the thermal noise and interference, the RSSI is equal to the total received signal power:

$$RSSI_{|\text{dBm}} = P_{R|\text{dBm}} = 61 - 134.72 = -73.72 \text{ dBm} \tag{11.15}$$

For $\Delta f = 15$ kHz and channel bandwidth of 20 MHz, the number of resource blocks $N_{RB} = 106$ and the RSRP is

$$RSRP_{|\text{dBm}} = RSSI_{|\text{dBm}} - 10\log_{10}(12 \times N_{RB}) = -104.76 \text{ dBm} \tag{11.16}$$

11.1 Cellular Network Performance Metrics and Modeling

and the RSQR value is

$$RSRQ_{|dB} = 10\log_{10}(N_{RB}) + RSRP_{|dBm} - RSSI_{|dBm} = -10.78 \text{ dB} \quad (11.17)$$

which is 1/12 in linear scale.

Let us now consider the other base station. It is located 2.5 km away from the UE location, and its pathloss is $PL(d = 2.5) = 143.06$ dB. The UE received two signals, one from the serving cell at the level of -73.72 dBm and the other from the neighboring cell at the level of $61 - 143 = -82$ dBm. Therefore, the RSSI at the UE location is

$$RSSI_{|dBm} = 10\log_{10}\left(10^{-\frac{73.72}{10}} + 10^{-\frac{82}{10}}\right) = -73.12 \text{ dBm} \quad (11.18)$$

The RSRP value is not affected by the interference, so the RSRQ value is

$$RSRQ_{|dB} = 10\log_{10}(106) - 104.76 + 73.12 = -11.38 \text{ dB} \quad (11.19)$$

Clearly, the RSRQ value we reduced with the addition of interference. In general, the more interference is added, the lower the RSRQ value becomes.

The same steps can be repeated for 5% load for both base stations, where the EIRP is reduced to $61 + 10\log_{10}(0.05) = 48$ dBm. The RSRQ value after calculation becomes 1.6 dB. ∎

11.1.4 Reference Signal SINR

This metric provides the serving cell's reference signal carrier to interference plus noise ratio S/(N+I), where signal and interference powers have bandwidths corresponding to one subcarrier (one RE). They are averaged and computed over RBs containing reference signals, so that the SINR expression is

$$SINR_{RS} = \frac{RSRP}{P_{I,\,RE} + P_{N,\,RE}} \quad (11.20)$$

where $P_{I,\,RE}$ is the reference signal interference power observed on reference signal resource elements in the serving cell and the noise power over one RE is calculated as

$$P_{N,\,RE} = k_B T_0 F \Delta f$$

where k_B is the Boltzmann constant, F is the noise factor, $T_0 = 290$ K, and Δf is the subcarrier spacing.

11.1.5 Synchronization Signal SINR

This metric provides the serving cell's synchronization signals carrier to interference plus noise ratio S/(N+I), including primary and secondary synchronization signals. Signal and interference powers are averaged and computed over RBs containing PSS and SSS signals.

11.1.6 PDSCH SINR

This following modeling provides the S/(N+I) of the PDSCH downlink data channel, under the assumption of 100% cell load.

$$SINR_{PDSCH} = \frac{P_{PDSCH}}{P_I + P_N} \quad (11.21)$$

where P_{PDSCH} is the received PDSCH signal power and P_I and P_N and the interference and noise powers at the receiver.

11.1.7 PDSCH Maximum User Achievable MCS and Spectral Efficiency

This metric provides information on the downlink maximum spectral efficiency that can be achieved for a given user. It provides information on the capacity offered on the downlink. The downlink maximum achievable spectral efficiency shows the number of bits per symbols associated with the downlink best available modulation and coding scheme.

The best available Modulation and Coding Scheme (MCS) is the MCS that achieves the target spectral efficiency for a given coverage.

11.1.8 PUSCH SINR

The SINR value in the UL, over the PUSCH channel is the ratio of received signal power over the interference plus noise, where the bandwidth for each component is equal is depends on the number assigned UL resourece blocks for a given user.

11.1.9 Maximum Achievable User Data Rate

This metric provides the maximum possible downlink (DL) or uplink (UL) user throughput that could be achieved with the best DL/UL modulation scheme. The maximum achievable data rate is the data rate provided by the computed best available modulation and coding scheme. The downlink maximum achievable data rate is directly proportional to the maximum achievable spectral efficiency. It also depends on the number of resource elements that can be used for data transmission. When computing the number of resource elements that can be used for data, control channels such as reference signal, PDCCH, and synchronization signal are accounted for.

While the actual peak data rates for UEs in the field depend on many parameters, including the network configuration, interference value, proximity to base station, etc., it is possible to estimate maximal achievable data rates in the laboratory using a simple methodology provided in the next example.

Example 3 Determine the peak DL data rate for an NR UE operating at 20 MHz channel bandwidth if $N_L = 2$ MIMO layers are transmitted; network is configured to operate using normal cyclic prefix and to use 256QAM. Assume the error control code rate of $R_c = 0.93$, subcarrier spacing of $\Delta f = 15$ kHz, and the control plane overhead of OH=15%.

Solution From TS 38.101-1 [3], we determine that the number of resource blocks for the given configuration is $N_R B = 106$.

Let us first look at the one slot time interval, which corresponds to 1 ms. Within one resource blocks, over the 1 slot duration, there are $12 \times 14 = 168$ REs, and over the entire bandwidth there are

$$168 \times N_{RB} = 168 \times 106 = 17,808 \qquad (11.22)$$

REs. Each RE caries $Q_m = \log_2(256) = 8$ bits, so that the total number of bits within one 1 ms slot is

$$17,808 \times 8 = 142,464. \qquad (11.23)$$

Therefore, the data rate per MIMO layer, excluding the code rate and control plane overhead, is 142.46 Mbps. Finally, if we account for the MIMO two-layer transmission and overheads, the peak data rate is

$$R_{peak} = 2 \times 142.46 \times 0.93 \times (1 - 0.15) = 225.23 \text{ Mbps.} \qquad (11.24)$$

■

While the method illustrated above can be used, for purpose of consistency, 3GPP has specified a method to calculate peak data rates in the DL and UL for FR1 and FR2 ranges in TS 38.306 [4]. A modified version, applicable to a single channel, is

$$R_{peak} = N_L Q_m f R_{\max} \frac{N_{PRB}^{BW,\mu} \times 12}{T_S^{\mu}} (1 - OH) \text{ bits/sec} \tag{11.25}$$

where N_L and Q_m are defined above; $R_{\max} = 948/1024 = 0.93$ is the maximal error control rate (same as R_c above); f is the scaling factor that can be 0.4, 0.75, 0.8, or 1, depending on device capabilities; μ is the configured numerology (subcarrier spacing); $T_S^{\mu} = 1/(14 \times 2^{\mu})$ is the average OFDM symbol duration in a subframe for the given numerology and normal cyclic prefix; and $N_{PRB}^{BW,\mu}$ is the maximum number of resource blocks for the given configuration and the overhead values are listed as follows:

$$OH = \begin{cases} 0.14, & \text{for FR1 range in the DL} \\ 0.18, & \text{for FR2 range in the DL} \\ 0.08, & \text{for FR1 range in the UL} \\ 0.10, & \text{for FR2 range in the UL} \end{cases} \tag{11.26}$$

Example 4 Determine the peak DL data rate using the TS 38.306 specified formula for an NR UE operating at 20 MHz channel bandwidth if $N_L = 2$ MIMO layers are transmitted; network is configured to operate using normal cyclic prefix and to use 256QAM. Assume the subcarrier spacing of $\Delta f = 15$ kHz.

Solution For the given subcarrier spacing of 15 kHz, we conclude that the numerology parameter $\mu = 0$. After substituting the values in the equation above, we get the peak data rate of

$$R_{peak} = 4 \times 8 \times 1 \times 0.93 \frac{106 \times 12}{71.43 \times 10^{-3}} (1 - 0.14) = 226.85 \text{ Mbps}. \tag{11.27}$$

∎

The peak spectral efficiency can be calculated based on the peak data rate and the channel bandwidth. For example, using the calculation from Example 4 for a 20 MHz channel and the subcarrier spacing of 15 kHz, the peak spectral efficiency is

11.1 Cellular Network Performance Metrics and Modeling

$$SE_{peak}^{NR} = \frac{R_{peak}}{BW} = 22.7 \text{ bps/Hz} \quad (11.28)$$

We can also look at the spectral efficiency per MIMO layer, in which case it is

$$SE_{peak/layer}^{NR} = 22.7/4 = 5.675 \text{ bps/Hz} \quad (11.29)$$

For LTE, the peak data rate calculation is more involved, but a simple formula used in the industry is

$$SE_{peak/layer}^{LTE} = 5 \text{ bps/Hz.} \quad (11.30)$$

For example, a 20 MHz LTE channel with 4 MIMO layers can reach a peak data rate of 400 Mbps. We can see that the peak LTE data rate is about 13% lower than the NR peak data rate for this specific example. This varies across different channel bandwidths. Note that the peak speeds and spectral efficiencies calculated in previous examples are based on 256QAM ($Q_m = 8$). This is a typical peak speed for most networks nowadays, but the 3GPP specifications allow for 1024QAM ($Q_m = 10$), in which case the values calculated above would scale by the 10/8 ratio.

Lastly, for TDD systems, the peak UE data rate is scaled by a duty cycle parameter $\alpha_{DL/UL}$, which depends on specific configuration. The general expression for the scaling parameter is

$$\alpha_{DL/UL} = \frac{\#DL/UL \text{ OFDM symbols in a frame}}{\#\text{total OFDM symbols in a frame}} \quad (11.31)$$

To illustrate the TDD scaling factor, let us consider the following example.

Example 5 Consider an NR TDD configuration with the slot pattern DFU-UDDDDDD, where D slot indicates a slot with 14 DL symbols, U slot is a slot with 14 UL symbols, and the flexible slot F is configured as F=DDDDGGGGGGUUUU (4 DL symbols, 6 guard time symbols, and 4 UL symbols). The slot pattern repeats. Calculate the DL and UL duty cycle.

Solution Within ten slots, there are seven DL slots, and the flexible slot has additional four symbols, so the DL duty cycle is

$$\alpha_{DL} = \frac{7 \times 14 + 4}{10 \times 14} = 0.7286, \quad (11.32)$$

and the UL duty cycle is

$$\alpha_{UL} = \frac{2 \times 14 + 4}{10 \times 14} = 0.2286. \tag{11.33}$$

∎

11.2 IEEE 802.11 Peak Data Rates

IEEE 802.11 or Wi-Fi data rates are calculated similarly to the peak data rates in 5G, with a few minor differences. Namely, the number of subcarriers is not calculated based on the number of resource blocks N_{RB}. Instead, we can use the number of data subcarriers parameter N_{SUB} directly. Also, there is no f scaling parameter. Lastly, the symbol duration is fixed (no need for averaging) and we will denote it T_{OFDM}. The peak Wi-Fi throughput is then calculated as

$$R_{peak} = N_L N_{SUB} Q_m R_c \frac{1}{T_{OFDM}} \tag{11.34}$$

where R_c is code rate and N_L and Q_m are the number of MIMO layers and the number of bits of digital modulation order (same as in 5G). Table 11.3 summarizes parameters for various generations of 802.11 technologies and the calculated respective peak data rates.

Note that these are maximal theoretical data rates. Also, for IEEE 802.11 ac we assumed a channel bandwidth of 160 MHz. For narrower channel bandwidths, the number of data subcarriers would be lower, and consequently the peak data rate would be lower. Also, the IEEE 802.11 ad data rate is calculated assuming OFDM implementation (there is also a single carrier mode).

Table 11.3 Peak data rates for various IEEE 802.11 generations of technology

Technology	N_L (max)	N_{SUB}	Q_m	R_c	T_{OFDM} (μs)	R_{peak}
IEEE 802.11 a/g	1	48	6	3/4	4	54 Mbps
IEEE 802.11 n	4	108	6	5/6	3.6	600 Mbps
IEEE 802.11 ac	8	468	8	5/6	3.6	6.93 Gbps
IEEE 802.11 ad	1	336	6	13/16	0.24242	6.76 Gbps

11.3 Network Simulations

11.3.1 Dynamic Cellular System Simulations

Simulation and evaluation of cellular systems is performed either by link-level simulations or system-level simulations. Link-level simulations are performed to evaluate the performance of a certain RF scheme on a single RF or cellular link, without considering the impact of interference or other system-level impacts. Typical evaluation metrics for link-level simulations include the block error rate (BLER), bit error rate, maximum capacity of certain physical layer channels, or required SNR levels needed to sustain a given evaluated scheme or modulation.

System-level simulations on the other hand are performed to evaluate the performance of cellular system on a holistic level, considering impacts of interference, cell loads, number of users, UE transmit power control, frame structure configurations, and impact of intra-channel and cross-channel configurations. Typical system-level evaluation metrics include user throughput, cell throughput, latency, cell resource load usage, spectral efficiency, or algorithmic percentage gains for certain system schemes. Performance of certain channels (e.g., PDCCH, PDSCH, PUCCH, or PUSCH) can be evaluated as well. The objective of a dynamic system-level simulator is to evaluate the impact of a given algorithm on the system performance.

The structure of a system-level simulator is typically based on two main loops that ensure gathering of sufficient and accurate channel statistics: a shadowing outer loop and a fast-fading time-slot inner loop. Each iteration of the shadowing loop drops the users uniformly across the simulation area, to model mobility of users and variations in large-scale fading statistics. Each iteration of the fast-fading loop updates the fast-fading parameters for each mobile to each cell in the network. Fast fading is generated according to one of various channel models, whereby multipath and propagation characteristics are modeled. Figure 11.1 illustrates an example of a drop of the shadowing loop for a heterogeneous network deployment.

A system-level simulator typically features multi-cell and multiuser support, whereby users are dropped randomly in simulation area and are uniformly distributed in the simulation area or within the coverage of a cell. System-level simulators feature wrap around modeling to reduce edge effects and accurately model interference of wide networks, whereby signals wrap around the simulation area edges to model a wide area network. Each cell is modeled with an antenna radiation pattern and directional antenna gain model.

11.3.2 Monte Carlo Simulations

Monte Carlo simulations rely on repeated random sampling of the cellular system to achieve deterministic evaluations. Unlike dynamic simulations, Monte Carlo simulations rely on evaluating multiple samples of the shadowing loop without

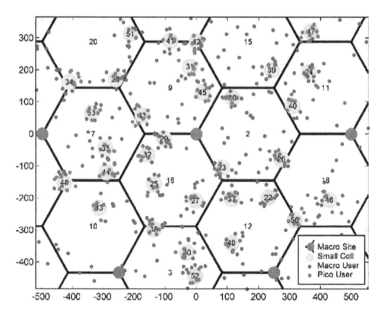

Fig. 11.1 An example of a simulated heterogeneous network deployment

running a time-domain fast fading loop, as fast fading averages over long periods of time. The intention of Monte Carlo simulations is to predict simulation results within a reasonable simulation run time and thus are typically used by radio network planning tools to evaluate the performance of large networks.

A Monte Carlo simulation also evaluate a number of runs in each simulated cell, whereby in each run a number of users are served, blocked, or not-assigned resources. Users are randomly distributed within the cell coverage, where users are dropped according to location-dependent traffic definitions. A traffic definition can drop users uniformly per simulation area, proportional to the cell's signal strength, or proportional to pre-generated traffic maps. User throughput is increased according to their channel conditions, distribution, path loss, and traffic profile.

Different scheduling algorithms can be used to assign resources to a given user. A scheduling algorithm cycles sequentially through users using such criteria until all users are scheduled or cell resources are depleted in the current scheduling run. The simulation can be configured with a minimum user throughput for a given service to simulate, which the scheduling algorithm will strive to meet at the minimum. With such minimum throughput defined, the scheduling algorithm may first go through a first pass to assign users resources necessary to meet the minimum throughput and then go through a second pass to assign remaining resources whereby each user k is assigned a share of remaining resources equal to

$$\text{Number of available cell PRB} \times \frac{PF_k(t)}{\sum_k PF_k(t)} \qquad (11.35)$$

where $PF_k(t)$ is the fairness factor assigned to user k, described in the next section for the proportional fair scheduling algorithm. In a more fairness-centric scheduling implementation, $PF_k(t)$ can be estimated at the inverse of the maximum uplink data rate of user k.

11.3.3 Scheduling Algorithms

Scheduling is the process of assigning physical cell resources to users in served by the cell. The use of OFDMA in NR and LTE systems facilitates the potential for scheduling cell users on orthogonal time-frequency resource blocks selectively. Channel-dependent scheduling is a commonly used technique in various cellular systems. Multiple access is achieved through OFDMA in the downlink and SC-FDMA or DFT-s-OFDM in the uplink, which allows scheduling to be performed in both the frequency and time domains.

Frequency-selective scheduling algorithms leverage the channel's frequency selectivity, while time-selective scheduling leverages the channel fast fading gain reported by different users in the cell [5]. Scheduling is a function of the available cell load, user mobility, the number of users per cell, data traffic characteristics, and the selected modulation and coding scheme (MCS). Channel reporting-dependent scheduling leverages the channel's time and frequency selectivity to allocate valuable radio resources in an optimal manner. Users periodically report CQI, which is exploited by the scheduler when allocating resource blocks and selecting suitable MCS. Frequency-selective scheduling provides benefits to both FDD and TDD networks. Downlink resource allocation relies on the CQI reported by the user. CQI must be reported at the sub-band level for frequency-selective scheduling. FDD has the advantage of having more instantaneous CQI feedback, as PUCCH CQI feedback can be instantaneous. This is more advantageous in the case of severe multipath or high user mobility. TDD includes a time split within the radio frame between the uplink and downlink, which results in a CQI reporting delay that depends on the TDD UL/DL split frame configuration.

Different scheduling algorithms can be evaluated in terms of their impact on the service packet delay, packet loss ratios, and throughput. The following scheduling algorithms are described herein along with benefits and an evaluation.

11.3.3.1 Proportional Fair Scheduling

Proportional fair scheduling provides a balance between increasing the cell capacity and the cell-edge user performance. The algorithm favors the following two groups of users equally: cell edge users with the worst radio and interference conditions and cell center users who contribute the best increase in spectral efficiency. The classification of these two groups can be performed based on the reported CQIs. Remaining users, including cell-edge users with decent radio conditions and cell-center users with limited spectral efficiency increase potential, are left with the remaining resource blocks. The algorithm determines the group of cell-center users providing maximal increase in spectral efficiency according to their potential to reach higher-order modulation and coding schemes when assigned better quality resource blocks. The spectral efficiencies of standardized MCSs are taken into considerations by the scheduler.

In one proportional fair scheduler model [6], the algorithm allocates a user k^*, which maximizes the proportional fairness metric for an assigned resource. The proportional fairness metric is described as the ratio of the user's instantaneous data rate $R_k(t)$ on RB r to the user's past data rate $T_k(t)$ filtered through a time window of memory length t_c

$$k^* = \arg \max_{1 \leq k \leq K} PF_k(t) = \arg \max_{1 \leq k \leq K} \frac{R_k(t)}{T_k(t)} \quad (11.36)$$

where

$$T_k(t) = \left(1 - \frac{1}{t_c}\right) T_k(t-1) + \frac{1}{t_c} R_k(t-1).$$

The algorithm cycles sequentially through users using such criteria until all users are scheduled, or cell resources are depleted in the current scheduling slot, whereby each user k is assigned a bandwidth equal to

$$\text{Number of available cell PRB} \times \frac{PF_k(t)}{\sum_k PF_k(t)} \quad (11.37)$$

11.3.3.2 Round Robin Scheduler

In a round robin scheduler, the algorithm assigns each user an equal share of resources. Each user is assigned a bandwidth equal to number of available cell PRB/Number of UEs, where the number of UEs in cell is the number of UEs with buffered data in the current scheduling slot.

In a simple system-level evaluation, the two scheduling algorithms can be compared. The following data rate results can be observed using a SISO antenna configuration. These figures demonstrate the downlink user and cell rate CDFs under the two scheduling algorithms (Figs. 11.2 and 11.3).

Simulation results are summarized in the Table 11.4.

11.3 Network Simulations

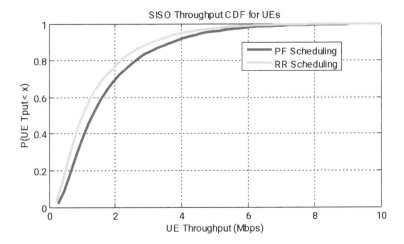

Fig. 11.2 CDF of user data rates in the cell using proportional fair and round robin scheduling

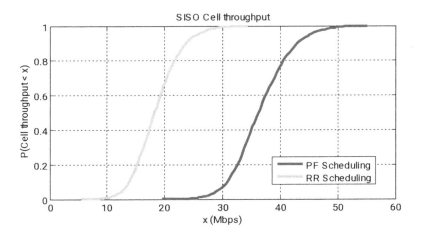

Fig. 11.3 CDF of cell data rates in the network using proportional fair and round robin scheduling

Table 11.4 Analysis of user and cell data rates using proportional fair and round robin scheduling

Evaluation	Mean user rate (Mbps)	Cell-edge user rate (Mbps)	Peak user rate (Mbps)	Mean cell rate (Mbps)
Proportional fair scheduler	3.66	0.38	28.01	36.72
Round robin scheduler	1.52	0.27	17.09	18.67

Problems

1. The RSSI range for the 5G NR is from $-120\,\text{dBm}$ to $-13\,\text{dBm}$. Based on the range of L3 SS-RSRP values defined by 3GPP and the table of bandwidths in FR1 range included below, determine the range of SS-RSRQ values. Determine the index values for these SS-RSRQ values (SS-RSRQ_xxx values).
   Make sure that the SS-RSRP and RSSI values are paired appropriately (e.g., SS-RSRP of $-31\,\text{dBm}$ cannot correspond to RSSI of $-120\,\text{dBm}$).

2. A UE is located between two base stations, at distance $d_{2D,1} = 300$ m from the serving base station and at distance $d_{2D,2} = 400$ m from the interfering base stations. If the UE reads the dBm scale RSRP level of $-81.311\,\text{dBm}$, and the RSRQ level of 0.14 (linear), determine the following:

   a. RSSI level, assuming an NR air interface and a channel bandwidth of 30 MHz.
   b. Signal level received from the serving cell as well as the signal level received from the interfering base station assuming that the serving base station is not fully loaded at transmits 32 watts of power at the antenna port (before antenna gain). Assume the following:
   $h_{BS} = 25$ m, $h_{UT} = 2.5$ m, base station antenna gain is 18 dBi and the carrier frequency is $f_c = 1930\,\text{MHz}$ and

   $$PL_{NLOS} = 13.54 + 39.08 \log_{10}(d_{3D}) + 20 \log_{10}(f_c) - 0.6(h_{UT} - 1.5).$$

   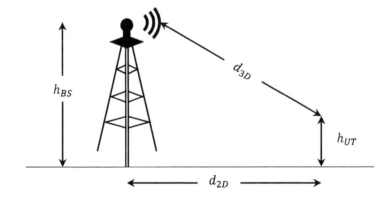

   The f_c in the pathloss formula above is expressed in GHz, d_{3D} is in meters and h_{UT} is a UE height in meters.
   What are the load and the nominal power level (Tx power at 100% load) of the serving base station?

   c. What is the load of the interfering base station if it has the same nominal power as the serving base station?

3. Given the table of 5G channel bandwidths expressed in terms of number of resource blocks:

11.3 Network Simulations

SCS (kH)	50 MHz	100 MHz	200 MHz	400 MHz	800 MHz	1600 MHz	2000 MHz
60	66	132	264	N/A	N/A	N/A	N/A
120	32	66	132	264	N/A	N/A	N/A
480	N/A	N/A	N/A	66	124	248	N/A
960	N/A	N/A	N/A	33	62	124	148

calculate peak speeds for each configuration. Assume a normal cyclic prefix, 64QAM modulation, and two MIMO layers ($N_L = 2$).

4. Peak NR data rates, expressed in Mbps, are listed in the table below and cover a subset of possible bandwidth and subcarrier spacing configurations:

SCS (kHZ)	5 MHz	10 MHz	15 MHz	20 MHz	25 MHz	30 MHz	40 MHz	50 MHz
15	107	222.57	338.14	453.70	569.27	684.83	924.53	1155.66
30	94.16	205.45	325.30	436.58	556.43	667.71	907.41	1138.54

Assuming:

- 256QAM modulation
- 4 MIMO layers
- code rate $R_{max} = 948/1024$
- scaling factor $f = 1$

calculate the number of resource blocks for each of the entries in the table above.

5. A UE located inside a building is receiving signal from 3 macro base stations, BS1, BS2, and BS3, as shown below.

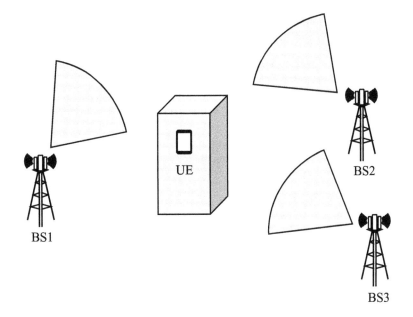

The location of the UE inside the building is at the (latitude, longitude) coordinates in decimal degrees UE = (43.65962, −79.3969), and geo-locations of the base stations are as follows:

BS1 = (43.658563, −79.398951),
BS2 = (43.66014, −79.392951),
BS3 = (43.661166, −79.386916)

The distance d between two points for a given latitude φ_i and longitude λ_i coordinates in radians ($i = 1, 2$) is calculated using the following haversine formula:

$$a = \sin^2\left(\frac{\Delta\varphi}{2}\right) + \cos(\varphi_1)\cos(\varphi_2)\sin^2\left(\frac{\Delta\lambda}{2}\right)$$

$$c = \tan^{-1}\left(\frac{\sqrt{a}}{\sqrt{1-a}}\right)$$

$$d = Rc$$

where $R = 6{,}371$ km is the earth's mean radius. The following parameters are known:

- Transmit power for each base station is 40 watts
- Antenna gains in the direction of the UE for BS1 and BS2 are 12.45 dBi and the antenna gain for BS3 = 17 dBi (given the sector orientation)
- Pathloss formula is $PL(d) = 128.1 + 10\log_{10}(d_{km}) + L_{wall}$
- Wall loss between BS1 and UE is $L_{wall} = 45$ dB, between BS2 and UE is $L_{wall} = 40$ dB and BS3 and UE is $L_{wall} = 35$ dB

Calculate the following assuming LTE technology (the N_{RB} value differs compared to NR, the other definitions are the same):

a. Distances between each base station and the UE using the haversine formula
b. Pathlosses between each base station and the UE
c. EIRP in the direction of UE and the received power from each BS
d. RSSI, RSRP, and RSRQ values at the UE receiver assuming full power at the transmitter, $\Delta f = 15$ kHz; system bandwidth of 20 MHz is used at all base station and that the serving base station is BS1
e. RSSI, RSRP, and RSRQ values at the UE receiver assuming that BS2 and BS3 are suddenly turned off.

6. In LTE system, the reference signal (RS) settings within a resource block (RB) configured as shown below.

11.3 Network Simulations

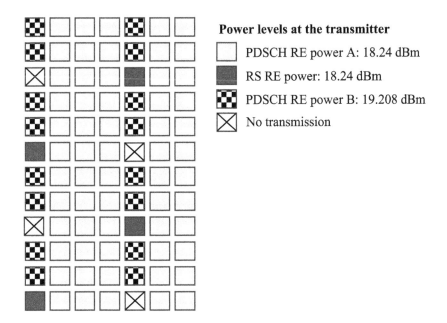

Power levels at the transmitter

- ☐ PDSCH RE power A: 18.24 dBm
- ■ RS RE power: 18.24 dBm
- ▦ PDSCH RE power B: 19.208 dBm
- ☒ No transmission

a. Determine RSRP, RSRQ, and RSSI values at the transmitter assuming there is no interference, and the channel bandwidth is 10 MHz. What is the transmitter power in watts?

b. Assume that such signal is received at some distance away from the transmitter such that the pathloss is 100 dB. The detected RSRP power at the receiver is −81.76 dBm, and the RSSI power is −50 dBm and includes a wideband uniformly distributed interference contribution. Determine:

   i. Total received power in dBm
   ii. Level of interference at the receiver in dBm
   iii. RSRQ value at the receiver

7. An ultra-reliable, low-latency communication system similar to 5G NR is used to transmit 32 bytes of information over a very short time duration. Assuming that a channel code with the code rate r = 1/6 is applied, determine the following:

   a. How many bits are transmitted over the air excluding the CRC (apply only the channel coding to the data stream)?
   b. If QPSK modulation is used, how many resource elements are required to transmit all the bits calculated in (a)?
   c. Ignoring the CRS signals and any control plane, if all the REs are being transmitted over one symbol duration, determine the required bandwidth to transmit all the bits.
   d. What is the instantaneous information data rate assuming the CP for the symbol has 160 samples (first symbol)?
   e. Repeat (c) and (d) assuming that all the bits are transmitted over 5 symbols, where the first symbol uses 160 samples for CP and the next 4 symbols each have 144 samples for CP.

References

1. NR, Physical layer measurements. https://www.3gpp.org/ftp/Specs/archive/38_series/38.215/38215-i30.zip Cited 21 Jul 2024
2. NR, Requirements for support of radio resource management. https://www.3gpp.org/ftp/Specs/archive/38_series/38.133/38133-i60.zip Cited 21 Jul 2024
3. NR, User Equipment (UE) radio transmission and reception, Part 1: Range 1 Standalone. https://www.3gpp.org/ftp/Specs/archive/38_series/38.101-1/38101-1-i60.zip Cited 21 Jul 2024
4. NR, User Equipment (UE) radio access capabilities. https://www.3gpp.org/ftp/Specs/archive/38_series/38.306/38306-i20.zip Cited 21 Jul 2024
5. F. Alfarhan, R. Lerbour, Y. Le Helloco, G. Donnard, *Analysis of Practical Frequency Selective Scheduling Algorithms in LTE Networks* (IEEE Vehicular Technology Conference (VTC), Vancouver, Canada, 2014)
6. J. Lim, H. Myung, K. Oh, D. Goodman, Proportional fair scheduling of uplink single-carrier FDMA systems, in *IEEE 17th International Symposium on Personal, Indoor and Mobile Radio Communications (PIMRC)* (2006), pp. 1–6

Chapter 12
Carrier Aggregation with Licensed and Unlicensed Bands

12.1 LTE Carrier Aggregation

As we recall from Chap. 4, one of the key targets in developing the IMT-Advanced specifications was a peak downlink data rate of 1 Gbps. On the other hand, the maximum LTE channel bandwidth of 20 MHz can support at most 400 Mbps using a spectral efficiency of 20 bps/Hz (assuming 256QAM modulation and a high coding rate). In 2012, 3GPP Rel-10 introduced the concept of carrier aggregation (CA), combining two or more carriers,[1] to meet (and even exceed) the IMT-Advanced target data rates of 1 Gbps. It was not an entirely new concept as a similar idea was introduced in 3G/UMTS, where two adjacent channels could be combined into a logical Dual Cell carrier that aggregates two channels to achieve higher data rates. The data rates in a carrier aggregation are additive, so if one carrier supports a peak rate of 400 Mbps, two will support a peak of 800 Mbps and so on. Thus, a 1 Gbps data rate in the DL can be achieved by combining three carriers.

What differentiates the LTE CA from the HSPA+ is the ability to aggregate noncontiguous channels, known as component carriers (CCs). Furthermore, these CCs may be from within the same band,[2] but CCs from different bands can also be aggregated. Another difference is that the aggregated LTE channels do not

---

[1] It is important to distinguish the combination of carriers from using multiple subcarriers in an OFDM/OFDMA setting. In carrier aggregation, we use multiple modulators and demodulators, i.e., we have multiple parallel OFDM/OFDMA transmissions.

[2] We note again that a band, as referenced here, usually consists of multiple LTE or HSPA+ channels. For example, a 3GPP band 7 is 70 MHz wide in the DL, whereas individual LTE channels can be up to 20 MHz.

Supplementary Information The online version contains supplementary material available at (https://doi.org/10.1007/978-3-031-76455-4_12).

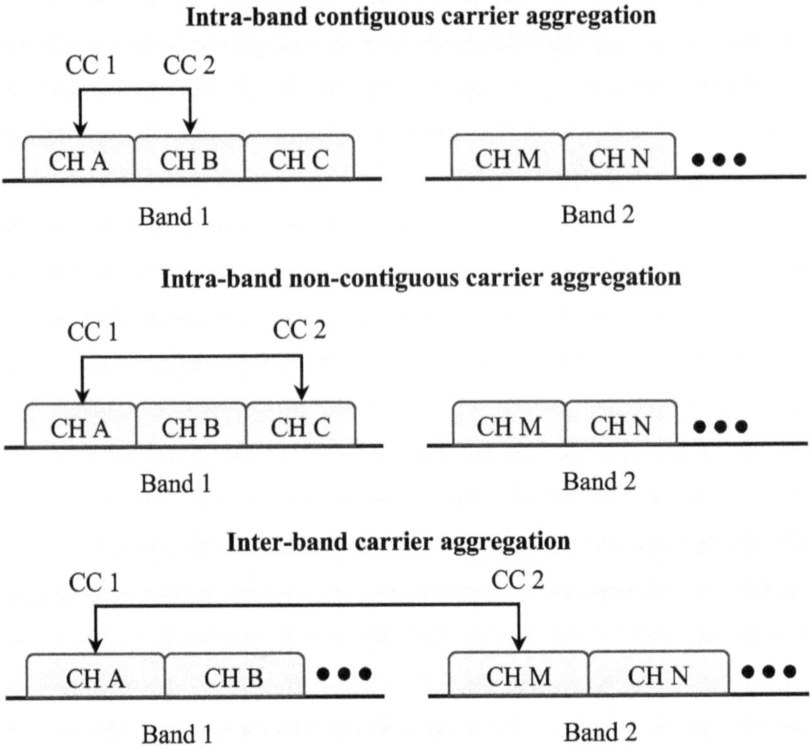

Intra-band contiguous carrier aggregation

```
     CC 1    CC 2
      ↓       ↓
  | CH A | CH B | CH C |        | CH M | CH N | • • •
         Band 1                        Band 2
```

Intra-band non-contiguous carrier aggregation

```
     CC 1           CC 2
      ↓              ↓
  | CH A | CH B | CH C |        | CH M | CH N | • • •
         Band 1                        Band 2
```

Inter-band carrier aggregation

```
     CC 1                         CC 2
      ↓                            ↓
  | CH A | CH B | • • •    | CH M | CH N | • • •
         Band 1                   Band 2
```

Fig. 12.1 Types of carrier aggregation supported by LTE

need to have the same bandwidth as was the case in HSPA+, where only two 5 MHz channels could be aggregated. An illustration of various options is shown in Fig. 12.1.

Inter-band CA is the most flexible option available as it allows different bands to be combined to achieve higher data rates while benefiting from the different propagation properties of different channels.[3] This is likely the most commonly used type of CA as most operators hold a single channel per band. Some operators, however, may hold multiple noncontiguous blocks within the same band, in which case different CCs can be combined using the intra-band CA option. Alternatively, operators may have spectrum holdings that exceed the maximum channel bandwidth of 20 MHz; in this case intra-band *contiguous* CA can be used.

---

[3] It is worth noting that there could be cases in which no benefit arises given different coverages for different bands.

12.1 LTE Carrier Aggregation

Table 12.1 CA bandwidth classes

CA bandwidth class	Aggregated transmission bandwidth	Number of contiguous CCs
A	$N_{RB,agg} \leq 100$	1
B	$25 < N_{RB,agg} \leq 100$	2
C	$100 < N_{RB,agg} \leq 200$	2
D	$200 < N_{RB,agg} \leq 300$	3
E	$300 < N_{RB,agg} \leq 400$	4
F	$400 < N_{RB,agg} \leq 500$	5
I	$700 < N_{RB,agg} \leq 800$	8

12.1.1 Bandwidth Classes

3GPP has introduced different bandwidth classes in order to differentiate between different CA scenarios. Table 5.6A-1, from TS 36.101 [1], is reproduced in Table 12.1.

As can be seen from Table 12.1, bandwidth class "A" represents a single contiguous CC whose channel bandwidth does not exceed 20 MHz, or in terms of number of resource blocks (RBs), it does not exceed the aggregated number of RBs $N_{RB,agg} = 100$. On its own, a single CC does not represent carrier aggregation, but multiple channels, each with bandwidth class "A," are typically aggregated. Bandwidth class "C" aggregates two channels that are up to 20 MHz each (20+5, 20+10, 20+15 combinations are also allowed,[4] so that the aggregated channel bandwidth is up 40 MHz or up to 200 resource blocks. Some LTE bands have channels that do not exceed 10 MHz of bandwidth, but the whole band is wider than 10 MHz (e.g., band 5 is 25 MHz wide in the DL and supports channels up to 10 MHz of bandwidth). These bands can still be aggregated, but the maximum combined bandwidth with two component carriers would equal to 20 MHz. For these bands, a bandwidth class (BC) "B" has been introduced (e.g., Table 5.6A.1–1 from the TS 36.101 [1] lists CA_5B, CA_12B, and other two CC intra-band contiguous CAs).

Wider aggregated channels, consisting of 60 MHz (BC "D"), 80 MHz (BC "E") and higher, are also possible, especially with a licensed TDD band 41 (e.g., CA_41E) or unlicensed band 46 (e.g., CA_46E).

12.1.2 LTE CA Notation and the Number of Supported CCs

Carrier aggregation notation involves a concatenation of band number and BC letters for each band, separated by a "-" between the bands. For example, inter-band CA involving two carriers from bands 2 and 5 is written as CA_2A-5A. If the CA applies

---
[4] But 10+10 is not since the number of aggregated RBs does not exceed 100.

E-UTRA CA configuration, 2 bands										
Configuration		Bands	Bandwidth (MHz)							BC set
DL	UL		1.4	3	5	10	15	20	Max Agg.	
CA_2A-5A	CA_2A-5A	2			Y	Y	Y	Y	30	0
		5			Y	Y				
		2			Y	Y			20	1
		5			Y	Y				
CA_2A-26A	-	2			Y	Y	Y	Y	35	0
		26			Y	Y	Y			

Fig. 12.2 Example of 3GPP specified CA scenarios, from TS 36.101 [1]

to the same band, we differentiate between noncontiguous (e.g., CA_7A-7A) and contiguous CA scenario (e.g., CA_7C).

The CA scenarios are specified in 3GPP in a such way that supported channel bandwidths are listed for each band, as shown in Fig. 12.2. While the 3GPP specifies also UL CA combinations for some DL CA combinations, in practice they are rarely used and only one CC is used (i.e., no CA in the UL), which results in a complete DL and UL asymmetry (e.g., five DL components and one UL component). The dash ("-") notation in the UL configuration indicates that any of the DL bands can be used for the UL, but only one of them at any given time. In the specific example below, for the CA combination CA_2A-26A, the UL is either provided by band 2 or by band 26, but not by both at the same time. However, the first DL combo CA_2A-5A does have UL aggregation of bands 2 and 5 in the UL, although band 2 or band 5 can be transmitted in the UL individually (i.e., in a single UL CC configuration).

The maximum aggregated bandwidth is determined as a sum of the maximum channel bandwidths per band. Sometimes, different bandwidth combinations sets (BCS) are defined as in the example of the CA_2A-5A combo, where band 2 can support up to 20 MHz (BCS 0) or up to 10 MHz (BCS 1). These BCS are normally requested by different operators and reflect their specific requirements.

Initially, 3GPP specified generic support for up to 5 CCs for CA, and over time that number has been extended up to 32 CCs (up to 640 MHz as stated in Section 5.5 of TS 36.300 [2]). However, in practice, LTE devices support up to five CCs, which normally involve four different bands (e.g., CA_2A-7A-7A-29A-66A). This device capability follows the specific combinations listed in TS 36.101 [1]. As of Rel 17, up to five bands are specified. Device limitations related to number of supported CCs come both from baseband processing capabilities, and from the RF capabilities limited by the number of receiver chains, antennas, duplexers, etc.

12.1.3 LTE CA and Channel Spacing

As introduced earlier, we can have contiguous and noncontiguous carrier aggregation within the same band. To differentiate between these two CA scenarios, 3GPP defines the nominal spacing between any two CC center frequencies. For example, contiguous 40 MHz spectrum block can be an aggregation of two contiguous 20 MHz channels, which is known as a CA scenario with a BC "C" notation, or can be considered a noncontiguous CA scenario with BC "A" notation depending on how far apart their center frequencies are. The nominal channel spacing between the CC center frequencies for contiguous CA setting is specified in TS 36.101 [1]

$$\text{Nominal CH spacing} = \left\lfloor \frac{BW_{CH(1)} + BW_{CH(2)} - 0.1 \left| BW_{CH(1)} - BW_{CH(2)} \right|}{0.6} \right\rfloor \times 0.3 \quad (12.1)$$

and for the example of two contiguous 20 MHz LTE channels, it means that the nominal channel spacing is 19.8 MHz carrier aggregation. The center frequency spacing for contiguous CA scenario can be adjusted to any multiple of 300 kHz less than the nominal channel spacing (e.g., 19.5 MHz, 19.2 MHz), as long the subcarriers from the adjacent channels do not overlap. Conversely, if the nominal spacing exceeds 19.8 MHz (e.g., we set it to 20.1 MHz), two component carriers can be aggregated using noncontiguous CA.

A close look at (12.1) say that the nominal channel spacing uses 300 kHz increments. The reason why the 300 kHz steps are adopted has to do with the center frequency raster of many devices that is 100 kHz and with the fact that orthogonality between subcarriers spaced 15 kHz apart needs to be maintained. Therefore, 300 kHz is calculated as the least common multiple between 100 kHz and 15 kHz. For the specific example of 20 + 20 MHz, noncontiguous CA can be set with the center frequency separation of 20 MHz as the inter-carrier interference is negligible given the 2 MHz of separation between the closest subcarriers (from different carriers). This is, in fact, the nominal channel spacing for two adjacent channels in a non-CA scenario.

12.1.4 Primary and Secondary Serving Cell

Different carriers within a CA combinations are effectively different cells, and they typically have different coverage, especially if they are selected from different bands. Normally, the cell with the largest coverage is declared as the primary serving cell (PSC) for far-away users, and the corresponding CC is called the primary component carrier (PCC). The primary cell operates at the primary frequency, that is, at the frequency at which the UE performs the initial access. For users that are closer to the base station, different PCC selection rules may apply, and typically the channel with the widest bandwidth would be selected as a PCC carrier, while load balancing among the carriers is taken into account. The PCC also determines the UL carrier (i.e., UL is serviced by a single carrier). Other carriers form secondary serving cells (SSC) and are called secondary component carriers (SCC). Secondary cells operate at secondary frequencies after the PCC RRC connection is established and are intended to provide additional radio resources (via CA). The PCC carries all the RRC and NAS signaling discussed in Chap. 11. The rest of the signaling, at the MAC and PHY layer (e.g., PDCCH, HARQ ACK/NACK) is carried out separately over each carrier, which is illustrated in Fig. 12.3.

The PCC remains active until a handover takes place, whereas SCC can be added or removed via RRC messages. Therefore, relative to a single CC, the LTE protocol stack is impacted at the RRC layer (addition, removal and reconfiguration of SCCs) and at the MAC layer, where separate HARQ process exist for each CC.

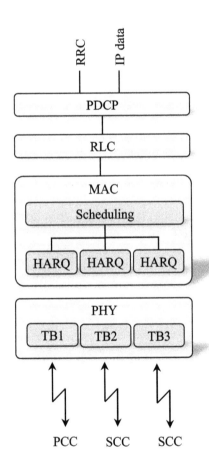

Fig. 12.3 LTE eNB protocol stack in CA scenario

12.1.5 *Peak Data Rate Calculations in LTE CA*

The peak data rate for a given CA scenario is simply calculated as a sum of data rates of individual carriers. For the purpose of illustration, we will consider a CA combination with five CCs, with the following parameters:

- Band 2, with the channel bandwidth of 20 MHz, and 4×4 MIMO support
- Band 7, with the channel bandwidth of 40 MHz, and 4×4 MIMO support
- Band 29, with the channel bandwidth of 10 MHz, and 2×2 MIMO support
- Band 66, with the channel bandwidth of 20 MHz, and 4×4 MIMO support

The best CA combination that captures these 4 bands is CA_2A-7C-29A-66A, with the aggregated bandwidth of $20 + 40 + 10 + 20 = 90$ MHz. Assuming 256QAM transmission, the spectral efficiency for mid-bands (bands 2, 7 and 66) is 20 bps/Hz, whereas the spectral efficiency for low bands (band 29 in this example) is 10 bps/Hz. Therefore, the peak data rate is

$$20 \times 20 + 40 \times 20 + 10 \times 10 + 20 \times 20 = 1700 \text{ Mbps}. \qquad (12.2)$$

Theoretically, if an operator had 5 mid-bands, each with 20 MHz of channel bandwidth, the peak DL data rate with 256QAM is $100 \times 20 = 2000$ Mbps, and using 1024QAM the max supported DL data rate is $100 \times 25 = 2500$ Mbps. This is often considered as transmission of 20 MIMO layers ($5 \times 4 = 20$, where 5 is the number of carriers, and 4 is the MIMO order per individual carrier). These 20 MIMO layers can be allocated differently. For example, it is possible to have three CCs, each with 4×4 MIMO layers, and four more CCs with 2×2 MIMO layers for the total of seven CCs, where the total number of MIMO layers is $3 \times 4 + 4 \times 2 = 20$. An example of such combination is CA_2A-7A-7A-46E combination that can combine up to 140 MHz of spectrum. This combination involves unlicensed band 46, which will be discussed more in a later section.

At this point, it is worth noting that the peak data rate calculation is only one possible metric; the actual data rates to individual are lower than the peaks that could be measured in a lab. Part of the reason is that users rarely experience ideal channel conditions, nor do users usually have all the resources available to them. Additionally, the peak data rate calculation assumes minimal signaling overhead, which is often not the case in practical networks. Finally, the peak data rate calculations assume the highest supported MIMO order, but channels rarely support the highest possible rank (e.g., 4×4 MIMO requires channel rank of 4, but that is not often observed in the field).

12.2 NR Carrier Aggregation

In 5G NR, carrier aggregation is based on the same principle as LTE, where there are intra-band contiguous, intra-band noncontiguous, and inter-band CA combinations. Technical specifications allow for up to 16 component carriers, but the actually specified combinations have a smaller number of CCs.

12.2.1 5G/NR Bandwidth Classes

The definition of bandwidth classes in NR is more complex when compared to LTE due to different maximum channel bandwidths associated with different subcarrier spacings (e.g., $\Delta f = 15$ kHz subcarrier spacing has a maximum channel bandwidth $BW_{Channel,max} = 50$ MHz, $\Delta f = 30$ kHz has a maximum channel bandwidth $BW_{Channel,max} = 100$ MHz, etc.). An example of a complete Rel-18 table for the FR1 range, specified in TS 38.101-1 [3], is shown in Table 12.2.

Table 12.2 NR CA bandwidth classes

NR CA BC	Aggregated channel BW	Number of contiguous CCs	Fallback group
A	$BW_{Channel} \leq BW_{Channel,max}$	1	1, 2, 3[4]
B	$20 \leq BW_{Channel,CA} \leq 100$ MHz	2	2, 3[4]
C	$100 \leq BW_{Channel,CA} \leq 2 \times BW_{Channel,max}$	2	1, 3[4]
D	$200 \leq BW_{Channel,CA} \leq 3 \times BW_{Channel,max}$	3	
E	$300 \leq BW_{Channel,CA} \leq 4 \times BW_{Channel,max}$	4	
G	$100 \leq BW_{Channel,CA} \leq 150$ MHz	3	2
H	$150 \leq BW_{Channel,CA} \leq 200$ MHz	4	
I	$200 \leq BW_{Channel,CA} \leq 250$ MHz	5	
J	$250 \leq BW_{Channel,CA} \leq 300$ MHz	6	
K	$300 \leq BW_{Channel,CA} \leq 350$ MHz	7	
L	$350 \leq BW_{Channel,CA} \leq 400$ MHz	8	
M[3]	$50 \leq BW_{Channel,CA} \leq 200$ MHz	3	3[4]
N[3]	$80 \leq BW_{Channel,CA} \leq 300$ MHz	4	
O[3]	$100 \leq BW_{Channel,CA} \leq 400$ MHz	5	

NOTE 1: $BW_{Channel,max}$ is maximum channel bandwidth supported among all bands in a release
NOTE 2: It is mandatory for a UE to be able to fallback to lower order NR CA bandwidth class configuration within a fallback group. It is not mandatory for a UE to be able to fallback to lower order NR CA bandwidth class configuration that belongs to a different fallback group.
NOTE 3: This bandwidth class is only applicable to bands identified for use with shared spectrum channel access in Table 5.2-1 in TS 38.101-1 [3].
NOTE 4: Fallback group 3 is only applicable to bands identified for use with shared spectrum channel access in Table 5.2-1 in TS 38.101-1 [3].

12.2.2 NR CA Notation

The CA notation in NR is similar to the notation in LTE. For example, for intra-band CA, if we have bands "nX" and "nY," the CA notation is CA_nXA-nYA. However, in intra-band CA combinations, especially in the FR2 range, where a lot of spectrum is available, it is possible to have a larger number of CCs, and it is not convenient to write the combinations using concatenations of band numbers and bandwidth classes. Instead, a shorter notation has been adopted in NR where the number of repetitions is indicated in braces. For example, CA_n257A-n257A-n257A-n257A-n257A-n257A-n257A-n257A is written as CA_n257(8A). Similarly, for intra-band, noncontiguous and contiguous combination CA_n257A-n257A-n257A-n257C, the short notation is CA_n257(3A-C).

12.2.3 Supplemental Uplink

Another difference with respect to LTE is the introduction of a supplemental UL (SUL) component. The main reason for introducing SUL combinations is the limited uplink link budget for mid- and high bands, especially for those above 3 GHz. For these bands, where the DL link budget exceeds the UL link budget by a wide margin, a low band component is added to provide UL coverage for the cell area where the high band UL coverage does not exist (Fig. 12.4).

Unlike in LTE, where the DL and UL components are always paired (except for asymmetric bands, where part of the UL may be missing), in NR we can have two bands providing the uplink for one DL band. This combination of two uplinks and one downlink is considered a single cell (logically) (Fig. 12.5).

Bands intended for SUL are indicated by their UL component only. For example, SUL band n89, listed in TS 38.101-1, Table 5.2-1 [3], covers the frequency range 824 MHz–849 MHz, which is the same as the regular NR band n5, as shown in the

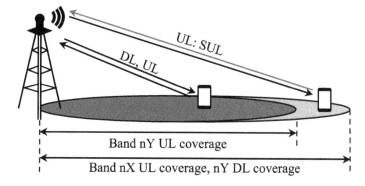

Fig. 12.4 Supplemental uplink illustration find a ZTE reference and reference it here

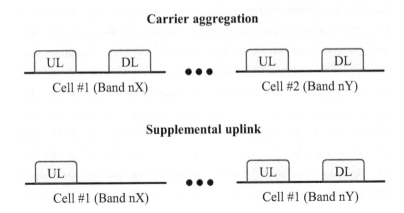

Fig. 12.5 Illustration of differences between a regular CA and supplemental UL in NR technology

Table 12.3 Selected NR bands operating in FR1

NR operating band	UL range (MHz)	DL range (MHz)	Duplex mode
n5	824–849	869–894	FDD
n89	824–849	N/A	SUL

short extract in Table 12.3. SUL tables are specified the same way as the regular CA combinations, with the difference being the prefix that explicitly includes the SUL notation (e.g., SUL_n78A-n89A).

It is worth noting, however, that, at the time of writing, it is not clear if operators are deploying this feature.

12.2.4 NR Channel Spacings for CA

Similarly to LTE, the nominal channel spacing for intra-band contiguous CA between two adjacent CCs is defined in TS 38.101-1, section 5.4A.1 [3] as

$$\text{Nominal CH spacing} = \left\lfloor \frac{BW_{CH(1)} + BW_{CH(2)} - 2\left|GB_{GB(1)} - BW_{CH(2)}\right|}{0.6} \right\rfloor \times 0.3 \tag{12.3}$$

provided that channels are from the bands with a 100 kHz channel raster. A number of NR channels have a channel raster of 15 kHz or 30 kHz (positions where an NR center frequency can be located are in discrete grid points). For these bands and their channels, the nominal channel spacing between adjacent intra-band contiguous carriers is (TS 38.101-1, section 5.4A.1 [3])

$$\text{Nominal CH spacing} = \left\lfloor \frac{BW_{CH(1)} + BW_{CH(2)} - 2\left|GB_{CH(1)} - GB_{CH(2)}\right|}{0.015 \times 2^{n+1}} \right\rfloor 0.015 \times 2^n \tag{12.4}$$

where $n = \mu_0$ is the largest subcarrier spacing parameter μ for the given channel bandwidths, and $GB_{CH(i)}$, $i = 1, 2$, is the minimum guardband for the given channel bandwidth and selected μ value. A smaller channel spacing can be configured as long as it is a multiple of the least common multiple of the channel raster and subcarrier spacing Δf. If the channel spacing is larger than the nominal channel spacing defined above, CA becomes intra-band noncontiguous.

12.2.5 Peak Data Rate Calculation in NR CA

The peak data rate in NR CA combinations is obtained by simply adding the peak speeds of individual carriers, using the formulas introduced in Chap. 11. As an example, let us calculate the peak data rate for the CA combination CA_n7A-n25A-n66A, where each channel is 20 MHz wide. Assuming that each CC operates in 4×4 MIMO mode, their individual peak data rates are about 454 Mbps, and the combined peak data rate is $454 + 454 + 454 = 1362$ Mbps. If we compare this to a similar LTE combination CA_7A-25A-66A, whose peak speed would be 1200 Mbps, we note that the NR peak speed is 13.5% higher when compared to LTE.

A more significant increase in peak data rates comes from wider NR channels, which are not supported in LTE. For example, with a three CC combination CA_n7A-n66A-n77A, whose individual channels are 40 MHz, 30 MHz, and 100 MHz for bands n7, n66, and n77, respectively, then the peak NR data rate is $924 + 685 + 1503 = 3112$ Mbps, which is significantly higher than the maximum of 1200 Mbps achievable with three CCs in LTE.

The peak data rate difference becomes even more pronounced if FR2 range bands are carrier aggregated, because individual channels in FR2 range can be up to 400 MHz wide for frequencies below 50 GHz. A simplified comparison of the widest channel support using an example of two CCs for CA with LTE, NR in FR1, and NR in FR2 is shown below (Fig. 12.6).

12.3 Dual Connectivity

Carrier aggregation discussed so far assumes that individual component carriers come from channels/bands that reside at the same base station. It is also possible to aggregate carriers that are not co-located physically, but are still within a relatively small distance (e.g., up to a few hundred meters from each other). A simple

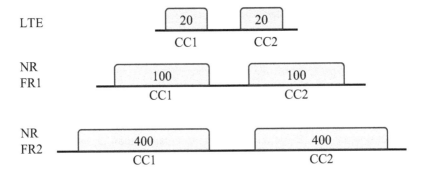

Fig. 12.6 Comparison of the widest channels supported in LTE and NR carrier aggregation using two component carriers

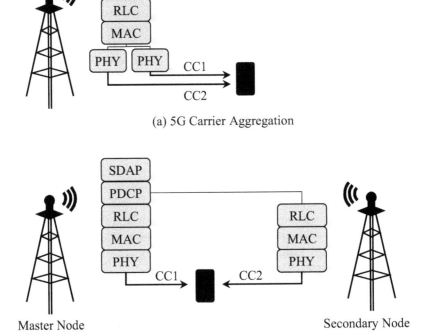

Fig. 12.7 5G carrier aggregation and 5G dual connectivity

illustration is shown in Fig. 12.7, where we see that in CA the traffic split between the carriers is at the MAC layer, whereas on dual connectivity (DC) the traffic split is at the PDCP layer.

12.3 Dual Connectivity

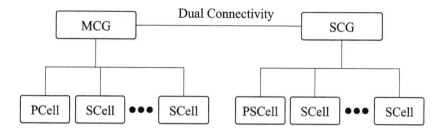

Fig. 12.8 Master and secondary cell groups

By aggregating resources from multiple cells, DC increases overall network capacity and spectral efficiency, allowing for better utilization of available spectrum and improved user experience, especially in dense urban areas or high-traffic locations where some small cells of varying radio conditions can be assisted by connectivity with a higher frequency cell for added reliability. DC also enables the network to perform load balancing and offload traffic, e.g., during congestion and peak usage hours.

In DC, the master node is in control as it connects the UE to the core network, and it is the first node to which the UE connects. The simplified illustration above shows only two CCs, but it is possible to have multiple CCs on both master and secondary nodes. For that purpose, multiple component carriers aggregated within a master node are called the master cell group (MCG), whereas the component carriers grouped within the secondary node are called the secondary cell group (SCG). Cells within an MCG that serve as primary serving cells are indicated as primary cells (PCells); secondary serving cells within both MCG and SCG are indicated as secondary cells (SCells), whereas within the SCG there is also the primary secondary cell (PSCell) which performs the initial access under SCG, as illustrated in Fig. 12.8.

The simplified explanation above does not capture all the elements of the network in NR-DC connection mode. 3GPP TS 37.340 [4] provides more details about the architecture and shows that the PDCP traffic split applies to split bearer (a radio bearer with RLC bearers both in MCG and SCG) only (Fig. 12.9).

In general, 3GPP supports Multi-Radio Dual Connectivity (MR-DC) where one node provides NR access, while the other node can be E-UTRA (LTE) or NR. The most widely deployed type of dual connectivity is the so-called E-UTRA-NR Dual Connectivity (EN-DC), which is also known as the non-standalone 5G architecture.

12.3.1 DC Notation and Peak Data Rate Calculation

In dual connectivity, the notation used to indicate carrier aggregation is inherited from the regular LTE and NR notation, with only one difference which is the "DC"

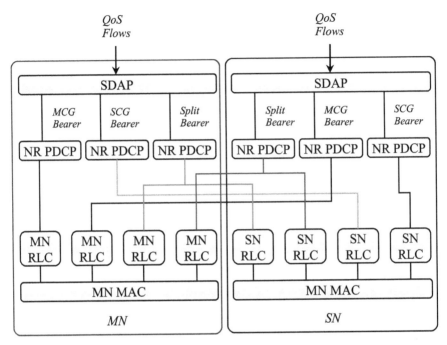

Fig. 12.9 Radio protocol architecture for master cell group (MCG), secondary cell group (SCG), and split bearers in MR-DC with 5GC (37.340 [4])

prefix. For example, the NR-DC notation for the carrier aggregation of bands n2 and n77, DC_n2A-n77A, provided that both use BC "A." In EN-DC, similarly, we use a mix of LTE and NR CA notation and again prefixed with "DC." An example is DC_2A-7A-7A-66A_n78(2A), where the master and secondary cell groups are separated by an underscore. The peak data rate is calculated the same way as in the LTE and NR carrier aggregation scenarios, that is, by adding peak data rates of individual component carriers.

12.4 License-Assisted Access

As introduced earlier, carrier aggregation has been first introduced in LTE starting from Rel-10. As the need for higher data rates increased over time, a low-hanging fruit was to incorporate LTE technology into *unlicensed* spectrum and to use it in combination with licensed spectrum. The rationale is that unlicensed spectrum in the 5 GHz range potentially provides a lot of bandwidth, typically in the excess of 500 MHz per region or country, and the mechanisms for coexistence with Wi-Fi technology were already well-developed. 3GPP conducted a global spectrum regulation analysis and summarized the results in TR 36.889 [5]. The conclusion

12.4 License-Assisted Access

Table 12.4 License-exempt spectrum in 5 GHz range (unlicensed spectrum) in Canada

Sub-band	5150–5250	5250–5350	5470–5600	5650–5725	5725–5895
EIRP	23 dBm	30 dBm	30 dBm	30 dBm	30 dBm/36 dBm
Deployment	Indoor	Indoor/outdoor	Indoor/outdoor	Indoor/outdoor	Indoor/outdoor

was that there is significant global spectrum overlap within the unlicensed 5 GHz band.

To use unlicensed spectrum, in 2016, within Rel-13, 3GPP introduced a type of CA solution that combines licensed and unlicensed bands, known as license-assisted access (LAA). The first specification allowed use of unlicensed spectrum in the DL direction only. The subsequent Rel-14 introduced enhanced LAA (eLAA), where the ability to support DL and UL operation using unlicensed spectrum was added. More enhancements were added in Rel-15, with the introduction of further enhanced LAA (FeLAA), where support for autonomous UL transmission (grant-free transmission that does not rely on a scheduling request) was added.

3GPP has defined an LTE TDD band 46 in TS 36.101 [1] and TS 36.104 [6] that operates in frequency range 5150–5925 MHz. This band is restricted to licensed-assisted operation and cannot be used as a standalone band. In Canada (and USA), the frequency range that overlaps with band 46 ranges from 5150 to 5850 MHz and is divided in several sub-bands, each with different transmit power and application rules, as specified in RSS-247 [7] (Table 12.4).

As an example, devices are not allowed to transmit in the 5600–5650 MHz sub-band because Environment Canada's weather radars operate in this frequency range. Also, the sub-band division of 5725–5850 MHz and 5850–5895 MHz is also defined, but for the purpose of deployment illustration and power levels, the same rules apply.

Aside from license-exempt type of access to this spectrum, a notable difference with respect to licensed bands is the limitation on the output power. While it is not uncommon to have the EIRP in excess of 60 dBm for licensed bands, unlicensed carriers are limited to 30 dBm for indoor use and to 36 dBm for outdoor use. This severely limits the LAA coverage, which is one of the main reasons why this technology has not seen significant commercial success.

The basic principle of CA operation with unlicensed band is illustrated in Fig. 12.10. The primary carrier is selected from licensed bands, and it carries all the signaling, while the secondary carrier is one or multiple unlicensed carriers. The key enabler of this type of CA is the fact that unlicensed bands are technology-agnostic. While most of the users use IEEE 802.11 family of solutions (a.k.a. Wi-Fi), there is nothing in spectrum regulation that prevents other technologies from being used in the unlicensed bands. However, any device operating in these bands must be a good "citizen" in a sense that it needs to share resources fairly with other users.

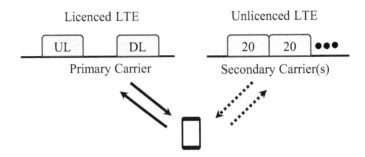

Fig. 12.10 Basic principle of LAA operation

12.4.1 Interference Mitigation

If interference cannot be prevented, the use of unlicensed spectrum is on a no-interference protection basis. There are several mechanisms that can be used to control or avoid interference. One of them is Dynamic Frequency Selection (DFS), a mechanism used to avoid interference with radar systems. In Canada, sub-bands 5250–5350 MHz, 5470–5600 MHz, and 5650–5725 MHz are subjected to DFS. The main idea is that the eNB detects signal levels in individual channels, and if the radar signal level exceeds -64 dBm (for devices with up to 1 Watt of transmission power), that channel cannot be used.

Another widely used mechanism is listen before talk (LBT). The main principle of LBT is the following: every time the LAA base station wants to transmit, it must "listen" to the channel to determine if it is in use—a so-called clear channel assessment (CCA). The CCA uses energy detection to determine the presence or absence of other signals on the channel. In case the channel is not in use, a transmission can ensue. The LBT algorithm has been defined to be in line with ETSI EN 301 893 [8] as of this writing, with the following steps [9]:

1. Transmitting device waits for the channel to be idle for 16 μs. Channel is considered to be *idle* if there is no signal detected above an energy detection threshold (-75 or -80 dBm/MHz, depending on the maximum transmit power of the coexisting device).
2. The device runs a CCA for 9 μs over several observation slots.
3. The next stage is called the back-off stage, where additional an N CCAs are performed.
4. Finally, if the channel is clear, the transmission can commence. The transmission duration is upper bounded by the maximum channel occupancy time (MCOT), which cannot exceed 10 ms (Fig. 12.11).

12.5 NR-Unlicensed (NR-U)

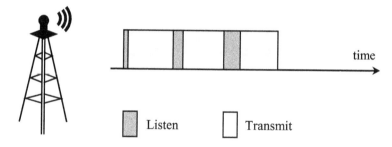

Fig. 12.11 Listen before talk and data transmission in LAA operation

E-UTRA CA configuration, 3 bands									
Configuration		Bands	Bandwidth (MHz)					Max Agg.	BC set
DL	UL		1.4	3	5	10	15	20		
CA_2A-7A-7A-46E	-	2			Y	Y	Y	Y	140	0
		7	See CA_7A-7A Bandwidth combination set 1 in table 5.6A.1-3						
		46	See the CA_46E Bandwidth combination set 0 in Table 5.6A.1-1						

Fig. 12.12 Example of 3GPP specified LAA-based carrier aggregation combination

12.4.2 LAA Combinations

The LAA-based carrier aggregation combinations follow the same notation as in the case of licensed LTE. Typically, 1, 2, or 3 licensed bands are aggregated together with several contiguous band 46 carriers, thus forming 6 or 7 component carriers per combination. An example of a seven-component carrier LAA combination is CA_2A-7A-7A-46E, listed in TS 36.101, Table 5.6A.1–2a [1], as shown below. Its maximum combined bandwidth is 140 MHz, and its peak data rate is calculated as $3 \times 400 + 80 \times 10 = 2000$ Mbps, where we assumed 4×4 MIMO in the licensed bands (each is 20 MHz wide) and 2×2 MIMO for the unlicensed carriers (Fig. 12.12).

12.5 NR-Unlicensed (NR-U)

After the 5G 3GPP Rel-15 was introduced for licensed bands, the idea of expanding the NR air interface to unlicensed spectrum was a natural choice.

In Rel-16, support for the NR-Unlicensed (NR-U) operation was introduced, with support for 5 GHz and 6 GHz unlicensed bands. NR-U supports radio access operating with shared spectrum channel access to operate in different modes where either PCell, PSCell, or Scells can be in shared spectrum. A Scell may or may not be configured with UL. The following NR-U deployment scenarios are supported:

- NR in unlicensed spectrum (standalone NR-U),
- Carrier Aggregation between NR in licensed spectrum and NR in unlicensed spectrum,
- Dual Connectivity between LTE in licensed spectrum and NR in unlicensed spectrum,
- NR cell in unlicensed spectrum and UL in licensed spectrum, and
- DC between NR in licensed spectrum and NR in unlicensed spectrum.

For standalone NR-U deployments, initial access supports PRACH with a larger PRACH transmission power under PSD limitation and/or to meet occupied channel bandwidth (OCB) requirements, longer PRACH sequences of length 1151 and 517 are introduced (applicable to both 15 and 30 kHz SCS), as explained in Chap. 8. The legacy length 839 PRACH sequence is not supported in a cell with unlicensed spectrum channel access. NR-U further supports new SSB[5] positions, up to 20 (30 kHz) and 10 (15 kHz) candidate SSB positions to allow for more transmission opportunities without requiring LBT prior to every SSB transmission.

In Rel-17, NR-U unlicensed channel access has been extended to include the frequency range above 52 GHz. NR-U has many similarities with LAA, including coexistence with WiFi in the 5 GHz band, LBT mechanism for interference mitigation, support for CA, and so on. However, there are some important differences that may make this technology potentially more successful than its predecessor. Among the chief differences are:

- Ability to support wider channels of up to 2000 MHz in FR2 (n263 band, specified in 57000–71000 MHz frequency range), and up to 100 MHz in FR1 (e.g., n46 and n96)
- Scalable numerology as part of the NR air interface
- More supported bands (e.g., n46, n96, n102, n263)
- More deployment scenarios (e.g., EN-DC, NR-DC, NR-CA, NR standalone)

The ability to support wider channel bandwidths is a natural consequence of NR technology, where up to 400 MHz channels have been supported from Rel-15. Since Rel-17, 3GPP has added support for 57–71 GHz unlicensed band n263, which adds support for individual channels being up to 2 GHz.

Within the FR1 portion of the spectrum, the unlicensed band 46 has evolved into n46 to include the 5 GHz band. At the same time, a new band near 6 GHz, n96, has been added that spans 1.2 GHz in total. There is also band n102 (5925–6425 MHz), which applies to markets outside of North America.

---

[5] The synchronization signal block (SSB) is discussed in Chap. 8.

12.5 NR-Unlicensed (NR-U)

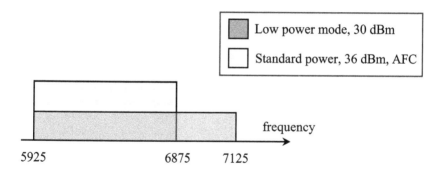

Fig. 12.13 Illustration of RSS-248 specified power and operation rules for 6 GHz band in Canada

In Canada, the 6 GHz band radio requirements are specified in RSS-248 [10]. The total range of 1.2 GHz can be used in low-power mode (30 dBm per access point), for indoors use. A subset of frequencies, ranging from 5925 to 6875 MHz, can also be used in a standard-power mode, where an NR base station or a Wi-Fi access point can transmit up to 36 dBm of EIRP. Current power rules applied to the 6 GHz band in Canada are illustrated in Fig. 12.13.

Standard power mode access points are subjected to Automated Frequency Coordination (AFC) system. The AFC is an Innovation, Science and Economic Development Canada (ISED) designated database system that maintains records of available frequencies and associated maximum power levels for use by a standard-power access point or a fixed client device. There records are linked to a specific time and geographic location.

Availability of large amount of spectrum, combined with the ability to deploy NR-U in a standalone mode (no need for licensed PCell), makes this technology potentially more interesting compared to LAA and a potential solution for private wireless networks.

12.5.1 NR-U Carrier Aggregation

NR-U can be deployed in variety of deployment scenarios, including the carrier aggregation. The CA notation is inherited from the NR and LTE, so it is easy to identify the type of technology. The table below highlights some of NR-U CA combinations and their applicable deployment scenarios (Table 12.5).

Peak speed calculations are done the same way as for any regular CA combination, by adding peak data rates of individual component carriers. However, as mentioned earlier, practical data rates are significantly lower than this calculation would show.

Table 12.5 Selected NR-U Carrier Aggregation combinations and their deployment type

CA combo	Frequency Range	Deployment type
CA_n46C	FR-1	NR standalone
CA_n96E	FR-1	NR standalone
CA_n263C	FR-2	NR standalone
CA_n46A-n96A	FR-1	NR standalone
CA_n7A-n46A-n78A	FR-1	NR CA
DC_2A_n46A	FR-1	EN-DC
DC_n7A_n46D	FR-1	NR-DC

Problems

1. An operator has different spectrum holdings in cities A and B as shown below.

   City A: n2:20 MHz | n7:20 MHz | n7:20 MHz | n29:10 MHz | n5:10 MHz
   City B: n7:40 MHz | n25:30 MHz | n5:10 MHz

   Small gaps between the blocks of spectrum indicate noncontiguity. Do the following for each city:

   a. Construct the best NR combos consisting of 3, 4, and 5 component carriers (CCs) that result in the highest bandwidth under the following UE limitations:

      i. For 3CC CA, 4×4 MIMO can be used for all three component carriers (blocks) provided that they are all mid-bands (i.e., bands n2, n7, and n25 are considered mid-band).
      ii. For 4CC CA, 4×4 MIMO is supported on mid-band carriers, a low band n5 (if needed) can only have 2×2 MIMO
      iii. For 5 CC CA, 4×4 MIMO can be used for mid-band carriers only, low bands support 2×2 MIMO only. For each band, supported channel bandwidths are provided in TS 38.101-1, Table 5.3.5-1 and should be taken into account. Specify individual carrier bandwidths and the aggregate bandwidth for each combination. Proper carrier aggregation notation should be used.

   b. Calculate peak speeds for the carrier aggregation combinations from a) based on the formulas provided in lecture 6.

2. For the same spectrum holdings as in Problem 1, determine the best EN-DC combinations in terms of peak data rates assuming that one n7 block is configured to use the NR technology and the remaining spectrum is configured to use the LTE technology.

3. Operator A has 3 noncontiguous blocks of n41 spectrum, each is 20 MHz wide, whereas operator B has one 60 MHz block on n41 spectrum. Compare the peak data rates for the two operators if operator A aggregates the three blocks. Assume

	Phone A	Phone B	Phone C
NSA (EN-DC)	TDD+TDD NR, up to 20 LTE layers	2 NR (TDD or FDD), up to 20 LTE layers	2 NR (TDD or FDD), up to 20 LTE layers
	FDD+TDD NR, up to 8 LTE layers		
SA (NR CA)	3 CCs, up to 4 layers/CC	4 CCs, up to 4 layers/CC	5 CCs, up to 4 layers/CC

DL combo	DL BWs
CA_n7A-n25A-n66A-n77(2A)	20-15-30-100-50
CA_n7A-n66A-n78(2A)	20-30-40-40
DC_2A-7A_n41(2A)	20-20-50-50
DC_2A-7A-7A-66A_n77(2A)	20-20-20-15-100-50
DC_2A-7A-7A-66A_n66A-n78A	20-20-20-15-30-40
CA_n66A-n78(2A)	30-40-40

that both operators configure their networks to use the subcarrier spacing of 30 kHz and that the TDD duty cycle in the DL is $\alpha_{TDD} = 0.75$.

4. Peak UE data rates are limited by various implementation constraints. Three different phones, implemented by different original equipment manufacturers (OEMs), have their carrier aggregation limitations listed in the table below: calculate the peak data rates for the following carrier aggregation combinations (individual channel bandwidths are specified in the DL bandwidth column):
For TDD NR bands you can use the TDD scale of $\alpha_{TDD} = 0.75$, and 256QAM modulation order for all bands. Assume that n77 and n78 bands are configured to use SCS=30 kHz, while all the other bands use SCS=15 kHz.
For LTE bands, use SE=5 bps/Hz per MIMO layer for all bands.

5. For the Problem 4, determine the combined aggregated bandwidth and the overall (combined) spectral efficiency of the aggregated carriers.

References

1. Evolved Universal Terrestrial Radio Access (E-UTRA), User Equipment (UE) radio transmission and reception. https://www.3gpp.org/ftp/Specs/archive/36_series/36.101/36101-i50.zip. Cited Apr 20, 2024
2. Evolved Universal Terrestrial Radio Access (E-UTRA) and Evolved Universal Terrestrial Radio Access Network (E-UTRAN), Overall description, Stage 2. https://www.3gpp.org/ftp/Specs/archive/36_series/36.300/36300-i10.zip. Cited Apr 20, 2024
3. User Equipment (UE) radio transmission and reception, Part 1: Range 1 Standalone. https://www.3gpp.org/ftp/Specs/archive/38_series/38.101-1/38101-1-i50.zip. Cited Apr 20, 2024
4. NR, Multi-connectivity, Overall description, Stage-2. https://www.3gpp.org/ftp/Specs/archive/37_series/37.340/37340-i10.zip. Cited Apr 20, 2024
5. Feasibility Study on Licensed-Assisted Access to Unlicensed Spectrum. https://www.3gpp.org/ftp/Specs/archive/36_series/36.889/36889-d00.zip. Cited Apr 20, 2024

6. Evolved Universal Terrestrial Radio Access (E-UTRA), Base Station (BS) radio transmission and reception. https://www.3gpp.org/ftp/Specs/archive/36_series/36.104/36104-i50.zip. Cited Apr 20, 2024
7. Digital Transmission Systems (DTSs), Frequency Hopping Systems (FHSs) and Licence-Exempt Local Area Network (LE-LAN) Devices. https://ised-isde.canada.ca/site/spectrum-management-telecommunications/en/devices-and-equipment/radio-equipment-standards/radio-standards-specifications-rss/rss-247-digital-transmission-systems-dtss-frequency-hopping-systems-fhss-and-licence-exempt-local. Cited Apr 20, 2024
8. 5 GHz RLAN, Harmonised Standard for access to radio spectrum. https://www.etsi.org/deliver/etsi_en/301800_301899/301893/02.02.00_20/en_301893v020200ev.pdf. Cited Jun 2, 2024
9. J. Wszolek, S. Ludyga, W. Anzel, S. Szott, Revisiting LTE LAA: Channel Access, QoS, and Coexistence with WiFi. IEEE Commun. Mag. **59**(2), 91–97 (2021)
10. Radio Local Area Network (RLAN) Devices Operating in the 5925-7125 MHz Band. https://ised-isde.canada.ca/site/spectrum-management-telecommunications/en/devices-and-equipment/radio-equipment-standards/radio-standards-specifications-rss/rss-248-radio-local-area-network-rlan-devices-operating-5925-7125-mhz-band. Cited Apr 20, 2024

Chapter 13
Cellular Network Planning, Design, and Optimization

13.1 Minimization of Drive Tests and Call Trace Data Collection

Signal strength predictions from RF propagation models may not be accurate enough at any given location. For the purpose of more accurate planning, drive tests are usually required. As its name suggests, a drive test involves driving through the service area with test equipment measuring signal strengths. These measurements are also used as input to calibrate signal propagation and prediction models. However, drives tests are often very costly and slow to produce test results. In addition, drive tests are limited to accessible roads and public areas and therefore cannot provide accurate estimates for the higher layers of 3D coverage.

Starting from Release 10, 3GPP has introduced the minimization of drive testing (MDT) framework [1] in 4G, which was later also added to 5G specifications within Release 16. The MDT allows for various RF measurements and user locations to be collected directly from the UEs connected to the network. This helps to minimize the cost and time involved in manual drive tests. MDT thus aims to optimize network performance and quality of service while reducing the need for physical drive tests. The primary goal of MDT is to gather network performance data and user experience metrics without conducting traditional, resource-intensive drive tests.

MDT measurements and data are typically used for network optimization, troubleshooting, and network planning. For example, MDT can be adopted into self-organizing networks (SON) as part of the network optimization process. Analyzing

---

Supplementary Information The online version contains supplementary material available at (https://doi.org/10.1007/978-3-031-76455-4_13).

the collected data helps in identifying and resolving network issues, optimizing network configuration, and improving overall performance. MDT data can also assist in identifying the root causes of network problems, such as coverage holes, interference, or capacity issues. In network planning and design, MDT data provides valuable insights for planning new network deployments and expansions, especially in complex environments like urban areas.

MDT uses mobile terminals to collect data, channel measurements, as well as call traces. MDT data includes various parameters related to signal quality metrics, network throughput, connection density, and UE performance. The collection process is designed to be non-intrusive and generally occurs in the background during normal device operation. Examples of collected MDT data in 5G networks include:

- Signal quality and measured channel conditions metrics, including RSRP, RSRQ, SNR, and CQI.
- Throughput data metrics, including actual downlink and uplink throughput achieved by the user, which is crucial for evaluating the network's data-handling capacity.
- Connectivity metrics, including:
  - Handover success and failure rates.
  - Cell ID and time spent in each cell, which can be useful for identifying coverage issues or turning off cells in low load conditions (e.g., to save network energy).
- Network latency metrics, including signal round trip time (RTT).
- QoS parameters, including:
  - Packet loss rate statistics, which indicate the percentage of packets lost in transmission, affecting the quality of voice and video calls.
  - Jitter: a measure of the variability in packet arrival time at the UE buffer, impacting real-time services like VoIP and video streaming.
  - Mobility patterns, including UE location and movement patterns. Logs may include location and movement direction, helping in understanding how users move within the network coverage area and creating accurate traffic maps for user demand in the network.
- Device performance metrics, including:
  - Battery usage: records battery consumption related to network activities, which can indicate network efficiency.
  - Device temperature: monitors the device's temperature during different network activities to ensure device safety and performance.
- Application-specific data, including app usage statistics. The UE collects data on application usage and performance over the network, helping in optimizing network resources for popular applications.

- Environmental factors, including reporting of physical obstructions and building materials. Logged data are reported in areas with specific environmental challenges that might affect signal propagation.

There are two main types of MDT: immediate and logged MDT. Immediate MDT involves real-time data collection for immediate analysis, often triggered by specific events or conditions in the network (e.g., a mobility event). Logged MDT involves storing the collected data in the UE for later retrieval and analysis. Logged MDT is typically used for less time-sensitive data collection. An example of logged MDT data is RSRP measurements from several surrounding cells, which can be used by the location and mobility function (LMF) entity in the network for postprocessing and UE positioning.

A typical MDT data collection process go through the following phases:

- Acquiring UE reports from the network
- Using the UE report for positioning of mobile terminals
- Mapping the mobile terminal measurements at the time of the report and likely locations
- Using the reported information for network optimization

The overall description of the MDT is provided in 3GPP TS 37.320 [1].

13.2 Positioning Methods in Cellular Networks

Positioning of UEs in the network serves as an important area for network optimization, giving operators the opportunity to optimize resources where most users are geo-located. Global Positioning System (GPS) and Global Navigation Satellite System (GNSS) services serve as important input, but are not always available for all call traces, available only for a subset of UEs capable of GNSS, or available only for a subset of measured locations. In the absence of GPS data, cellular network traffic positioning becomes an important source of user geo-localization, where user positions are estimated using cellular radio signals. Further, GPS estimates are not accurate in dense urban 3D settings.

The main methods for UE positioning using cellular radio rely on either Observed Time Difference of Arrival (OTDOA), Enhanced Cell ID, Uplink TDOA (UL-TDOA), or RF fingerprinting pattern matching. These methods are explained further herein.

Downlink-based positioning methods typically involve using downlink reference signals such as positioning reference signals (PRS). The UE receives multiple reference signals from different transmission-reception points (TRPs) and measures DL reference signal time difference (RSTD) or RSRP, where the TRP is defined as a set of geographically co-located antennas that perform transmission and reception functions (basically, a base station). Examples of DL positioning methods are Downlink Angle of Departure (DL-AoD) or DL-TDOA positioning [2].

DL-TDOA (Downlink Time Difference of Arrival) relies on measuring the time difference in the arrival of signals from multiple base stations or TRPs. In this method, the UE receives signals transmitted from different cells at slightly different times, due to the varying propagation distances. By calculating the time difference of these arriving signals, the network can estimate the relative distance of the device from each TRP. Using the data from multiple TRPs, the position of the device can be triangulated. This triangulation is based on the principle that if you know the distance of an object from several known points, you can accurately determine its location by probabilistically predicting the intersection between the predicted distance to each TRP. DL-TDOA is particularly effective in urban areas where GPS signals may be obstructed, and it is a key component in LTE and 5G networks, including for emergency response and location-based applications.

DL-AoD is a positioning method used in 5G networks. This technique involves determining the direction from which signals are sent from a base station to the UE. By using multiple antennas at the base station, the network can accurately calculate the angle at which the signal is transmitted toward the device. This angle, known as the Angle of Departure, is crucial for pinpointing the device's location. The method becomes especially effective in environments with multiple TRPs, as the intersection of the angles from different stations can be used to triangulate the UE's position with a high degree of precision. DL-AoD leverages massive MIMO and beamforming capabilities to enhance location accuracy and provide reliable positioning, especially in urban environments where traditional GPS might be less effective.

Uplink-based positioning method include positioning methods that use uplink reference signals such as sounding reference signals (SRS) for positioning. In UL positioning, the UE transmits SRS to multiple TRPs and the TRPs measure the Uplink Round Trip Time of Arrival (UL RTOA) and RSRP. Examples of UL positioning methods include UL-TDOA or Uplink Angle of Arrival (UL-AoA) positioning [2]. As with DL-TDOA, UL-TDOA relies on measuring the time differences in signals transmitted from the UE to multiple base stations or TRPs. Each base station receives these signals at slightly different times due to the varying propagation distances. By analyzing these time differences, the network calculates the distance between the device and each base station. Triangulating these distances from several base stations allows the system to accurately pinpoint the device's location. Similarly, UL-AoA relies on measuring the angle at which signals arrive at a base station from the device. By using antenna arrays at the base station, UL-AoA calculates the angle based on the phase difference of the incoming signal across these antennas. By collecting angle information from multiple base stations, the network can triangulate the device's position.

DL and UL positioning methods include positioning methods that uses both uplink and downlink reference signals for positioning. In one example, a UE transmits SRS to multiple TRPs and gNB measures Rx-Tx time difference, the difference between arrival time of the reference signal transmitted by the TRP (e.g., PRS), and transmission time of the reference signal transmitted from the UE. The gNB can measure RSRP for the received SRS. The UE measures the Rx-Tx time difference for the PRS transmitted from multiple TRPs. The UE can measure RSRP for the received PRS. The Rx-Tx difference and possibly RSRP measured at UE and gNB are used to compute the round trip time.

An example of DL and UL positioning method is multi-RTT positioning [2]. Multi-RTT positioning involves measuring the time it takes for a signal to travel from the device to multiple base stations and back again. Each round-trip time measurement gives an estimate of the distance between the device and a specific base station. By collecting RTT data from several base stations and applying advanced algorithms, the system can triangulate the device's position. Multi-RTT's effectiveness increases with the number of base stations involved in the measurement, offering a robust solution for urban and indoor positioning.

13.2.1 Positioning by RF Finger Printing and Geolocations of Collected MDT Data

RF fingerprinting pattern matching relies on collecting measurements and comparing them with a training database to find the best fingerprint match, which then is used as the UE's position estimate. Fingerprinting positioning algorithms typically rely on either deterministic or probabilistic estimation. The deterministic approach relies on considering measurements by their value only, while probabilistic approach relies on considering probabilistically matching measurements to their best likelihood position match. A training dataset is usually used; the training data is based on generated propagation signal strength predictions or from external measurements over a set of various locations. Each logged measurement contains measurements to the serving cell and a number of its neighbor cells (e.g., RSRP measurements). RF fingerprinting then localizes each logged measurement or trace to a location estimate.

Logged MDT measurements can be used to aid operators with positioning of logged call traces. An example framework for postprocessing of geolocalized calls in [3, 4] is explained herein. Given a set of possible locations $\mathbf{V}$, the probability of an MDT measurement being located at location v_i conditioned by the set of measurement reports $\mathbf{M}$ can be expressed as

$$P(v_i|\mathbf{S}) = p(v_i|\mathbf{S}) \, p(v_i|\text{TA}) \, p(v_i|\text{cell ID}) \quad (13.1)$$

where $\mathbf{S}$ is a vector of measurement values reported for the serving cell and a number of its neighbors for a single MDT measurement report, where $\mathbf{S} \subseteq \mathbf{M}$. TA is the logged uplink timing advance value to the serving cell.

A reported timing advance value represents the time offset that compensates for the round-trip propagation delay between the UE and the base station. The term $p(v_i|\text{TA})$ filters out the locations that are outside of the ring provided by the propagation distance computed from the reported timing advance value, with an added margin for error. The third term, $p(v_i|\text{cell ID})$, filters out locations outside of the likely coverage area of the reported serving cell in the MDT report.

$p(v_i|S)$ represents the probability that the UE is in location v_i given the set of measurements reported in a measurement report S and a set of training data. Using Bayes equality, $p(v_i|S)$ can be represented as

$$p(v_i|S) = \frac{p(S|v_i)p(v_i)}{p(S)} \qquad (13.2)$$

where $p(v_i)$ is the probability that any given UE is located at location v_i. This can be modeled as a uniform distribution in areas where UEs can be located and otherwise 0 in locations not accessible. $p(S)$ can be modeled as a constant, assuming measurements can vary equally among all locations.

The term $p(S|v_i)$ is the probability of having the logged measurements array at the given location v_i, which can be estimated as the probabilistic average of a Gaussian probability function over the logged measurement samples; input samples to the Gaussian distribution function can be taken as the difference between the logged measurement and the training data sample for a given cell; the standard deviation for the distribution function used can be an environment specific value that accounts for possible RSRP variations due to fading and measurement errors. $p(S|v_i)$ can be denoted as

$$p(S|v_i) = \left(\prod_{j=1}^{N} p_j(s|v_i)\right)^{\frac{1}{N}} \qquad (13.3)$$

$$p_j(s|v_i) = e^{-\frac{RSRP^s_{v_i,j} - RSRP^t_{v_i,j}}{2\sigma^2}} \qquad (13.4)$$

where $RSRP^s_{v_i,j}$ s the logged measured RSRP value from report $s \in S$ from cell j at location v_i and $RSRP^t_{v_i,j}$ is the RSRP value from the training data from cell j and at location v_i (e.g., based on signal strength propagation modeling at location v_i). σ denotes the standard deviation of the large-scale fading on a log-normal distribution at the location of the serving cell.

For a given measurement report, the likely location is the one that maximizes $p(v_i|S)$. Using the computed $p(v_i|S)$, an operator can thus generate a positioning likelihood map from logged MDT measurements, which can then be used as an input to optimize network parameters and operation.

13.2.2 Generation of Traffic Demand Maps from Geolocated MDT Data

A traffic map in the area of collected MDT data can be generated by averaging the positioning probability maps of all measurement reports, where reports are weighted by the amount of data volume d_S that form part of the same MDT report traffic

$$\mathit{Traffic\ map}(v_i) = \frac{\sum_{S \in \mathbf{M}} d_S P(v_i|S)}{\sum_{S \in \mathbf{M}} P(v_i|S)} \tag{13.5}$$

where the units of *Traffic map*(v_i) are either bits or bits/s, depending on whether data volume or throughput is used as input for ds. The primary objective of traffic demand map generation is to identify cellular traffic hotspots rather than pinpointing user and demand locations. However, one drawback of this traffic estimation method is that it assumes traffic demand based on reported actual data transfer volumes, which, in reality, may be limited by coverage. Traffic maps can thus be enhanced by combining multiple sources, in addition to MDT data.

For each reporting time in the MDT report, a *Traffic map*(v_i) value can be generated. The operator can thus create traffic maps per time period in the day. Geolocated MDT data can thus provide more realistic generation for traffic maps for user demand per time period, which enables the creation of different traffic maps for different periods of the day or week. This enables the network optimization process to add more capacity cells during periods of the day where demand and cell loads are higher and turn off capacity cells when demand and cell loads are low.

13.3 Small Cell Placement

Small cell placement and dimensioning are critical aspects of modern cellular network design, particularly in dense urban environments where demand for high-speed data services is continuously growing. Small cells are low-power, short-range wireless transmission systems (like femtocells, picocells, and microcells) that complement the traditional macrocell networks by providing enhanced coverage and capacity in targeted areas.

The primary goals of small cell placement are to enhance network capacity, improve coverage in areas with weak signals, and provide better service quality in high-traffic areas. However, challenges include interference management with macrocells and other small cells, ensuring adequate backhaul and backhaul connectivity and capacity, cost of site ownership, and dealing with varied urban topographies and building densities. Practical considerations such as adhering to local regulations and the availability of physical sites for installation and access to power and backhaul also influence the placement.

Capacity and coverage dimensioning for small cells involves determining the optimal size and capacity of a small cell to adequately serve the demand within its coverage area. Backhaul requirements ensure that each small cell has adequate backhaul support, either through wired or wireless connections, to handle the data traffic.

Cellular network planning is an optimization task involving base station quantity, configurations (like power, antenna type, tower height), and geographic placement. While network planning tools simulate and optimize the performance of deployed radio access networks, they do not always provide precise knowledge of traffic demand patterns. Network-collected MDT data enhances these tools' accuracy. Small cell placement location optimization, crucial for maximizing network capacity or coverage and determining optimal antenna deployment radii, is one example showing the advantages of such optimization in cellular systems. Understanding mobility and user density patterns helps in placing small cells in locations where users frequently gather or pass through, like shopping centers, business districts, or public transport hubs. Further, the small cell placement optimization process also involves ensuring that small cells do not cause excessive interference with other existing cells sharing the same carrier, which requires careful planning of power settings and handover parameters.

13.4 Automatic Cell Planning and Optimization

Automatic cell planning (ACP) and optimization focuses on optimizing the quality of a cellular telecommunications network. Optimization of cellular networks involves assessing if the network meets predetermined quality criteria and adjusting parameters for optimized performance. Common criteria include coverage, interference, capacity, and non-coverage, typically defined in relation to performance thresholds, which can be static or user-adjustable. However, relying on such thresholds for optimization criteria can be limiting and does not guarantee overall network performance.

ACP tackles automation of the first stage of deployment of wireless networks. For any service provider, the process of network planning and initial commissioning can be simplified substantially with ACP. ACP is considered a semi-automated tool and is not considered a real-time autonomous optimization tool. It is however useful for initial planning and optimization of RF coverage and capacity.

ACP aims to optimize coverage, interference, capacity, quality of service, and exposure to electromagnetic fields across the different radio access technologies in the network. The optimization parameters include the antenna type, antenna height, antenna azimuth and tilt, antenna beamwidth, candidate site selection, and the positioning of sites. The optimization procedure is then started after compiling necessary data inputs and creating optimization clusters. The optimization could also be controlled through a user-defined cost function with criteria weighting. A combination of key performance indicators for different networking layers is then used to observe and control the optimization output.

ACP further aims to improve cellular network quality without relying on defining threshold levels. ACP software can estimate at least one performance or quality metric (e.g., coverage probability or a minimum bit rate) at a matrix of various locations V covered by a number of TRPs, based on a matrix of associated antenna parameters P. Each location v_i can be a position of a terminal in the network and includes data regarding the performance of at least one terminal or more generally a bin in the coverage area offered by the operator. Performance indicators are essential

13.4 Automatic Cell Planning and Optimization

data for measuring, estimating, or representing network performance levels, e.g., per bandwidth, carrier, or frequency, which may not be quantified by thresholds. Optimization parameters in **P** can include antenna transmit powers, antenna tilts, antenna azimuths, MIMO smart antenna gain patterns, number of antenna ports, number of antenna elements used, and/or number of carriers operated. **P** may also include other configuration parameters, such as frequency planning (in case frequency reuse of 1 is not used), code planning (e.g., including PCIDs or tracking area codes), and/or PRACH sequence parameters.

ACP aims to estimate an overall quality level (e.g., RSRP, RSRQ, SINR, or QCI) of the network from local quality levels at each point, depending exclusively on the performance indicator. This is based on the assumption that each location v_i is associated with at least one serving cell or TRP.

ACP involves executing an optimization process, which can involve a number of iterations and can be run with or without a heuristic process. A convex optimization problem can be, for instance, executed without involving heuristics; an example of which is provided in the next section. Throughout the optimization, the ACP process adjusts and stores parameters **P** corresponding to the best overall quality level for chosen optimization objective. The ACP process can iteratively modify antenna parameters **P** throughout the optimization while estimating performance indicators and storing optimal parameters. Throughout the optimization, the ACP process may require recomputing signal strength predictions using the new assumptions of the optimization process in the current iteration, such as to test the coverage after changing certain parameters (e.g., antenna azimuth, tilt, or gain pattern).

An ACP process can be used also to determine the optimal locations for new placement of new antennas (e.g., small cells) within an existing network (e.g., a macro-cell network). Such a process can also be used to determine the optimal locations for placement of cells from scratch (e.g., a greenfield deployment in a new frequency or a new geographic area). An ACP process can also be used to optimize various objectives, possibly in combination, including coverage-based objectives, capacity-based objectives, minimum throughput objectives, which cells or carriers to turn on or off, where to place small cells, etc. Often, generated traffic maps or collected logged MDT data can be used as input to the ACP optimization process, in order to select parameters in **P** with optimal parameters at the identified hotspots. The ACP outcome can also provide a number of optimal configuration parameters **P** for different deployment strategies, different periods of the day/night, or different assumptions.

13.4.1 Optimization and Planning of Network Configuration Parameters by Mixed Integer Programming Optimization

Optimization of wireless access networks relies on the configuration of key parameters that guarantee quality of operation and service. Poorly planned network access parameters, such as channel assignments or cell identifying codes, result in increased interference on system-critical control channels, network access failures, and poor quality of service. Therefore, the optimization of such network parameters is a vital task for network planning and optimization for the operator. Such an optimization can be carried by modeling the optimization problem as a mixed integer quadratic program (MIQP) [5]. The optimization of frequency channel assignments, tracking area codes, physical cell identifiers, and coordination clusters are possible applications.

Self-organizing wireless networks rely on a number of vital configuration parameters that guarantee quality of service and reliable network access. LTE and NR systems use physical cell identifiers (PCIs) for cell identification during the cell search and network access procedures and for the configuration of the system-critical reference signals. This need for cell planning is true for all wireless networks. 3G networks use orthogonal scrambling codes to distinguish cells and separate the serving cell's data signals from neighboring cells. Wi-Fi networks rely on channel separation between neighboring cells to reduce inter-cell interference. Poorly planned Wi-Fi channel assignment thus results in increased inter-cell interference. Since network configuration parameters are assigned to cells from a pool of limited resources, channels, or codes, the optimization of parameter assignment is a mandatory task for network planning and self-organizing wireless networks. A suboptimal parameter plan translates to increased dropped calls, system access failures, lower data rates, and degraded overall system performance.

Critical network parameter optimization problems encountered by modern self-organizing wireless networks can be modeled using MIQPs. Such problems is characterized by optimization objective functions and constraints. The engineering problems at stake are complex, thus requiring long computation times to return acceptable solutions, which are frequently coupled with satisfactory optimization convergence levels. However, mature mixed integer program (MIP) solvers are readily available in the literature; background information on MIPs and MIQPs is explained herein followed by exemplary applications for optimization for the majority of channel or code assignment problems using MIQPs, followed by an example of optimization of tracking area codes (TAC).

The following matrix notation is used throughout this section. Uppercase bold letters are used to denote matrices, while lowercase bold letters are used for vectors. For a matrix $\mathbf{A}$, $\mathbf{A}^T$ is its transpose of $\mathbf{A}$. $\mathbf{A}[i, j]$ is the (i, j)-th entry of matrix $\mathbf{A}$. Subscripts added to notations of vectors and matrices are used for additional description. $Tr(\mathbf{A})$ is the trace of matrix $\mathbf{A}$. the operation repmat($\mathbf{A}, N, M$) repeats copies of a two-dimensional matrix $\mathbf{A}$ N times on the first dimension and M times on the second dimension. $\mathbf{A} \times \mathbf{B}$ denotes element-wise multiplication, while $\mathbf{AB}$ denotes the inner product of $\mathbf{A}$ and $\mathbf{B}$. $\mathbf{1}$ denotes a vector of ones having a length that is dependent on the inner product involved.

13.4.1.1 Mixed Integer Programming

Integer programming is a branch of mathematical optimization programs, where some or all of the variables are restricted to integer values. More specifically, a mixed integer linear program (MILP) is a program, whereby the objective function is linear. An MIQP is an integer program with a quadratic objective function. A general MIP optimization problem is formulated as

$$\min \; \mathbf{x}^T \mathbf{Q} \mathbf{x} + \mathbf{q}^T \mathbf{x}$$

s.t. $\mathbf{A}_{eq}\mathbf{x} = \mathbf{b}_{eq}$ (linear equality constraints)

$\mathbf{A}\mathbf{x} \leq \mathbf{b}$ (linear inequality constraints)

$\mathbf{x}^T \mathbf{Q}_i \mathbf{x} + \mathbf{q}_i^T \mathbf{x} \leq \mathbf{b}_i$ (quadratic constraints)

$\mathbf{I} \leq \mathbf{x} \leq \mathbf{u}$ (bound constraints)

Some or all $\mathbf{x}$ must take integer

values (integrality constraints),

where $\mathbf{x}, \mathbf{q}, \mathbf{q}_i, \mathbf{I}$, and $\mathbf{u}$ are vectors of dimension N. $\mathbf{A}$ and $\mathbf{A}_{eq}$ are $(M \times N)$ matrices. $\mathbf{Q}$ and $\mathbf{Q}_i$ are $(N \times N)$ matrices. $\mathbf{b}, \mathbf{b}_i$, and $\mathbf{b}_{eq}$ are length-M vectors, where M is the number of equality or inequality constraints, respectively. Since some MIQP solvers require the $\mathbf{Q}$ matrix in the objective to be symmetric, in these cases, $\mathbf{Q}$ could be replaced by $(\mathbf{Q} + \mathbf{Q}^T)/2$ without altering the objective value.

The integrality constraints allow the MIP to capture the discrete nature of some decisions. For example, a binary variable of values restricted to 0 or 1 can be used to decide whether or not a particular assignment is made, such as assigning a certain wireless channel or resource to a cell or scheduling an agent to perform a specific job.

The integer nature of MIPs makes them non-convex optimization problems. From a complexity point of view, MIPs are generally NP-hard, as discussed in [6, 7]. However, there are well-developed algorithms and solvers specific to MIPs. An MIP solver that uses the branch-and-bound mixed-integer optimization algorithm [8] can be used, coupled with the cutting planes method when needed [9]. In one example realization, the branch-and-bound algorithm relies on initially relaxing the MIP problem to a non-integer linear program (LP), then adding additional constraints on the solution space according to the solution of the relaxed LP. The additional constraints branch the solution space of x to exclude the region between nearest integer approximations of the LP solution. Each additional branching constraint thus creates an additional further restricted MILP, which is also solved in the same manner. The best of the optimal solutions of the branched MIPs is also optimal to the original root MIP. This creates a search tree, where MIPs generated via the search procedure are called nodes of the tree. In general, the original MIP is optimally solved if a point at which all leaf nodes are solved or discarded is reached [8].

13.4.1.2 General Network Channel or Code Assignment Optimizations

In a self-organizing wireless network of N cells and R resources (carriers, channels, scrambling codes, or PCIs), cells using the same resource or channel assignment interfere with each other, and self-organizing wireless networks characterized by high levels of inter-cell interference are considered to be poorly designed. The aim of self-organizing network planning is to minimize channel re-use as much as possible, especially if mutual interference between cells using the same channel exceeds target levels. Generally speaking, assigning the same resource to two cells, A and B, has a design cost of the interference estimate of cell A on cell B plus the interference estimate of cell B on cell A. This is an optimization problem that could be solved through mixed integer programming. More specifically, the problem could be formatted as an MIQP. Values of the inter-cell interference values can be obtained from MDT logged measurements or from predictions generated from a planning software tool.

Assuming **X** is the binary ($N \times R$) optimization variable representing the channel/code assignment matrix. $\mathbf{X}[i, j]$ is equal to one if cell i is assigned to channel j or zero otherwise. Performing a matrix-to-vector transformation to **X** (by concatenating matrix columns) yields vector **x** of length NR. The transformation is not necessary, but is performed for all optimization models to conform to the standard MIQP form, with **Int** as the ($N \times N$) cell-to-cell interference matrix, which is typically generated by a radio planning and propagation tool or from MDT measurements. **Int** quantifies the average interference strength between all cells in the network. For instance, $\mathbf{Int}[B, A]$, the interference of cell B on cell A, is estimated by a radio propagation tool as

$$\mathbf{Int}[B, A] = \frac{\sum_{bins \in \text{Optimization area}} P_{bin}^{Rx}(B) Pr_{bin}^{service}(A)}{\sum_{bins \in \text{Optimization area}} Pr_{bin}^{service}(A)} \quad (13.6)$$

where $Pr_{bin}^{service}(A)$ is the probability of having the bin covered and served by cell A, which depends on the predicted received powers from all cells and the shadowing profile at the bin meeting a minimum threshold. $Pr_{bin}^{service}(A)$ could be simplified to be 1 if the signal received from cell A is the largest and 0 otherwise. $Pr_{bin}^{service}(B)$ is the received reference signal power at the bin from cell B. It is also common to use relative interference matrices, where the interference is relative to the received signal power of the serving cell. For relative interference matrices, the interfering power $Pr_{bin}^{service}(B)$ in (13.6) is divided by $Pr_{bin}^{service}(A)$. The diagonal elements of **Int** are zeros. **Z** is a ($N \times N$) matrix of zeros. Let $\mathbf{Q}_{int}$ be the ($NR \times NR$) matrix generated in the following manner:

$$\mathbf{Q}_{int} = \begin{pmatrix} \mathbf{Int} & \mathbf{Z} & \cdots & \mathbf{Z} \\ \mathbf{Z} & \mathbf{Int} & \cdots & \mathbf{Z} \\ \vdots & \vdots & & \vdots \\ \mathbf{Z} & \mathbf{Z} & \cdots & \mathbf{Int} \end{pmatrix} \quad (13.7)$$

13.4 Automatic Cell Planning and Optimization

where the number of $\mathbf{Z}$ matrices inserted per dimension is equal to $R - 1$ and **Int** matrices only exist along the diagonal of $\mathbf{Q}_{int}$. The objective function is then simply $\mathbf{x}^T \mathbf{Q}_{int} \mathbf{x}$. It is therefore expected that $\mathbf{Q}_{int}$ is sparse and large for a large number of cells and resources but is created to conform to standard MIQP form. In non-standard MIQP form, the objective function is $\text{Tr}\left(\mathbf{X}^T \mathbf{Int} \mathbf{X}\right)$. The additional constraints are thus as follows:

- All elements of $\mathbf{x}$ are binary integers, meaning a resource is either assigned to a cell or not assigned. This constraint is formulated as $\mathbf{l} \le \mathbf{x} \le \mathbf{u}$, where $\mathbf{l}$ is a vector of zeros and $\mathbf{u}$ is a vector of ones of length $N \times R$.
- A cell can only be assigned to one resource (channel, carrier, or code), thus requiring $\sum_{j=1}^{R} \mathbf{X}[i,j] = 1$

Since the optimization variable $\mathbf{x}$ is the vector version of $\mathbf{X}$, this equality constraint could be written as $\mathbf{A}_{eq}\mathbf{x} = \mathbf{b}_{eq}$, where $\mathbf{A}_{eq}$ is the $N \times N \times R$ matrix constructed as repmat($\mathbf{I}$, 1, R) and $\mathbf{I}$ is the $(N \times N)$ identity matrix. $\mathbf{b}_{eq}$ is a length-N vector of ones.

With the objective and constraints defined, the general network channel/code optimization problem is the MIQP:

$$\min \quad \mathbf{x}^T \mathbf{Q}_{int} \mathbf{x}$$
$$\text{s.t.} \quad \mathbf{A}_{eq}\mathbf{x} = \mathbf{b}_{eq}$$
$$\mathbf{l} \le \mathbf{x} \le \mathbf{u}$$

all $\mathbf{x}$ must take integer values

13.4.1.3 Tracking Area and Paging Area Code Optimization

A tracking area in LTE or a RAN paging area in 5G serves the high-level function of pinpointing the locations of mobile terminals by encompassing the cells' coverage area. They are integral for manageable user paging, effectively decreasing the volume of cells paged within the network. In tracking area provisioning, it is important to avoid having areas of increased mobility at the edges of tracking areas since moving across these boundaries necessitates terminal-specific tracking area update (TAU) signaling. Thus, it is imperative to optimize tracking area code (TAC) allocations on a per-cell basis to reduce the network's paging and TAU signaling burdens.

A tracking area blankets a group of cells and aids the network's self-organizing mobility management entity, such as the MME in an LTE network, in high-level mobile terminal tracking. The network tracks terminals to page them for incoming calls or data. To conserve battery life, mobile terminals typically enter a standby sleep-cycle, waking at specific times to check for relevant paging messages. Without knowing a terminal's exact cell, the MME pages every cell within the terminal's tracking area, leading to an increased number of paging messages with a larger tracking area. Smaller tracking areas, while paging-efficient, necessitate frequent TAU signals when mobiles cross tracking area borders, which is signal-intensive. One of the objectives of self-organizing wireless networks is to define tracking areas of minimal paging and TAU signaling in the network. High levels of paging and TAU loads could lead to MME failures, which should be avoided at all costs. However, these two objectives conflict with each other, as one aims to reduce the size of tracking area while the other aims to reduce tracking area boundaries by using fewer tracking areas with larger boundaries. This creates a challenging optimization problem to arrive at an optimal design.

The optimization cost function must jointly incorporate paging and TAU signaling costs. In order to estimate the paging cost, cell loads are used to estimate the number of paged terminals per cell. On the other hand, the number of inter-cell handovers is used in the computation of the TAU signaling cost. Cell load and handover statistics are available to operators through logged MDT data reports and network performance management tools. TAC optimization could be performed via MIQPs as well. In this case, the binary optimization variable $\mathbf{X}[i, j]$ is used to indicate whether a cell i is assigned to tracking area j or not. R represents the number of tracking area codes available for assignment in the network.

The number of paged terminals in a certain cell i could be estimated as $AF_{paging} \times MaxUEs_{paging} \times \mathbf{Loads}[i]$. $MaxUEs_{paging}$ is maximum number of paged terminals per cell, an input from the network dimensioning process. AF_{paging} is the terminal's estimated paging activity factor, which indicates the paging frequency averaged over the subscriber types that exist in the network. $\mathbf{Loads}[i]$ is the downlink time-frequency resource load of cell i. Since the paging cost is the sum of the paged terminals served by cells in the same tracking area, the paging cost of a tracking area plan is modeled as

$$Cost_{paging} = AF_{paging} MaxUEs_{paging} \mathrm{Tr}\left(\mathbf{X}^T \mathbf{L} \mathbf{X}\right) \qquad (13.8)$$

where $\mathbf{L}$ is a $(N \times N)$ matrix constructed as repmap($\mathbf{Loads}, 1, N$) and $\mathbf{Loads}$ is the length-N vector of cell loads. The paging cost in standard MIQP form is $\mathbf{x}^T \mathbf{Q}_{paging} \mathbf{x}$, where $\mathbf{Q}_{paging}$ is the $(NR \times NR)$ matrix generated in the following manner:

$$\mathbf{Q}_{paging} = AF_{paging} MaxUEs_{paging} \begin{pmatrix} \mathbf{L} & \mathbf{Z} & \cdots & \mathbf{Z} \\ \mathbf{Z} & \mathbf{L} & \cdots & \mathbf{Z} \\ \vdots & \vdots & \ddots & \vdots \\ \mathbf{Z} & \mathbf{Z} & \cdots & \mathbf{L} \end{pmatrix} \qquad (13.9)$$

where the number of $\mathbf{Z}$ matrices inserted per dimension is equal to $R - 1$ and $\mathbf{L}$ matrices only exist along the diagonal. Since the amount TAU signaling is equal to the sum of handovers occurring at TA boundaries, the TAU signaling cost for a certain TA design is

13.4 Automatic Cell Planning and Optimization

$$\text{Cost}_{\text{TAU}} = \mathbf{1}^T \left[\mathbf{H} \times (\mathbf{1} - \mathbf{X}\mathbf{X}^T) \right] \mathbf{1} = \mathbf{1}^T \mathbf{H} \mathbf{1} - \text{Tr}(\mathbf{X}^T \mathbf{H} \mathbf{X}) \quad (13.10)$$

where $\mathbf{H}$ is the $(N \times N)$ cell-to-cell handover matrix, which has zeros along its diagonal. The TAU cost in standard MIQP form is $\mathbf{1}^T \mathbf{H} \mathbf{1} + \mathbf{x}^T \mathbf{Q}_{\text{TAU}} \mathbf{x}$, where $\mathbf{Q}_{\text{TAU}}$ is the $(NR \times NR)$ matrix generated in the following manner:

$$\mathbf{Q}_{\text{TAU}} = \begin{pmatrix} -\mathbf{H} & \mathbf{Z} & \cdots & \mathbf{Z} \\ \mathbf{Z} & -\mathbf{H} & \cdots & \mathbf{Z} \\ \vdots & \vdots & \ddots & \vdots \\ \mathbf{Z} & \mathbf{Z} & \cdots & -\mathbf{H} \end{pmatrix} \quad (13.11)$$

where the number of $\mathbf{Z}$ matrices inserted per row and per column is equal to $R - 1$, and $-\mathbf{H}$ matrices only exist along the diagonal. Since $\mathbf{1}^T \mathbf{H} \mathbf{1}$ is a constant scalar that is independent from the optimization variable x, it is dropped from the cost function. Hence, the overall paging plus TAU cost of the TA plan is $\text{Tr} \mathbf{X}^T (\mathbf{L} - \mathbf{H}) \mathbf{X}$. In MIQP form, it is expressed as

$$\mathbf{x}^T \mathbf{Q}_{\text{paging}} \mathbf{x} + \mathbf{x}^T \mathbf{Q}_{\text{TAU}} \mathbf{x} = \mathbf{x}^T \mathbf{Q}_{\text{TA}} \mathbf{x} \quad (13.12)$$

where $\mathbf{Q}_{\text{TA}} = \mathbf{Q}_{\text{TAU}} + \mathbf{Q}_{\text{paging}}$. Since this is a binary TAC assignment problem, $\mathbf{u}$ and $\mathbf{l}$ are defined as per the MIQP optimization model. Given a cell could only be assigned one TAC at once,

$$\sum_{j=1}^{R} \mathbf{X}[i, j] = 1$$

holds and $\mathbf{A}_{eq}$ and $\mathbf{b}_{eq}$ are defined as per the MIQP optimization model. With clear objectives and constraints, the TAC optimization problem is thus the MIQP:

$$\begin{aligned} \min \quad & \mathbf{x}^T \mathbf{Q}_{\text{TA}} \mathbf{x} \\ \text{s.t.} \quad & \mathbf{A}_{eq} \mathbf{x} = \mathbf{b}_{eq} \\ & \mathbf{l} \leq \mathbf{x} \leq \mathbf{u} \end{aligned}$$

all x must take integer values

It is worth mentioning that it is also common practice in NR and LTE networks to solely optimize the TAU cost while adding paging capacity constraints, which are typically defined by the LTE MME and eNode-B paging capabilities. In this case, the paging cost is moved from the objective and placed as an optimization constraint instead.

13.4.1.4 PCI Code Optimization

Assigning physical cell identifiers (PCIs) is a crucial component in the architecture of LTE and NR self-organizing networks. A PCI emerges from the combination of an orthogonal sequence, the primary synchronization signal (PSS), and a pseudo-random sequence known as the secondary synchronization signal (SSS). Each pseudorandom sequence is linked to 1 of 168 unique cell identity groups, and each orthogonal sequence relates to a specific trio of cell identities within these groups. With three orthogonal sequences available, LTE has 504 unique PCIs, for instance. When a mobile terminal initiates a network connection following a power-off cycle, it utilizes the PSS and SSS for the cell search procedure. Identifying the cell occurs in two stages: the PSS determines the cell identity (0, 1, or 2), and the SSS identifies the cell identity group. Minimizing disruption to the PSS and SSS is critical to ensure the cell search is successful. Additionally, the PCI configures the cell's reference signals (CRS) and orchestrates their placement within a time-frequency resource block. These signals are indispensable for channel estimation, coherent demodulation, and handover processes. Consequently, devising a strategic PCI network plan is of paramount importance. A good PCI design should avoid the following factors as much as possible:

- PCI duplication in interfering cells: PCI collisions should be avoided for most interfering cells. This is ensured by assigning different PCIs to cells of large mutual interference.
- Inter-cell PSS interference: In the LTE example, PSS sequences are generated using Constant Amplitude Zero Autocorrelation (CAZAC) orthogonal sequences that are based on three root indices, 25, 29, and 34. These root indices are determined by the PCI mod (3) criterion, meaning cells with the same PCI mod (3) share the same CAZAC root index and thus suffer PSS interference.
- Inter-cell downlink RS collisions: In the LTE scenario, reference signals are mapped to resource elements within an LTE resource block based on the PCI mod (6) criterion for single-port antennas, or PCI mod (3) for multi-port antennas. Inter-cell RS collisions infer pilot interference, whereby terminals receive different reference signals from different sectors on the same resource elements. RS degradation impacts channel estimation, resulting in increased demodulation errors and a reduced downlink throughput.
- Inter-cell uplink demodulation RS collisions: LTE uplink demodulation reference signals (DM-RS) are designed using 30 different Zadoff-Chu sequences with low cross-correlation properties. Since DM-RS sequences are assigned to cells based on the cell PCI, cells with the same PCI mod (30) suffer from uplink DM-RS collisions, thus leading to uplink channel estimation errors.

Since PCI duplication is addressed by the MIQP optimization model, other factors are modeled by adding their PCI mod (D) criteria in the cost function. For a network with R available PCIs and N cells, the PCI mod (D) cost is modeled as follows: let $\mathbf{C}_D$ be a ($D \times R$) matrix. The first row of $\mathbf{C}_D$ is defined as a repeated sequence of a 1 followed by $D - 1$ zeros. The sequence is repeated until the number

13.4 Automatic Cell Planning and Optimization

of elements in the first row is R. Each subsequent row of $\mathbf{C}_D$ is a circular shift of the previous row, one position to the right. For example, for a network of six available PCIs, $\mathbf{C}_3$ for the PCI mod(3) criterion is [1 0 0 1 0 0; 0 1 0 0 1 0; 0 0 1 0 0 1]. Let $\mathbf{Q}_{\text{mod } D}$ be the $(NR \times NR)$ matrix generated in the following manner:

$$\mathbf{Q}_{\text{mod } D} = \text{rempat}\left(\mathbf{C}_D^T \mathbf{C}_D, N, N\right) \times \text{rempat}(\mathbf{Int}, N, N) \tag{13.13}$$

where **Int** is the matrix containing the average interference strength between all cells in the network. The cost of assigning PCIs with a PCI mod (D) criterion is then simply $\mathbf{x}^T \mathbf{Q}_{\text{mod } D} \mathbf{x}$. In non-standard MIQP form, the cost is $\text{Tr}\{\mathbf{C}_D \mathbf{X}^T \mathbf{Int} \mathbf{X} \mathbf{C}_D^T\}$. Since designing an optimal PCI plan should take into account all the factors outlined above, the costs of the individual objectives should be combined into one objective. For a single-port antenna network, the combined objective is

$$\mathbf{x}^T \mathbf{Q}_{\text{int}} \mathbf{x} + \mathbf{x}^T \mathbf{Q}_{\text{mod } 3} \mathbf{x} + \mathbf{x}^T \mathbf{Q}_{\text{mod } 6} \mathbf{x} + \mathbf{x}^T \mathbf{Q}_{\text{mod } 30} \mathbf{x} \tag{13.14}$$

This objective, however, assumes the same importance for all factors. The per-factor $\mathbf{Q}$ matrices should be scaled by importance weights, where the weights depend on the factor's collision severity. An overall $\mathbf{Q}$ matrix, $\mathbf{Q}_{\text{PCI}}$, could be used to sum the scaled per-factor Q functions:

$$\mathbf{Q}_{\text{PCI}} = \alpha_{\text{int}} \mathbf{Q}_{\text{int}} + \alpha_{\text{PSS}} \mathbf{Q}_{\text{mod } 3} + \alpha_{\text{DL-RS}} \mathbf{Q}_{\text{mod } 6} + \alpha_{\text{UL-RS}} \mathbf{Q}_{\text{mod } 30} \tag{13.15}$$

For example, the per-factor weights could be set to $\alpha_{\text{DL-RS}} = 2$, $\alpha_{\text{UL-RS}} = 2$, $\alpha_{\text{int}} = 10$, $\alpha_{\text{PSS}} = 10$, since downlink RS collisions are more expensive on the network's performance than uplink DM-RS collisions and since DM-RS collisions are more adverse than PSS collisions.

With the objectives and constraints identified, the PCI optimization problem is the following MIQP:

$$\min \quad \mathbf{x}^T \mathbf{Q}_{\text{PCS}} \mathbf{x}$$
$$\text{s.t.} \quad \mathbf{A}_{eq} \mathbf{x} = \mathbf{b}_{eq}$$
$$\mathbf{l} \leq \mathbf{x} \leq \mathbf{u}$$

all $\mathbf{x}$ must take integer values

where $\mathbf{A}_{eq}$, $\mathbf{b}_{eq}$, $\mathbf{u}$, and $\mathbf{l}$ are defined as in the MIQP optimization model.

13.4.1.5 Optimization of Coordination Clusters

The general code optimization model can also be specialized and detailed for LTE or NR coordination cluster optimization. Coordination clusters are used by many self-organizing inter-cell interference mitigation techniques. The use of coordination clusters is necessary in self-organizing networks as network-wide coordination is typically infeasible or too costly.

LTE incorporates advanced techniques to enhance network performance and efficiency, such as Coordinated Multi Point (CoMP), Inter-cell Interference Coordination (ICIC), and enhanced ICIC (eICIC). CoMP involves multiple cell sites working together to improve signal quality and reduce interference at the cell edges, enhancing user experience and network capacity. ICIC aims to manage and reduce interference between cells by coordinating resource allocation and power control among neighboring cells. Building on ICIC, eICIC introduces additional mechanisms like time-domain techniques and the use of almost blank subframes to further mitigate interference, particularly in heterogeneous networks where macrocells and small cells coexist.

Coordination is limited to a finite number of cells for many interference mitigation techniques such as CoMP, ICIC, eICIC, and could-RAN. The operation of CoMP, for example, is typically limited to a set of cells known as a cooperation set or CoMP cluster in which coordination is viable and only cells in the same cluster coordinate signal transmission and reception. Such cells typically require more involved inter-node connections to support low-latency exchange of interference information across the coordinated cells. In CoMP, a cooperation set size defines the maximum number of cells that could belong to a certain cluster. CoMP clusters could be static, where cells could only belong to a certain cluster, or dynamic, where cells could belong to multiple clusters, depending on the interference profile of the served user. The optimization of coordination clusters could be performed via MIQPs as well. Using the general network parameter optimization model, the binary variable **x** is used in this case to indicate whether a cell is assigned to a coordination cluster or not. R, the number resources, represents the number of coordination clusters available in the network. A cluster could contain a number of cells between zero and the maximum cooperation set size. Since the objective is to add the most interfering cells in the same cluster, the optimization objective function is to maximize $\mathbf{x}^T \mathbf{Q}_{int} \mathbf{x}$. In standard MIQP form, the objective function is to minimize $-\mathbf{x}^T \mathbf{Q}_{int} \mathbf{x}$. The constraints that must be accounted for are as follows:

- All elements of **x** are binary integers, meaning a cell is either assigned to a cluster or not.
- For static clustering, a cell can only be part of one cluster, thus requiring $\sum_{j=1}^{R} \mathbf{X}[i, j] = 1$. This constraint is written as $\mathbf{A}_{eq}\mathbf{x} = \mathbf{b}_{eq}$, where $\mathbf{A}_{eq}$ and $\mathbf{b}_{eq}$ are identical to the matrices explained in detail for the MIQP model. For dynamic coordination clusters, the equality constraint is changed to an inequality constraint and **b** is set as a length-N vector, where all elements are equal to the maximum number of clusters a cell could belong to at any time.
- The number of cells in any cluster must be less than or equal the maximum cooperation set size:

$$\sum_{i=1}^{N} \mathbf{X}[i, j] \leq \text{Cooperation set size}$$

This constraint is written as $\mathbf{A}_{CC}\mathbf{x} \leq \mathbf{b}_{CC}$, where $\mathbf{A}$ is an $(R \times NR)$ matrix. Initially, $\mathbf{A}$ is populated with zeros; elements $(i-1) \times N - toi \times N$ in each row are then replaced with ones, where i is the row index. $\mathbf{b}_{CC}$ is a length-R vector, where all elements are equal to cooperation set size.

With the objective and constraints defined, the self-organizing LTE coordination cluster optimization problem is formulated as the following MIQP:

$$\min \quad -\mathbf{x}^T \mathbf{Q}_{int} \mathbf{x}$$
$$\text{s.t.} \quad \mathbf{A}_{eq}\mathbf{x} = \mathbf{b}_{eq}$$
$$\mathbf{l} \leq \mathbf{x} \leq \mathbf{u}$$

all $\mathbf{x}$ must take integer values

13.5 Self-Organizing Wireless Networks

With the increase of demand for wireless broadband, service provides must respond effectively to satisfy the consumers' needs while simultaneously driving down the cost per bit. Complexity of modern cellular and wireless networks has grown to a level that is not tolerable to manage efficiently by manual planning and operations. The number of configuration parameters, UE features, and UE capabilities is also increasing exponentially with each RAN generation. SON is a solution to automate most granular process of wireless networks by autonomous frequent optimizations. As a result, SON significantly reduces capital expenditures (CAPEX) and operational expenditures (OPEX) of operators. Functions including cell configuration, mobility, coverage, load balancing, and interference management can be automated or at least aided by SON solutions.

The procedures of cellular and wireless networks have become more complex than those of initial generations cellular networks. The amount of equipment, signaling, and control parameters have been increasing consistently in order to increase the system's capabilities in terms of data rates. With this increase comes a high demand of automation of manual processes to reduce effort and cost and save time. The amount of capital expenditure and operations expenditures for a wireless service provider could be reduced by about 25% through automatic configuration and fast self-configuration. Automation, however, is not a new concept to wireless networks, as it is already used in some procedure such as scheduling or power control. The benefit from self-organization in those procedures is self-evident.

Moreover, the large number of parameter and configurations defining procedures in wireless networks such as handover or load balancing gives a substantial opportunity for self-optimization. Those processes are often too fast, granular, or complex for repetitive manual configuration. Through self-optimization, networks will efficiently utilize the scarce spectrum resource in order to provide better quality of service and experience to end users and increase system capacity.

SON was initiated by the Next-Generation Mobile Networks Alliance (NGMN) in 2006 and is also standardized by 3GPP, though some parts are left unspecified for the implementation. Here are some of the main goals of SON explained herein [10, 11]:

- Plug and play installation and self-configuration
- Automatic neighbor relation configuration
- Physical cell ID auto-configuration
- Mobility and handover optimization
- Minimization of drive tests
- QoS optimization
- Autonomous optimal load balancing
- Random access channel optimization
- Capacity/coverage compensation and optimization

The architecture of self-organizing networks could either be centralized, distributed, or a hybrid of those two. In a centralized SON structure, SON algorithms run on a separate SON server that is responsible for handling many cells. The output parameters of a centralized SON are then sent periodically with a variable frequency as needed. A centralized structure has the benefit of scalability and interaction between different SON algorithms across different cells, possibly from various network infrastructure vendors. However, centralized SON has the disadvantage of the delay of fast updates when they are required, as the processing and transmission of KPIs to the SON server incurs some delay, causing a slower response time. A centralized structure is more prone to outage and single point of failure issues in the backhaul links and thus is not typically used for dynamic optimizations.

In distributed SON architectures, SON algorithms run on each base station in the network and are cell specific. This allows for fast autonomous decision-making based on measurements made by users in the cell. A distributed structure has the benefit of simplicity of deployment for different radio access technologies and different manufacturing vendors. But the downside of this is the need to monitor KPIs across different RAN technologies and vendors to eliminate potential network instabilities. In a hybrid SON architecture, values of initial SON parameters could be done in a centralized manner on a SON server; then finer updates and parameters could also be done in a distributed manner on each base station.

13.6 Self-Configuration and Installation

With plug and play self-configuration and installation, a new base station is delivered with initial basic software to allow basic bootstrapping and basic connectivity. Once a new base station is powered up, it is allocated an IP address, and a connection to the backhaul and the core network is established. The base station is then authenticated and connected via secure upper layers connections. Once the connection to the core network is done, the base station establishes connections to its neighbors.

Self-configuration has the benefits of fast deployment and reduction in manual initial installation costs. It eliminates the need for multiple site visits, on-site configuration, and maintenance training cost for engineers.

13.6.1 Automatic Physical Cell ID Configuration

In cellular networks, a physical cell ID corresponds to the scrambling code used in the cell that is available on the pilot channel. A physical cell ID should be confusion- and collision-free, which is satisfied by having some uniqueness in its area of coverage. This condition is usually maintained by assigning orthogonal codes to different cells. In modern cellular networks, a physical cell ID is mapped to a cell global ID and an IP address.

Physical cell IDs are usually assigned manually by a network planner at the deployment stage or optimized by an ACP tool. Automatic physical cell ID configuration configures newly deployed cells with a unique cell ID that are confusion- and collision-free. Adjustments to physical cell IDs happen automatically when needed as network entities are added. This automation process reduces the pre-planning time at the initial deployment stage and reduces the re-planning time required after additional cells are deployed. This effectively reduces the CAPEX and OPEX a service provider encounters for this function.

13.6.2 Autonomous Load Balancing in Wireless Networks

In any shared network, network elements share the traffic load. Load balancing of users' traffic usually happen between network elements (base stations or core network elements) in order to increase the system capacity and the users' quality of service. This feature is very desirable for areas where periodic or scheduled concentrations of users regularly occur, conventions or sporting events, for example.

Load balancing optimization aims to increase system capacity while maintaining or increasing the QoS of users in the balancing area. As a result, load balancing improves end-user experience by reducing call drop rate, handover failure rate, and unnecessary redirection caused by unbalanced load. This objective is met by intelligently spreading the user traffic across the system's radio resources. The optimization could be modeled by a utility function to achieve this goal

$$F = \max(\text{System capacity}), \text{ while for each user } i \ QoS_i^{new} \geq QoS_i^{old} \tag{13.16}$$

where $QoS = f$ (requested rate, current rate, continuity of service). Load balancing may also be used by a service provider to shape the system load according to an energy savings policy or to achieve network energy savings at low cell loads. In this case, the aim is to minimize the energy consumption of the network by spreading the user traffic around while maintaining the QoS of users. The optimization cost function (energy function) here is

$$F = \min \text{(Network consumed energy)}, \text{ while for each user } i \ QoS_i^{new} \geq QoS_i^{old}$$

The benefit of automatic optimization of the load balancing feature here is substantial when compared with manual human intervention in the network management and optimization tasks.

Mobility load balancing refers to load balancing between cells and base stations. The load balancing here could include intra-carrier, inter-carrier, or inter-technology resources, as long as the load balancing optimization software is intelligent and aware of radio admission and continuity of services. The information required is exchanged between neighboring cells in order to adjust their handover thresholds. Once neighboring cells adjust their handover thresholds, users at the boundary are forced to change their serving cells accordingly.

High load fluctuations are usually accounted for by over-dimensioning the network during planning phase. A load imbalance condition is triggered when a cell edge user in a highly loaded cell cannot be served with guaranteed QoS due to lack of resources on the serving cell.

For example, if a load imbalance condition is triggered for a given UE and the number available resources of a neighboring cell is greater than the number of available resources of serving cell, the SON algorithm may adjust the load balancing handover (HO) offset of the UE's serving cell according to the difference in received signal strengths of both cells to force the UE to handover. The UE is then reallocated from the overloaded cell to a less loaded neighbor cell. After load balancing, the less loaded neighboring cell coverage increases due to the load balancing HO offset adjustment and it becomes the serving cell for users that have triggered a load imbalance condition. Consequently, the original serving cell becomes less loaded and the QoS for users increase as well as the overall system capacity.

13.6.3 Mobility Robustness Optimization

Optimization of mobility robustness aims to automatically detect and correct errors caused by incorrect mobility configuration and miss-timed handovers that could lead to radio link failures. Radio link failure is usually caused by too late handover, too early handover, or handover to the wrong cell. A SON algorithm aims to find the optimal mobility configuration for the parameters at a given time in order to reduce or eliminate unnecessary handovers. Suboptimal settings could lead to poor performance issues such as the ping pong effect, where the user is handed over between neighboring cells back and forth.

During the optimization, base stations collect user mobility information (e.g., MDT data) and transmit this information to neighboring cells for further analysis. This saves the unnecessary cost of manually optimizing the mobility parameters, thus improving user quality of experience by reducing the number of dropped calls or packets due to handover failures.

13.6.4 *Automatic Neighbor Relations*

Automatic neighbor relations (ANR) aims to avoid the manual work required to configure neighboring base stations for a newly deployed base station. The ANR process involves the following steps: users within the area of a newly deployed cell are instructed to report neighboring elements to their serving base station, e.g., part of MDT data. Users measure the physical cell ID of cells that are detected but have not been defined in their predefined neighbor list and report them to their serving base stations. If the serving base station receives a report by the user that indicates an unknown physical cell ID, it orders the user to read the global cell ID from the broadcast channel of the detected cell. The base station then requests the IP address corresponding to the cell's global ID from the core network, then updates its neighbor list, and establishes a connection with the new peer base station in order to exchange configuration data.

Automatic neighbor relations has the benefits of fast establishment of neighbor relations and saving manual work at the early stages of the network deployment. It also improves the quality of service by reducing the handover failures cause by missing neighbor relations.

13.6.5 *Coverage Optimization and Self-Healing*

Self-healing is a mechanism used to recover from a cell outage and cell failures and is especially effective when the base station equipment is unable to notify the service provider's operations management center about the failure.

Self-healing includes detection and compensation. A cell outage detection mechanism is designed to detect cell outages. An example of this detection is "sleeping cell" detection, where the outage is not reported to the operations center. A cell outage compensation mechanism is designed to extend the cell coverage of neighboring cells to compensate for the outage of a cell. This is usually done through adjusting the antenna tilt and azimuth. This is typically used when an equipment soft reboot does not solve the problem. Self-healing and coverage compensation improve operations and maintenance efficiency. It also improves the quality of experience of customers by increasing the network's stability.

13.6.6 PRACH Optimization

The RACH is the uplink random access unsynchronized channel, used for initial access and uplink synchronization. The random-access procedure performance influences call setup delay, handover delay, call setup success rate, and handover success rate. Therefore, RACH optimization is very important. Possible RACH optimization can include RACH physical resources, RACH preamble allocation (e.g., PRACH root sequence allocation, sample shift used to generate and separate preambles from the selected root sequence), RACH persistence level and backoff control, and RACH transmission power control.

Problems

1. Explain the benefits of automated MDT for an operator compared to collecting physical drive test data.
2. What are the benefits of using DL positioning method over UL positioning methods and vice versa?
3. a. What is the significance of a RAN paging area in 5G networks?
   b. Explain the relation between the number of cells in the paging area and the amount of paging.
   c. What does the UE do once it performs cell re-selection to a cell outside of its current RAN paging area?
4. Describe a few use cases for which SON can be useful for an operator.

References

1. Radio measurement collection for Minimization of Drive Tests (MDT), Overall description, Stage 2. https://www.3gpp.org/ftp/Specs/archive/37_series/37.320/37320-i20.zip Cited 28 Jul 2024
2. NG Radio Access Network (NG-RAN), Stage 2 functional specification of User Equipment (UE) positioning in NG-RAN. https://www.3gpp.org/ftp/Specs/archive/38_series/38.305/38305-i20.zip Cited 28 Jul 2024
3. X. Jiang, Y. Lui, X. Wang, An enhanced location estimation approach based on fingerprinting technique, in *International Conference on Communications and Mobile Computing* (2010)
4. R.F. Trifan, R. Lerbour, Y. Le Helloco, Enhanced 3D geolocation algorithm for LTE call traces, in *2016 IEEE 84th Vehicular Technology Conference (VTC-Fall), Montreal, QC, Canada* (2016), pp. 1–5
5. F. Alfarhan, A. Alsohaily, Self-organizing wireless network parameter optimization through mixed integer programming, in *2017 IEEE International Conference on Communications (ICC), Paris, France* (2017), pp. 1–6
6. C.H. Papadimitriou, K. Steiglitz, *Combinatorial Optimization: Algorithms and Complexity* (Prentice Hall, Englewood Cliffs, NJ, 1982)

7. C. Bliek, P. Bonami, A. Lodi, Solving mixed-integer quadratic programming problems with IBM-CPLEX: a progress report, in *Proceedings of 26th RAMP Symposium* (2014), pp. 171–180
8. A.H. Land, A.G. Doig, An automatic method of solving discrete programming problems. Econometrica **28**(3), 497–520 (1960)
9. J.E. Kelley, Jr., The cutting-plane method for solving convex programs. J. Soc. Indust. Appl. Math. **8**(4), 703–712 (1960)
10. NGMN Alliance, NGMN Recommendation on SON and O&M Requirements. NGMN White Paper (2008). https://www.ngmn.org/wp-content/uploads/NGMN_Recommendation_on_SON_and_O_M_Requirements.pdf Cited 28 Jul 2024
11. NGMN Alliance. NGMN Informative List of SON Use Cases. NGMN White Paper (2007). https://ngmn.org/wp-content/uploads/NGMN_Informative_List_of_SON_Use_Cases.pdf. Cited 28 Jul 2024

Index

A
Access stratum (AS), 240, 244, 329, 332–356
Alamouti scheme, 180–184, 193
Antenna efficiency, 135, 149
Antenna noise, 86, 94
Antenna polarization, 31–33
Array factor, 20, 21, 25, 49, 50
Automatic cell planning (ACP), 410–421, 423
Average power, 1–9, 32, 48, 50, 64, 72, 73, 77, 88, 209, 226, 227

B
Backhaul, 252–253, 257, 313, 409, 422
Bandwidth classes (BC), 383, 385, 388, 394
Beam failure recovery (BFR), 320, 321
Broadcasting, 298
Buffer status reporting (BSR), 304, 312–316

C
Carrier aggregation (CA), 297, 303, 304, 317, 329, 346, 350, 381–401
Carrier frequency offset (CFO), 229–232
Cell selection and reselection, 329, 340–344, 359
Channel (communications channel, channel model), 53–84, 133, 135, 136, 371
Channel spacing, 385, 390–391
Coherence bandwidth, 78–79, 81, 82, 211, 217
Coherence time, 79, 183
Common Public Radio Interface (CPRI), 243–249, 251, 256
Component carrier (CC), 229, 323, 381–394, 397, 399–401

Conditional handover (CHO), 347–350, 355, 357
Conducted power, 102, 146, 149, 361, 362, 364
Core network, 132, 239–242, 244, 325, 326, 331–333, 347, 356, 393, 422, 423, 425
COST 231, 58–60, 135
Cyclic prefix (CP), 219–225, 232, 235–237, 260–265, 271, 275, 283, 287, 289, 293, 367, 368, 377, 379

D
Directivity, 23–26, 49
Discontinuous reception (DRX), 318–319, 331, 333, 352, 353
Diversity gain, 180, 191, 195–197
Diversity order, 171–177, 180–182, 184, 185, 193, 195–198, 209
Dual connectivity (DC), 242, 246, 248, 266, 316, 325, 329, 347, 349, 391–394, 398

E
Effective antenna size, 55
Effective isotropic radiated power (EIRP), 33–34, 101–103, 118, 119, 121, 141, 149, 150, 160, 364, 365, 378, 395, 399
Energy, 1–7, 10, 12, 13, 15, 16, 18, 21, 32, 33, 38, 50, 53, 76, 136, 151, 156, 166, 191, 192, 208, 255, 267, 317, 341, 352, 396, 404, 424
Enhanced universal terrestrial radio access network (E-UTRAN), 239, 240, 242, 244
Equal gain combining (EGC), 176–177, 207

F

Foliage loss, 62–63
Fourier transform (FT), 5–8, 47, 73, 78, 215
Frame, 43, 44, 131, 260, 264, 265, 272–275, 286, 293, 306–308, 336, 353, 357, 369, 371, 373
Free space pathloss, 55, 57, 63, 64, 80, 121, 161
Frequency ranges: FR1, FR2, 133–135, 147, 148, 153–155, 229, 270–272, 290, 291, 293, 316, 317, 337, 342, 344, 351, 360, 361, 368, 376, 388–392, 398
Friis' formula, 55–58
Fronthaul, 249–251, 253–257

H

Half-power beamwidth (HPBW), 27–31
Handover, 240, 299, 325, 329, 332–334, 344–350, 355–357, 386, 404, 410, 416–418, 421–426
Hata model, 58–61, 80, 135

I

Initial access procedure, 299, 323, 336–344
Inter-carrier interference (ICI), 218, 229–231, 385
Interference, 44, 80, 85, 126, 185, 211, 240, 266, 362, 385, 404
International Mobile Telecommunications (IMT), 128–131, 136, 381
International Telecommunication Union (ITU), 30, 125, 127–131, 141
Inter-symbol interference (ISI), 72, 73, 75, 211, 212, 216, 218, 219, 221, 224, 236, 271
Isotropic antenna, 18, 19, 24, 33, 55

K

Key performance indicators (KPIs), 131, 359, 362, 410, 422

L

License assisted access (LAA), 394–399
Licensed spectrum, 126, 127, 141, 146, 394–398
License exempt spectrum, 395
Link budget, 61, 62, 85, 99–106, 150, 205, 389
Localized subcarrier mapping, 233–234, 237
Logical channel prioritization (LCP), 298, 308–314, 327
Logical channels, 295, 296, 298–299, 308–313, 315, 324
L1-RSRP, 360
L3-RSRP, 360

M

Massive MIMO, 165, 203–207, 249, 257, 406
Maximum achievable user data rate, peak data rate, 367–370
Maximum Allowable Pathloss (MAPL), 100–104, 205
Maximum ratio combining (MRC), 173–177, 179–181, 191, 199, 203, 204, 207
Measurements, 3, 58, 146, 171, 211, 282, 303, 330, 359, 403
Medium Access Control (MAC), 244, 245, 247, 248, 280–282, 295–323, 327, 328, 386, 392
Midhaul, 247–248, 253
Minimization of drive testing (MDT), 403–405, 407–411, 414, 416, 425, 426
Mobility, 240, 241, 244, 329, 331–334, 337, 340, 342, 344–351, 355, 357, 359, 371, 373, 404, 405, 410, 415, 421, 422, 424–425
Multiple input multiple output (MIMO) systems, 46, 165–210, 243, 245, 246, 249, 251, 252, 256, 257, 276, 279–280, 328, 367–370, 377, 387, 391, 397, 400, 401, 406, 411
Multiplexing gain, 196–198

N

Noise factor, 85, 88–93, 105, 116, 365
Noise figure, 85, 93–94, 98, 101, 104, 105, 116–118, 120, 155, 161
Noise rise, 85, 99–100
Non-access stratum (NAS), 240, 241, 244, 245, 329, 330, 332, 341, 352, 356, 386
NR-Unlicensed (NR-U), 397–401

O

Occupied bandwidth, 13, 35, 98, 118, 154, 160, 161, 237, 294
Open RAN, 253–257
Optimization, 173, 186, 200, 254, 255, 359, 403–426
Orthogonal frequency division multiplexing (OFDM), 43, 211–232, 235–237, 246, 251, 259–261, 263, 267, 283, 309, 336, 368–370, 373

Index

P

Packet Data Convergence Protocol (PDCP), 244, 245, 247, 248, 295, 310, 324, 325, 327, 328, 356, 392, 393

Paging, 144, 241, 244, 298, 299, 324, 329, 331–333, 335, 352, 353, 355–357, 415–417, 426

Passive Inter-Modulation (PIM), 107, 111–116, 122

Peak data rate in CA, 387, 391

Peak to average power ratio (PAPR), 225–228, 232, 235, 236, 284

Phase noise, 228–230, 281

Physical channels, 279–282, 289, 356

Physical signals, 87, 280–294

Positioning, 405–410, 426

Power delay profile, 75–77, 81

Power flux density (PFD), 14–18, 26, 55–57, 80, 120, 161

Power headroom reporting (PHR), 315–317

Power spectral density (PSD), 4, 7–12, 48, 57, 80, 86, 87, 93, 116, 120, 152, 153, 283, 398

Primary serving cell (PSC), 353, 385, 393

R

Radiation pattern, 18, 20–25, 27, 30, 34, 371

Radio Access Network (RAN), 132, 155, 239–257, 329, 333, 341, 352, 356, 410, 415, 421, 426

Radio bearers, 244, 295, 298, 303, 304, 325, 326, 329, 334

Radio Link Control (RLC), 244, 245, 247, 295, 309, 324, 325, 328, 355, 392, 393

Radio link failure (RLF), 302, 330, 331, 348, 349, 354, 355, 424

Radio link monitoring (RLM), 282, 320, 353–355

Radio resource control (RRC), 244, 245, 247, 275, 298, 302, 305, 308, 309, 312, 318, 321, 322, 329–335, 339–341, 344, 347, 348, 352–356, 386

Radio resource control states (connected, idle and inactive), 321

Random access (RA), 273, 282–289, 291, 299–303, 315, 320, 332, 344, 355, 422, 426

Rate matching, 276, 278

Receive diversity, 166–178, 181, 182, 208

Received Signal Strength Indication (RSSI), 362–365, 376, 378, 379

Receiver sensitivity, 85, 98, 99, 101, 120, 155

Reference signal received power (RSRP), 206, 300, 303, 342, 343, 347, 349, 350, 354, 357, 359–365, 376, 378, 379, 404–408, 411

Reference Signal Received Quality (RSRQ), 342, 347, 349, 354, 362–365, 378, 379, 404, 411

Remote radio head (RRH), 243–245, 247, 256

Resource block, 118, 185, 198, 266–269, 282, 285, 290, 293, 303, 361, 364, 367, 368, 370, 373, 374, 376–378, 383, 418

Resource element (RE), 245, 266–269, 281, 318, 359–361, 365, 367, 379, 418

S

Scheduling, 203, 240, 244, 266, 275, 282, 303–308, 310, 312, 313, 315–319, 322, 323, 334, 338, 372–375, 413, 421

Scheduling algorithms, 372–375

Secondary serving cell (SCS), 270, 271, 290, 291, 363, 377, 385–386, 398

Selection combining, 170–173, 175, 176, 206–208

Self-organizing networks (SON), 403, 414, 418, 420–422, 424

Service Data Adaptation Protocol (SDAP), 244, 247, 295, 325–328

Shadowing, 53, 54, 63–67, 70, 72, 95, 96, 101, 103, 371, 414

Shadowing, fade margin, 63–67

Signal bandwidth, 4, 9–14, 75, 87, 96, 125, 220

Signal to interference plus noise ratio (SINR), 85, 95–100, 116–120, 160, 185, 353, 365, 366, 411

Signal to noise ratio (SNR), 85, 88–91, 96–98, 104, 116, 118, 120, 121, 155, 156, 158, 166–177, 179, 181, 186, 187, 196, 197, 201, 206–210, 218, 247, 279, 371, 404

Single carrier FDMA (SC-FDMA), 232–234, 237, 238

Slow fading, flat fading, frequency selective, 68–69

Space-time coding, 183–185, 193–196

Spectrum, 13, 74, 95, 125, 218, 252, 264, 321, 355, 382, 421

Spectrum auctions, 126, 136–144, 146

Subcarrier spacing (SCS), 155, 156, 215, 218, 224, 228, 229, 235–237, 249, 259, 262, 263, 265, 268–272, 275, 285, 289, 290, 292, 306, 309, 323, 326, 328, 336, 349, 361, 364, 365, 367, 368, 377, 388, 391, 398, 401

Subframe, 260, 264, 265, 317, 351, 353, 368, 420
Supplemental uplink (SUL), 133, 303, 389–390
Synchronization signal block (SSB), 281–283, 301, 302, 320, 336–338, 340, 344, 351, 353, 355, 360, 361, 398
System information (SI), 244, 282, 283, 324, 329, 331, 337–340, 344, 352, 353
System temperature, 94, 104, 161

T
Thermal noise, 85–94, 119, 120, 157, 166, 360, 364
3rd Generation Partnership Project (3GPP), 4, 61, 94, 125, 232, 239, 262, 325, 332, 360, 381, 403
Third order intercept (TOI), 111–113, 121

Time domain numerology, 259–272
Timing advance, 286, 291, 299, 300, 303, 321, 351, 407
Traffic map, 372, 404, 408, 409, 411
Transmit diversity, 178–185
Transport block (TB), 275, 277–279, 297, 303–305, 307–309, 311, 312, 322, 327

U
Universal Terrestrial Radio Access Network (UTRAN), 132, 239
Unwanted emission, 17, 50, 151–155, 161

W
World Radiocommunication Conference (WRC), 130, 137

www.ingramcontent.com/pod-product-compliance
Lightning Source LLC
Chambersburg PA
CBHW070253180225

22050CB00008B/49